DIESEL ENGINE MANUAL

by Perry O. Black
revised by William E. Scahill

THEODORE AUDEL & CO.
a division of
THE BOBBS-MERRILL CO., INC.
Indianapolis/New York

Fourth Edition
First Printing

Published by The Bobbs-Merrill Company, Inc.
Indianapolis/New York

Manufactured in the United States of America

Library of Congress Cataloging in Publication Data

Black, Perry O.
Diesel engine manual.

Rev. ed. of: Audels diesel engine manual. 3rd ed. 1966
Includes index.
1. Diesel motor—Handbooks, manuals, etc.
I. Scahill, William F. II. Title.
TJ795.B534 1983 621.43'6 82-20635
ISBN 0-672-23371-1

FOREWORD

Diesel engines have been a major source of power in the industrial fields for many years. This trend has continued and has been even more pronounced in recent years with the increased use of diesel engines to reduce operating costs in trucks, buses, farm tractors, marine, and industrial power units. The diesel engine has proven itself to be a dependable source of economical power wherever engines high in horsepower and reliability are required.

This book compares and points out the advantages and disadvantages of gasoline and diesel engines and two-stroke-cycle and four-stroke-cycle engines; and it discusses practically all types and designs of diesel engines in use today. An attempt has been made to clearly present and explain the theory, operation, and maintenance of all types of diesel engines including stationary power units and truck, bus, tractor, and marine engines.

The question-and-answer method, along with many pertinent and informative illustrations, is used in presenting all the important aspects of theoretical and practical operation as well as maintenance of diesel engines.

The *Diesel Engine Manual* will be valuable to students, mechanics, operators of diesel-powered vehicles and stationary engines, and all others who are interested in or require a knowledge of diesel engines. The book will be especially valuable to the individual who is already working with diesel engines and desires to increase his knowledge or understanding of diesel engines in order to become more proficient in their maintenance or operation.

PERRY O. BLACK
WILLIAM E. SCAHILL

Contents

CHAPTER 1

CHAPTER 2

CHAPTER 3

CHAPTER 4

CHAPTER 5

CHAPTER 6

CHAPTER 7

CHAPTER 8

CHAPTER 9

CHAPTER 10

CHAPTER 11

CHAPTER 12

CHAPTER 13

CHAPTER 14

CHAPTER 15

CHAPTER 16

CHAPTER 17

CHAPTER 18

CHAPTER 19

CHAPTER 20

CHAPTER 21

CHAPTER 22

CHAPTER 23

CHAPTER 24

CHAPTER 25

CHAPTER 26

CHAPTER 27

CHAPTER 28

CHAPTER 29

CHAPTER 30

CHAPTER 31

CHAPTER 32

CHAPTER 33

APPENDIX

CHAPTER 1

The Diesel Principle

What is a diesel engine?

Answer: In the diesel engine, air alone is compressed in the cylinder; then, after the air has been compressed, a charge of fuel is vaporized by injection into the cylinder, and ignition is accomplished by the heat of compression.

Where did the diesel engine get its name?

Answer: Dr. Rudolf Diesel, a German inventor, patented the principle of the diesel engine in 1893.

Where can a diesel engine be used?

Answer: Diesel engines can be found in practically all heavy industry and in any place that needs tremendous power. The diesel engine has been in use since 1900.

What are the main differences between a diesel engine and a gasoline engine?

Answer: The diesel engine may use a low-grade fuel oil, which is ignited in the cylinder by the heat of air compression. The gasoline engine requires high-grade gasoline for fuel, which is ignited by an electric spark after the gasoline has been mixed with air in a carburetor, injected into the cylinder, and the mixture compressed.

How does a carburetor in a gasoline engine vaporize the fuel?

Answer: By blowing a stream of air over or through the fuel, sending it out in the form of a mist or a gas. This makes it easier to ignite than if it were in the liquid state. Also it mixes air with the fuel, making it into a high explosive. It is the explosion of the mixture in the cylinder that drives the piston and gives it power.

What is a true diesel engine?

Answer: One which ignites and burns the market grade of diesel fuel oil without preheating either the oil or any part of the cylinder to get the required heat. It uses neither outside igniting nor mixing devices.

What is a semi-diesel engine?

Answer: One which works in part on the diesel principle—that is, using some of the features of the diesel engine.

What is a hot-air engine?

Answer: One in which the expansion of heated air is used as the driving force. Air at high pressure is admitted to a cylinder wherein on expansion it moves the piston. The air is then exhausted to the atmosphere or recompressed and reused. Engines of this type were bulky for the amount of power they gave and were used mainly where a small power output was required and plenty of cold water existed for cooling.

How does the efficiency of a diesel engine compare with that of other types of heat engines?

Answer: The diesel engine is the most efficient type of engine known from a thermal (heat) point of view and is the true internal-combustion engine. It depends solely upon its own heat, generated in the cylinders, for igniting and burning the fuel.

How is heat for ignition obtained in the diesel engine cylinders?

Answer: By compressing the air in the cylinders until it has reached a temperature high enough to ignite the injected fuel.

What is the purpose of the carburetor on a gasoline engine?

Answer: Its purpose is to mix gasoline with air for ignition. Carburetors are made in different forms with various control devices to control the quantity of air and the quantity of gasoline mixed. The mixture is compressed in the upper part of the cylinder by the upward movement of the piston. Then a spark is sent into the cylinder to explode the mixture and send the piston downward to provide the power stroke for that cylinder (Fig. 1-1).

In the diesel engine, a carburetor is not needed to mix air with the fuel oil and a spark is not needed to ignite it. Air, alone, is compressed in the cylinder, and the fuel oil is sprayed into the compressed air. The heat of the compressed air ignites the fuel oil and pushes the piston downward. Therefore, all the energy is produced directly inside the cylinder without outside sources.

How does compressed air heat the cylinders?

Answer: When a volume of air is forced into a small space, it is heated in proportion to the amount to which it has been compressed.

What is ignition?

Answer: The act of setting on fire. In a gasoline engine, the electric spark sets the gasoline mixture on fire (Fig. 1-2). In the diesel engine, the heat of compression in the cylinders sets the fuel oil on fire (Fig. 1-2). In both instances, the fuel is "ignited."

What is meant by vaporization?

Answer: Vaporization is the changing or converting of a liquid, such as oil, into a vapor. This is done by blowing a stream of air over or through the oil, sending it out in the form of a vapor, a mist, or a fine spray. The purpose of vaporization is to change the liquid to the gaseous or vaporous state so that it will ignite more easily than if it were in the purely liquid state.

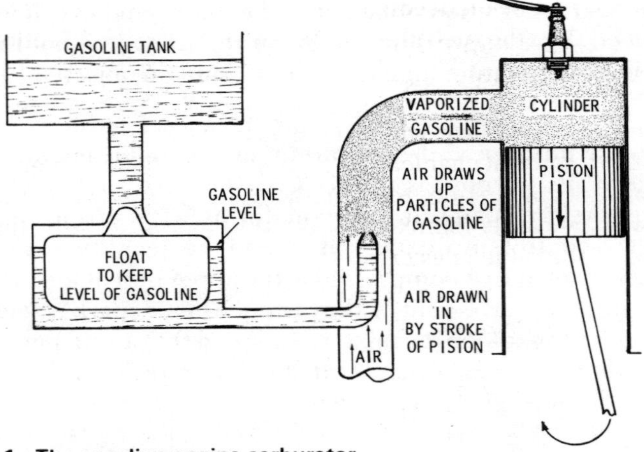

Fig. 1-1. The gasoline engine carburetor.

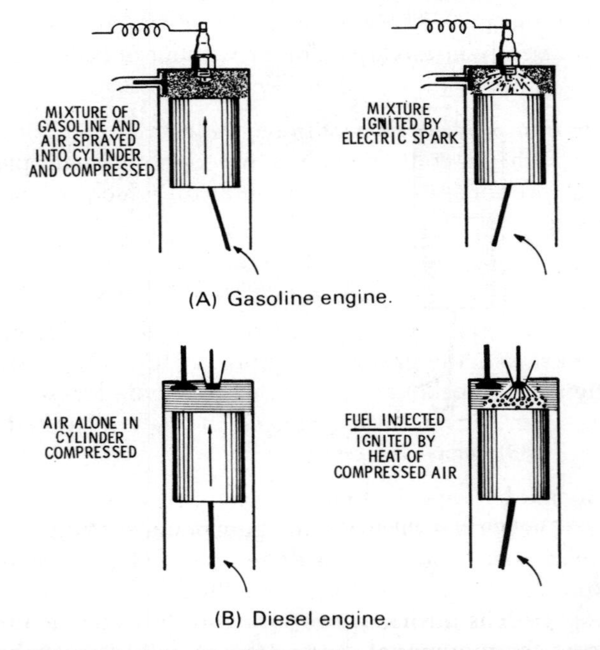

Fig. 1-2. Difference in fuel ignition in the gasoline engine and the diesel engine.

Is vaporization of oil necessary in the diesel engine?

Answer: No, the fuel oil is vaporized at the time of injection into the cylinder.

What causes air to heat when compressed?

Answer: The mechanical force pressing against the air particles, or molecules, squeezes them into a smaller space (Fig. 1-3). This pressure causes them to move or vibrate very rapidly, thus the friction of the molecules creates heat.

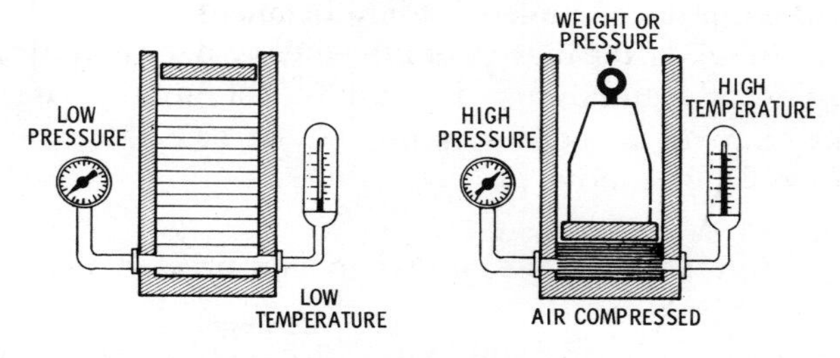

(A) Compression of air increases its temperature.

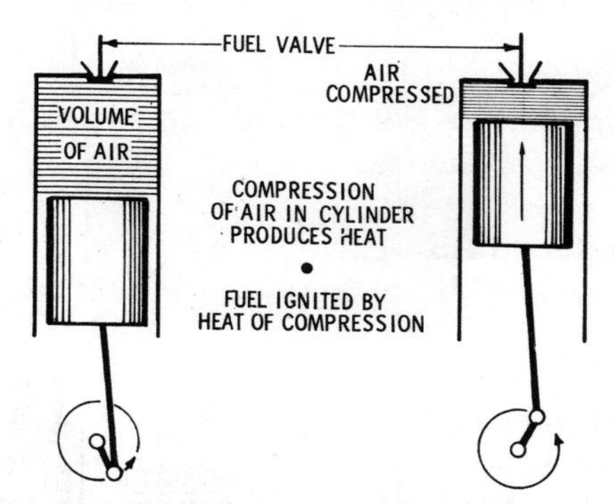

(B) Compression of air in the diesel cylinder.

Fig. 1-3. Illustration of temperature rise when air is compressed.

What is heat?

Answer: Heat is a form of energy and a source of power. It results from the movement of the molecules within a substance. All substances are made of molecules or tiny particles. These molecules are always in motion and the measure of their motion is heat.

What is a molecule?

Answer: A molecule is the smallest particle of which a substance is composed. There are smaller particles, atoms—and still smaller particles, but the molecule is the smallest particle that identifies or characterizes a substance.

Describe the motion of molecules in substances.

Answer: In solids, the motion of molecules is back and forth like tiny shuttles or pendulums. In liquids, the motion is in all directions—having no particular movement or arrangement. In gases, the molecules move in straight lines.

At what temperature is the air when compressed in a diesel cylinder?

Answer: At between 800 and 1,000 degrees Fahrenheit, depending upon the compression.

What is temperature?

Answer: Temperature is a term or name used to indicate the degree of heat of a substance.

How is temperature measured?

Answer: In general, by a thermometer. Other ways will be explained later.

Describe a thermometer.

Answer: A thermometer is an instrument for measuring temperature—the degree of heat or the degree of cold. Cold is the absence of heat. The thermometer is usually a glass tube with a bulb at the bottom that contains either mercury or colored alcohol.

Describe the principle upon which the thermometer works.

Answer: The principle of the thermometer is based upon the expansion of a substance when heated. Heat causes a substance to expand. If a substance such as mercury is placed in a tube and the base of the tube is heated, expansion will cause the mercury or liquid to rise in the tube. If the tube is marked off into divisions forming a scale, the amount of heat can be easily determined by the height to which the liquid rises.

Why is mercury used in most industrial thermometers?

Answer: Mercury is used in thermometers because it does not freeze until about 38 degrees below zero (−38°F.) and does not boil until about 675 degrees (675°F.) above zero. Therefore, it remains a liquid over a large temperature range and indicates temperatures quite accurately.

Name the three types of thermometers in use.

Answer: Fahrenheit, Celsius, and Reaumur (Fig. 1-4).

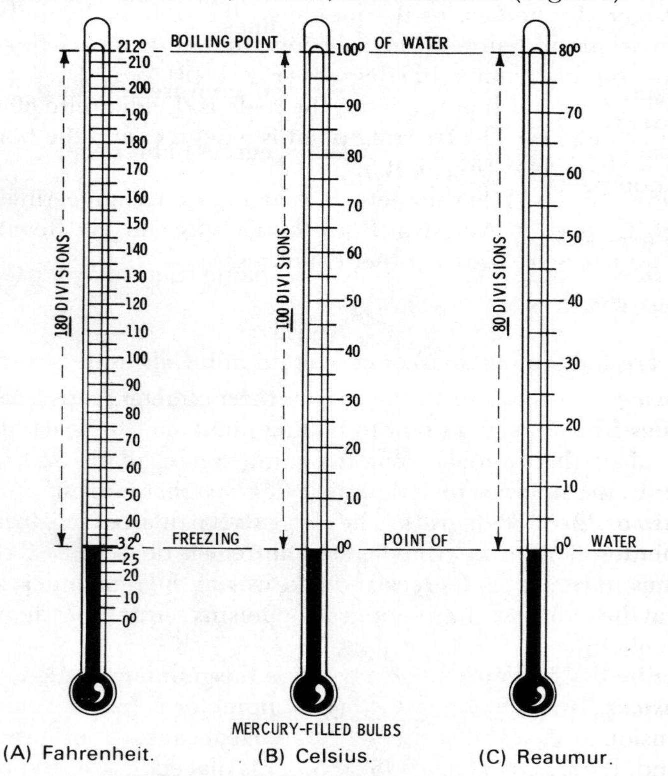

Fig. 1-4. Three types of thermometer scales.

Describe the Fahrenheit scale.

Answer: A scientist named Fahrenheit proposed that a thermometer scale be divided so that the freezing point of water would read 32 degrees above zero and the boiling point 212 degrees above zero (Fig. 1-4A). This scale has become the popular thermometer scale for commercial use in English-speaking countries, and temperature in general is now measured by the Fahrenheit (F.) thermometer.

What are the scale divisions of the Celsius and Reaumur thermometers?

Answer: In the Celsius thermometer, the scale is divided into 100 divisions or degrees. The freezing point of water is 0 degrees and the boiling point is 100 degrees (Fig. 1-4B).

In the Reaumur thermometer, the scale is divided into 80 divisions or degrees. The freezing point is 0 degrees and the boiling point is 80 degrees (Fig. 1-4C).

The Celsius (C.) thermometer is used mainly in engineering and scientific work in America. Both the Celsius and the Reaumur thermometers are used in other countries.

How are Fahrenheit degrees converted into Celsius degrees?

Answer: As the scale of the Celsius thermometer reads 0 to 100 degrees from freezing point to boiling point and the scale of the Fahrenheit thermometer, for the same range, reads 32 to 212 degrees, the number of Fahrenheit degrees that correspond are 212 minus 32 or 180 degrees. The range of 100 degrees is 5/9 of the remainder. Therefore, to convert Fahrenheit degrees to Celsius degrees, subtract 32 degrees and take 5/9 of the remainder.

Example: How many degrees Celsius equal 77 degrees Fahrenheit?

$$77 - 32 = 45$$

$$5/9 \text{ of } 45 = \frac{45 \times 5}{9} = \frac{225}{9} = 25°\text{C.}$$

How are Celsius degrees converted into Fahrenheit degrees?

Answer: Fahrenheit degrees equal 9/5 Celsius degrees; therefore, multiply Celsius degrees by 9, divide by 5 and add 32.

Example: 25 degrees Celsius equal how many degrees Fahrenheit?

$$\frac{25 \times 9}{5} = \frac{225}{5} = 45 + 32 = 77°F.$$

How are Reaumur degrees converted into Fahrenheit degrees? Into Celsius degrees?

Answer: As the Reaumur scale is divided into 80 parts or divisions between the freezing point and the boiling point, and the Fahrenheit thermometer has 180 divisions between these two points, 180 becomes 9/4 of 80 degrees. Therefore, Fahrenheit degrees are 9/4 of Reaumur degrees plus 32 degrees.

Example: 20 degrees Reaumur equal how many degrees Fahrenheit? Multiply by 9, divide by 4 and add 32

$$\frac{20 \times 9}{4} = \frac{180}{4} = 45 + 32 = 77°F.$$

Fahrenheit to Reaumur—subtract 32 and take 4/9 of the remainder.

Reaumur to Celsius—Reaumur reading times 5/4.

Celsius to Reaumur—Celsius reading times 4/5.

What is the absolute zero temperature?

Answer: Absolute zero is the temperature calculated to be the true or the absolute zero; the lowest possible temperature imaginable. It is the temperature at which gas, when cooled, would disappear entirely. This fact is determined on the following basis: A gas contracts as it cools; its volume becomes less as the temperature becomes less. If the cooling were continued until the gas disappeared entirely—that temperature would be the absolute zero. The absolute zero temperature is calculated to be 460° degrees below zero Fahrenheit (−460°F.) or 273 degrees below zero Celsius (—273°C.).

How are thermometer readings converted to absolute temperature readings?

Answer: By adding to the thermometer reading, 460 degrees for Fahrenheit or 273 degrees for Celsius.

What is the absolute temperature of the freezing point of water in Fahrenheit degrees?

Answer: Freezing point of water = 32°F.; plus 460° equals 492° absolute temperature.

What is the freezing point of water in Celsius degrees?

Answer: Freezing point of water = 0°C.; plus 273° equals 273° absolute temperature.

What is the highest temperature that can be measured by a mercury thermometer?

Answer: Normally about 500 degrees Fahrenheit because mercury boils at 674 degrees. It may be made to read 1,000 degrees or over by slowing down the expansion of the mercury by placing it under pressure. This is based upon the principle that the boiling action of a liquid can be delayed by exerting a force upon the liquid. Therefore, if the space in the tube above the mercury is filled with a gas, such as nitrogen or carbon dioxide, it will exert a pressure which will hold back the rising of the column of mercury and allow higher temperature readings. A temperature of 1,000 degrees F. is about the maximum temperature to which a thermometer can be used; the softening of the glass then becomes a factor.

How are temperatures above 1,000 degrees Fahrenheit measured?

Answer: By a pyrometer.

What is a pyrometer?

Answer: A pyrometer is an instrument used to measure high temperatures. It is made in a number of forms and can indicate temperatures to over 3,000 degrees Fahrenheit. There are means for determining still higher temperatures.

What is a mechanical pyrometer?

Answer: A mechanical pyrometer is one that works on the principle of expansion of metals by heat. If a metal rod, connected to a movable pointer, is heated, the pointer will move according to the amount that the metal has lengthened or expanded; and this, in turn, is according to the amount of heat applied. The pointer can be made to move over a scale that reads directly in degrees of heat.

Name two forms of electrical pyrometers.

Answer: Resistance pyrometers and thermoelectric or thermocouple pyrometers.

Describe a resistance pyrometer.

Answer: A resistance pyrometer measures temperatures by using the principle that heat increases the electrical resistance of metals. The more heat applied, the higher the resistance becomes. Therefore, if the resistance of a strip of metal is measured both before and after heating, the amount of heat applied can be determined from the difference in resistance. High-resistance metal strips or wire, such as platinum, are used and the instruments are made to read directly in degrees of heat.

Describe an electrical thermocouple pyrometer.

Answer: A thermocouple pyrometer (Fig. 1-5) is an instrument that measures temperatures by the use of the principle of an electrical thermocouple; that is, a heat couple. If two different kinds of metals are joined together and heated at the joint, an electrical current is generated in the metals. If the open ends are connected to an electrical meter, the amount of current generated can be determined.

This current is proportional to the amount of heat applied. The two metals are metals that can withstand intense heat without melting. Usually they are of platinum combination—one being pure platinum and the other an alloy of platinum and rhodium. They have a very high melting point and can withstand heat of more than 3,000 degrees Fahrenheit. This type of pyrometer can be used for measuring very high temperatures by direct contact.

There are other types of metal combinations for lower temperature requirements. Two of these combinations are iron and con-

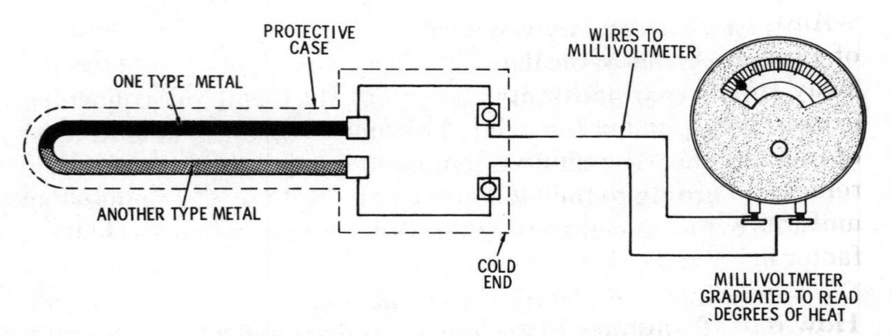

Fig. 1-5. Principle of a thermocouple pyrometer. A thermocouple is used for measuring high temperatures. When heat is applied to the junction of two dissimilar metals, a current of electricity begins to flow in proportion to the amount of heat applied. This current can be brought to a meter and translated in terms of heat. The thermocouple for diesel engine use is inserted in an opening in the cylinder exhaust, and by its temperature reading, denotes the burning condition of the fuel. It is possible to compare the actions in the various cylinders by use of the thermocouple.

stantan and copper and constantan; constantan is an alloy of nickel and copper.

What general advantage has the electrical pyrometer over the mechanical pyrometer?

Answer: The electrical pyrometer or thermocouple can be used to indicate temperature at a distant point from where the heat exists. For instance, the heat in an engine cylinder can be indicated on an instrument at a distance from either the engine or the engine room.

What is a recording pyrometer?

Answer: A recording pyrometer is one that records or registers temperatures over a period of time. It generally consists of a spring-wound clock mechanism that turns a paper or chart upon which rests an inked pen connected with the pyrometer. As the paper moves, an inked line that shows the degrees of heat at various times is marked on the paper. The paper is usually marked off into units of time and temperature.

Why is measurement of heat, and a knowledge of heat effect, important to the proper operation of a diesel engine?

Answer: Because heat is involved in all phases of the operation of a diesel engine. Heat ignites the fuel, heat must be conducted away from the engine to permit proper lubrication, heat in the exhaust tells whether the cylinders are getting the correct amount of fuel, heat must be removed from air compressors, heat must be removed from the cooling water, and heat must be maintained uniformly in all cylinders for equal power and efficiency. Heat is a factor in all parts of the engine; it must be known and controlled.

How is heat conducted from, or to, a body?

Answer: Heat travels in three different ways—by conduction, by radiation, and by convection. It is conducted from a body in the same way that it is conducted to a body.

Describe the way in which heat is transferred from one body to another by conduction.

Answer: As was previously stated, all bodies are composed of small particles, or molecules, which are always in a state of motion. Some molecules are more affected by heat than are others. Some substances, like metals, have the property of taking up heat and transmitting it from molecule to molecule more readily than others, such as wood or water. Metals will pass heat along or, as is said, will conduct it more quickly, by conduction, from which comes the term conductor, either a better or a poorer conductor of heat. Iron is a better conductor than wood. The degree of conduction by different substances or their ability to take up heat and pass it along is known. Silver is the best conductor, copper the next best, etc.; and when a substance is a very poor conductor of heat, it is said to be a heat insulator and is used as such.

Scale from impure water depositing inside either a water jacket or a boiler will form an insulation or coating and then require more heat to increase the temperature of the water than if the scale were not there. The scale does not permit the iron to conduct the heat of the water. For that reason, clean water must be used in a water jacket or in a heat chamber. Conduction and heat transfer can now be understood.

Describe transfer of heat by radiation.

Answer: The sun heats the earth. There is no direct contact between the sun and the earth, yet the effects of its heat are felt. You can warm your hands by holding them over a hot stove. You can heat a substance by having a flame near it—not touching it. Heat passes through the air; heat rays or heat waves pass from the hot substance—they radiate from it—and radiate in all directions. This is heat transfer by radiation. A cold body near a hot body will get its heat by radiation—the amount depending upon conditions.

Describe transfer of heat by convection.

Answer: Heat transfer by convection is based upon another principle. Heat applied to the bottom of a vessel containing water cannot heat the full quantity at one time, but it heats the portion closest to the flame. This causes the water to expand. When it expands it becomes lighter. Lighter substances rise so that the hot water rises to the top and is displaced by the cold water until the entire quantity is heated. This form of heat transfer is called "convection" or "convection currents." The changing of positions of the warm water and the cold water, until all is at a uniform temperature, is called heating by convection.

What is sensible heat?

Answer: Sensible heat is the heat that can be indicated either by contact or by a thermometer.

What is latent heat?

Answer: Latent heat is the opposite of sensible heat and can neither be indicated by contact nor by a thermometer or other instrument.

What is the unit of heat?

Answer: The amount of heat required to raise the temperature of one pound of water one degree Fahrenheit at its greatest density.

If a pound of water were put over a flame to be heated, the quantity of heat it would take to heat that water just one degree Fahrenheit would be what has been taken as the standard for heat measure or for the heat unit and is called the *British thermal unit*, abbreviated Btu.

If one pound of water at the freezing point is heated to the boiling point (32° to 212° F.), the amount of heat consumed is 180 Btu.

Assuming no loss of heat, it will require 180 heat units to raise the temperature of one pound of water from 32° to 212° Fahrenheit. If the transfer of heat takes place at a uniform rate and it requires six minutes to increase the temperature of the water from 32° to 212°, then one heat unit will be transferred to the water in (6 - 60) ± 180, or two seconds.

What is meant by specific heat?

Answer: Specific heat is the ratio of the amount of heat, or the number of heat units, or the number of Btu required to raise the temperature of a substance one degree to the amount of heat required to raise the temperature of an equal weight of water one degree.

Some substances heat more quickly than others. Metals, for instance, get hot much quicker than glass, wood, or air. If a particular substance requires 1/10 the amount of heat to bring it to a definite temperature as for an equal weight of water, the number

Table 1-1. Specific Heats of Some Common Substances

Solids		Liquids	
Copper	0.0951	Water	1.0
Wrought iron	0.1138	Sulfuric acid	0.335
Glass	0.1937	Mercury	0.0333
Cast iron	0.1298	Alcohol	0.7
Lead	0.0314	Benzine	0.95
Tin	0.0562	Ether	0.5034
Steel, Hard	0.1175		
Soft	0.1165		
Brass	0.0939		
Ice	0.504		

Gases

	At constant pressure	*At constant volume*
Air	0.23751	0.16847
Oxygen	0.21751	0.15507
Hydrogen	3.409	2.41226
Nitrogen	0.2438	0.17273
Ammonia	0.508	0.299
Alcohol	0.4534	0.399

of heat units, or the number of Btu becomes 1/10; and its specific heat is then 1/10 (0.1).

What is the relationship of heat to work?

Answer: As heat is a form of energy, it can do a number of things; it can expand metals, fuse metals, burn substances, evaporate liquids, produce steam which creates pressure and does work, cause air to expand and create pressures, etc. Therefore it has a direct relationship to work—the more work needed, the more heat needed; the more heat needed in a substance, the more work or mechanical energy is needed to produce that heat. The relationship between heat and work is called the "mechanical equivalent of heat."

What is the unit of the mechanical equivalent of heat?

Answer: 778 foot-pounds. The quantity of mechanical energy or work which would have to be expended in lifting a weight of 778 pounds to a height of 1 foot (or a weight of 1 pound to a height of 778 feet) is equivalent to the amount of heat required to raise the temperature of 1 pound of water 1 degree Fahrenheit. In other words, 1 Btu or 252 calories of heat energy expended is equal to 778 foot-pounds of work or mechanical energy expended. Therefore, a definite force is required to produce a definite amount of heat, and a definite amount of heat is required to produce a definite amount of force or work.

What is a foot-pound?

Answer: A foot-pound is the unit or measure of work; it is the amount of work done in lifting a one-pound weight through a distance of one foot.

What is the unit of power?

Answer: The unit of power is the horsepower. It is equivalent to the amount of work or energy required to lift a weight of 33,000 pounds through a distance of one foot in one minute (or a lesser or greater weight a proportional distance in a proportional time—the total equaling 33,000 foot-pounds per minute) and is defined as 33,000 foot-pounds per minute. The horsepower is the unit used for measuring the power of motors or engines.

What is pressure?

Answer: Pressure is a force exerted against anything that resists or opposes that force. Working against that force requires power. The load on an engine resists the turning of that engine, therefore power is required to overcome the load—hence the reason for pressure in the cylinders.

What is absolute pressure?

Answer: Absolute pressure is the pressure measured from the point of no pressure or true zero. Air exerts a pressure of 14.7 pounds upon every square inch of surface; therefore, a pressure gauge shows pressures above that of the air or the atmosphere. In order, therefore, to get absolute pressure, it is necessary to add 14.7 (roughly 15) pounds to the reading indicated by the gauge.

What is a pressure gauge?

Answer: It is an instrument with a dial for registering the pressure of a fluid or liquid. The usual gauge operates on the principle that a bent oval tube tends to straighten itself when under pressure.

What is meant by atmospheric pressure?

Answer: Atmospheric pressure is the pressure of the atmosphere, or air. It is equal to 14.7 pounds per square inch at sea level. Weights and pressures are given at sea level because air becomes rarefied (thinner) as height or altitude increases—therefore its weight becomes less. The decrease in weight is equivalent to about one-half pound per square inch for each 1,000 feet above sea level.

How is cylinder pressure measured in a diesel engine?

Answer: By an engine indicator which measures cylinder pressure (Fig. 1-6).

What is normal air compression in a diesel cylinder?

Answer: Between 350 and 500 pounds of air pressure per square inch, depending upon the method of fuel injection.

Why is high air pressure needed in a diesel engine cylinder?

Answer: To obtain the required amount of heat with which to ignite the fuel oil.

What is diesel fuel oil?

Answer: Diesel fuel oil is a refined distillate of crude oil prepared for use in diesel engines.

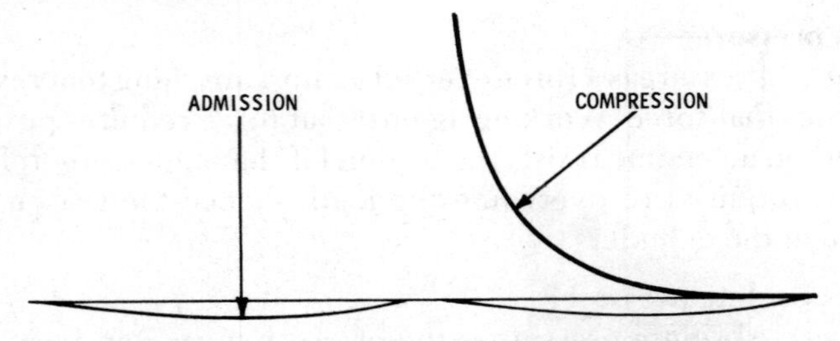

(A) In the admission stroke, the charge comes in below the atmospheric pressure line.

(B) In the compression stroke, heat is generated for ignition.

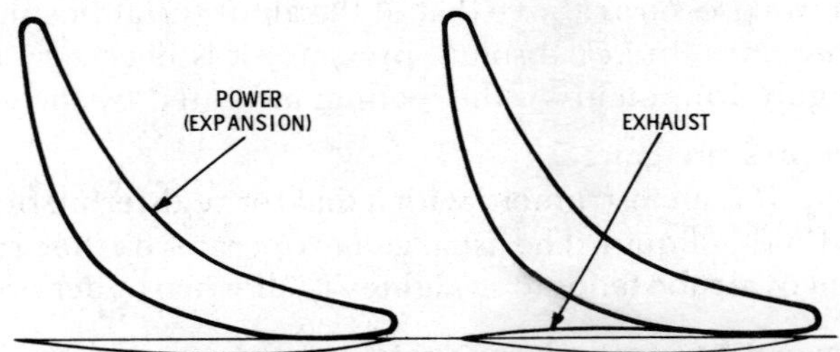

(C) In the power stroke, combustion products cause expansion.

(D) In the exhaust stroke, all the burned gases are ejected above atmospheric pressure.

Fig. 1-6. The diesel cycle presented progressively by an indicator card.

What makes fuel oil burn?

Answer: The hydrocarbons in the oil—the carbon and the hydrogen.

What is carbon and what is hydrogen?

Answer: Carbon is a substance, found in nature, which will burn. It is the basis of lampblack, of charcoal and other coals, and of liquid fuels.

Hydrogen is a gas found in nature and in many substances, which will burn easily.

What elements are found in fuel oil?

Answer: Roughly about 85 percent of carbon and 15 percent of hydrogen. It contains other ingredients, such as water, sulfur, and ash, which are a hindrance or drawback to burning and are obstacles to lubrication.

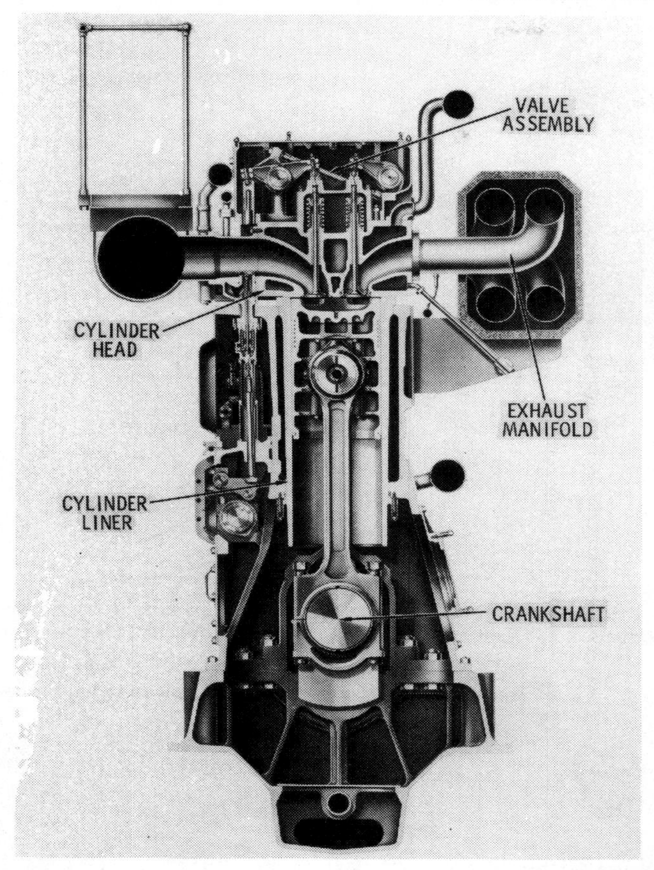

Courtesy Worthington Corp.

Fig. 1-7. Transverse section of four-stroke-cycle Worthington SEH diesel engine.

Why is air needed to make fuel burn?

Answer: Because burning or combustion cannot occur without oxygen; air supplies the oxygen.

What is air?

Answer: Air is a mechanical mixture of two gases—oxygen and nitrogen—with about one part oxygen and four parts nitrogen; by volume, actually 21 parts oxygen and 79 parts nitrogen; by weight, 23 parts oxygen and 77 parts nitrogen.

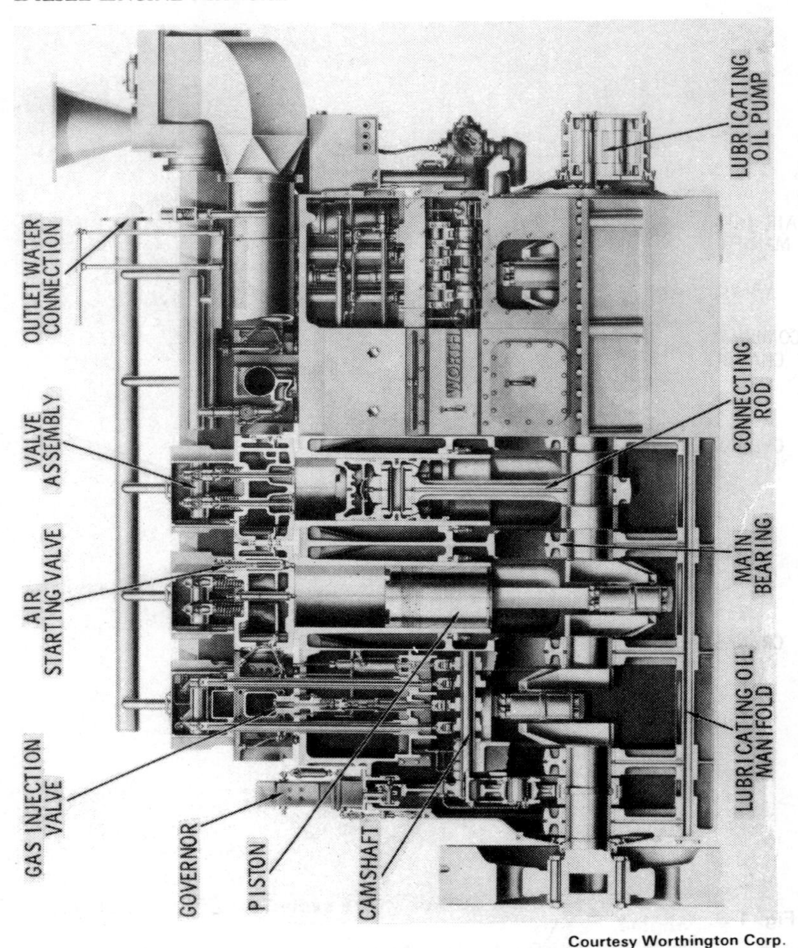

Courtesy Worthington Corp.

Fig. 1-8. Longitudinal section of four-stroke-cycle Worthington SEH diesel engine.

Does air have weight or pressure?

Answer: Yes. Air has weight which causes a pressure of 14.7 pounds per square inch at sea level.

Does air have a relationship of volume to weight?

Answer: Yes. At 32 degrees Fahrenheit, a pound of air has a

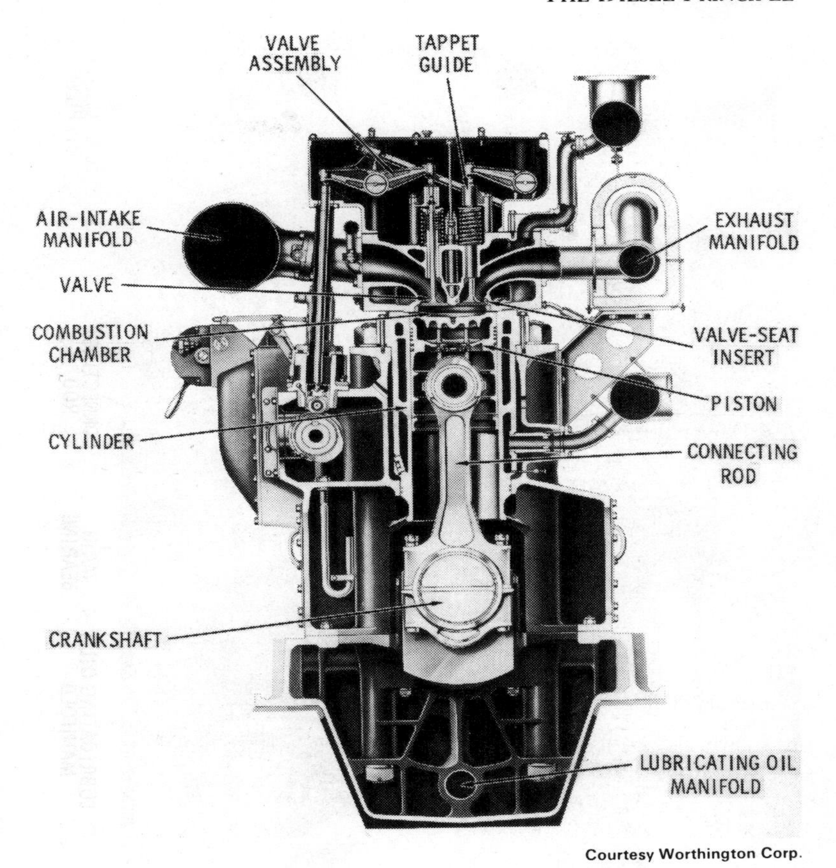

Courtesy Worthington Corp.

Fig. 1-9. Transverse section of four-stroke-cycle Worthington SW diesel inline engine.

volume of 12.38 cubic feet; at 62 degrees Fahrenheit, a pound of air has a volume of 13.14 cubic feet. There is also a relationship to temperature. For every degree in temperature rise, there is an increase in volume because of expansion.

What is meant by combustion?

Answer: Combustion is the chemical combination of oxygen with other substances to cause a flame or burning.

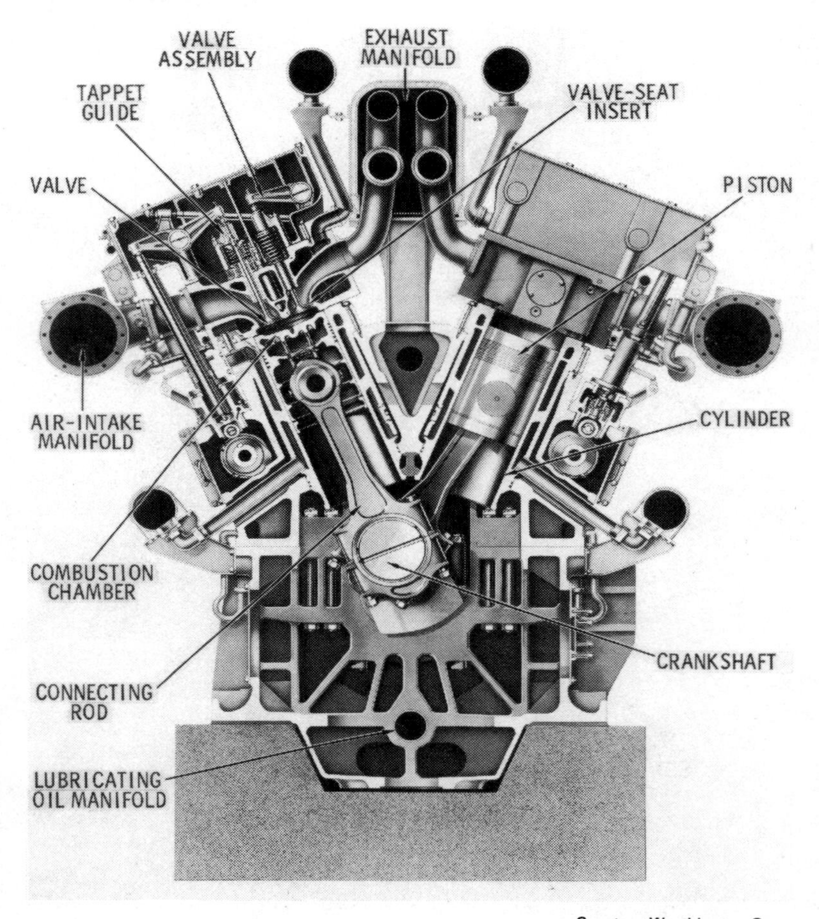

Courtesy Worthington Corp.

Fig. 1-10. transverse section of four-stroke-cycle Worthington SW diesel vee type engine.

What is meant by internal combustion?

Answer: Burning the fuel within the cylinder of the engine.

What is an internal-combustion engine?

Answer: An engine in which the fuel is burned directly within the working cylinder. Both gas and diesel engines are examples of internal-combusiton engines.

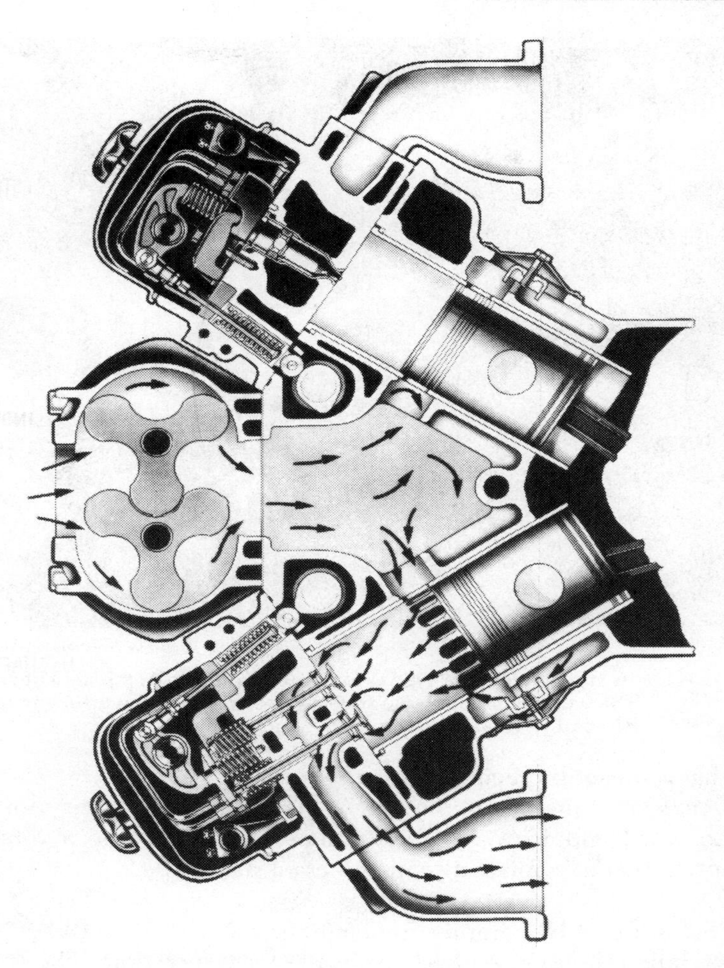

Courtesy General Motors

Fig. 1-11. Air-intake system through blower and engine.

What is an external-combustion engine?

Answer: An external-combustion engine is one in which the fuel is burned outside of the power cylinder. For example, in a steam engine the fuel is burned and heats the water in a boiler which produces the steam that is sent into the cylinder.

Courtesy Worthington Corp.

Fig. 1-12. A 16-cylinder, vee-type Worthington SW diesel engine to power the world's largest radio transmitter, which is powerful enough to send its message around the world.

What is meant by reciprocation motion?

Answer: A to-and-fro motion. As the piston in a cylinder moves backward and forward, this motion is converted to a rotary motion by the connecting rod and crankshaft.

What is meant by compression ignition?

Answer: Ignition caused by heat of compression. The heat resulting from compressed air in a diesel cylinder ignites the fuel. A diesel engine is a compression-ignition engine.

How is air obtained for scavenging in the diesel engine?

Answer: In the scavenging process employed in GM 71 Series two-stroke-cycle engines, air is forced into the cylinders by a blower, which thoroughly sweeps out all the burned gases through the exhaust valve ports (Fig. 1-11). This action also helps to cool the internal engine parts, particularly the exhaust valves. At the beginning of the compression stroke, therefore, each cylinder is filled with fresh, clean air, which permits highly efficient combustion.

The cylinder of a diesel engine must have air for compression to create the heat necessary to ignite the charge of fuel oil; therefore, it is necessary to cleanse the cylinder of all burned gases resulting from the combustion of the previous charge and to send clean air into it for the next compression and charge. In the four-stroke-cycle engine this is done by two of the four strokes—one stroke forces out the remaining burned gases and another stroke draws air into the cylinder. In the two-stroke-cycle engine, the two events practically take place in one operation. An air supply, under slight pressure, two to six pounds from some source, enters the cylinder, blows out the burned gases, and leaves clean air for compression. Crankcase-compression engines supply the air from the crankcase.

CHAPTER 2

Operating Principles

Upon what basic principle does the diesel engine operate?

Answer: A charge of fuel oil is injected and vaporized into heated air in the cylinder. The heat ignites the fuel oil, causing compression, which in turn moves the piston.

What creates the heat in the cylinder?

Answer: Air is compressed until it attains a temperature high enough to ignite the fuel oil.

What temperature is required to ignite fuel oil?

Answer: The temperature required depends upon the kind and the density of the oil used and the design of the engine; but ignition generally occurs at between 800 and 1,000 degrees Fahrenheit.

What air pressure is required to produce the required temperatures?

Answer: From 350 to 500 pounds per square inch.

How does the principle of diesel engine operation differ from that of the gasoline engine?

Answer: It differs in that, in the gasoline engine, a charge of gasoline and air is mixed in a carburetor, injected into the cylinder, compressed in the cylinder and then ignited by an electric spark; while in the diesel engine, a charge of air alone is compressed in the cylinder, fuel oil is injected and ignited by the heat of the compressed air.

What are the four strokes in the four-stroke-cycle diesel engine?

Answer: In the four-stroke-cycle diesel engine (Fig. 2-1), four strokes of the piston are required to complete one cycle or series of events which must take place, in regular order, to operate the engine. The four strokes are as follows:

1. First stroke draws air into the cylinder.
2. Second stroke compresses the air.
3. Third stroke is produced by the burning gases. It is the power stroke.
4. Fourth stroke expels the burned gases.

List the series of events in the four-stroke-cycle diesel engine.

Answer: The series of events taking place in a cylinder of a

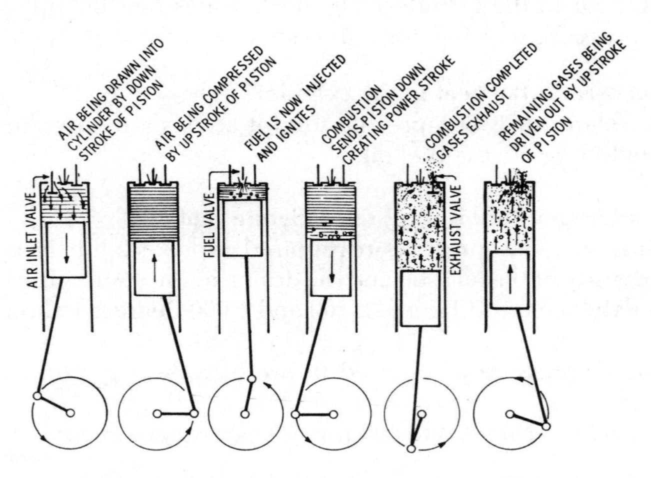

Fig. 2-1 Operating principle of a four-stroke-cycle diesel engine.

four-stroke-cycle engine, and making up one complete cycle, are shown in Fig. 2-1 as follows:

1. The air-inlet valve opens, permitting air to be drawn into the cylinder by the downward stroke of the piston.
2. The valve is closed and the piston starts upward, compressing the air for producing the heat needed to ignite the fuel oil.
3. The fuel valve opens, and the fuel oil is injected into the hot air, where it ignites.
4. The burning fuel forms gases that create pressure and send the piston downward; this is the power stroke of the engine.
5. When the piston has completed its power stroke, the exhaust valve opens and permits the burned gases to escape.
6. The piston returning upward forces out the remaining gases in the cylinder. The exhaust valve closes and the cycle repeats.

What is the two-cycle principle?

Answer: In the two-cycle engine, intake and exhaust take place during part of the compression and power stroke. In contrast a four-cycle engine requires four piston strokes to complete an operating cycle; thus, during half of its operation, the four-cycle engine functions merely as an air pump. See Figs. 2-2 through 2-5.

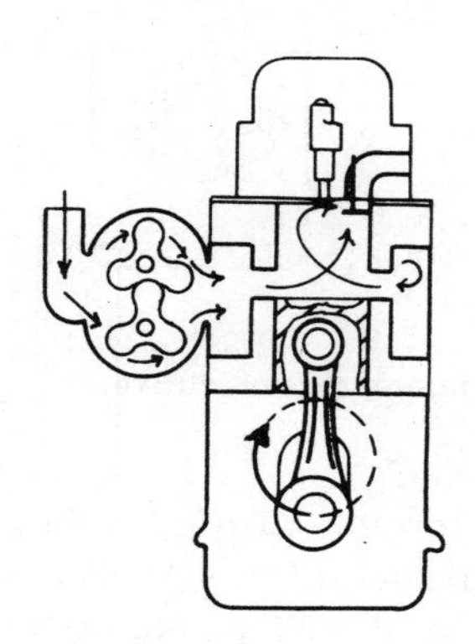

Courtesy General Motors

Fig. 2-2. Cross section of scavenging air.

Why is a blower used?

Answer: A blower, which is mounted on the side of the engine, forces air into the cylinders to expel the exhaust gases and supply the cylinders with an abundance of fresh air for combustion. A series of ports cut into the cylinder wall, which are above the piston when it is at the bottom of its stroke, admit the air from the blower into the cylinder as soon as the top face of the piston uncovers the ports (Fig. 2-2). The undirectional flow of air toward the exhaust valves produces a scavenging effect, leaving the cylinders full of clean air when the piston again covers the inlet ports.

What is the compression stroke?

Answer: As the piston continues on the upward stroke, the exhaust valves close and the charge of fresh air is subjected to the final compression (Fig. 2-3).

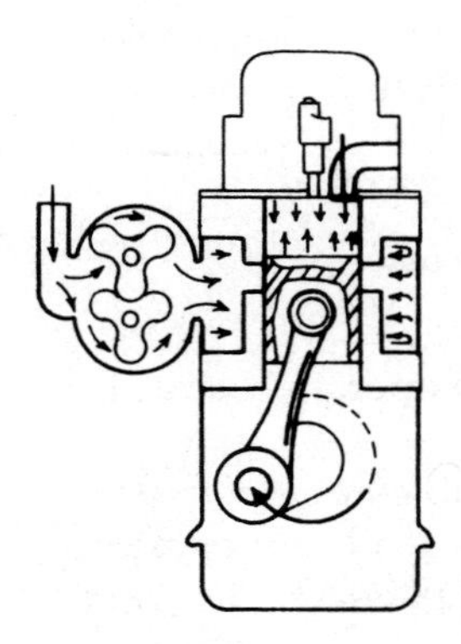

Courtesy General Motors

Fig. 2-3. Cross section of compression stroke.

What is the power stroke?

Answer: Shortly before the piston reaches its highest position, the required amount of fuel is sprayed into the combustion space by unit fuel injector (Fig. 2-4). The intense heat generated during the high compression of the air ignites the fine spray immediately, and the combustion continues as long as the fuel spray lasts. The resulting pressure forces the piston downward on its power stroke until the exhaust valves are again opened (Fig. 2-5). The burned gases escape into the exhaust manifold as the downward moving piston is about to uncover the inlet ports. When these ports are uncovered, the cylinder is again swept with clean scavenging air (Fig. 2-2). This entire combustion cycle is complete in each cylinder for each revolution of the crankshaft, or in other words, in two strokes; hence, the "two-stroke cycle."

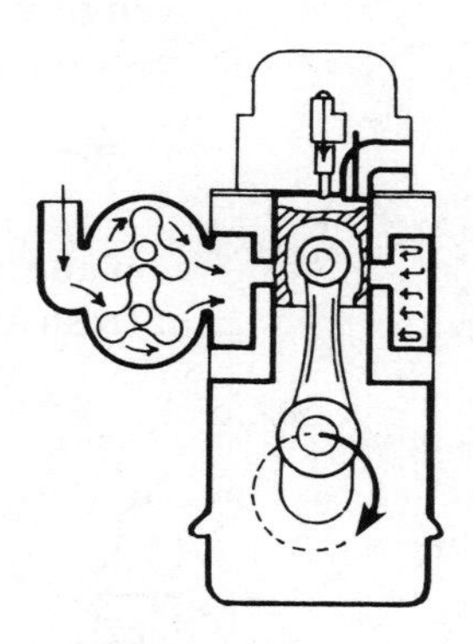

Courtesy General Motors

Fig. 2-4. Cross section of power stroke.

What is the principle of operation of the double-acting two-stroke-cycle diesel engine?

Answer: In the double-acting two-stroke-cycle diesel engine, the number of power strokes is doubled in each revolution. A two-stroke cycle, single-acting engine produces one power stroke in each revolution; a double-acting, two-stroke-cycle engine produces two power strokes; a four-stroke-cycle, single-acting engine produces one power stroke in two revolutions; a double-acting, four-stroke-cycle engine produces two power strokes in two revolutions. Power acts first on one end of the piston, then on the other; it is a reciprocating action with power each way.

The engine shown in Fig. 2-6 is a double-acting, two-stroke-cycle engine. It is similar to two single-acting engines working together. Each requires two movements of the piston to complete one cycle or series of events as follows:

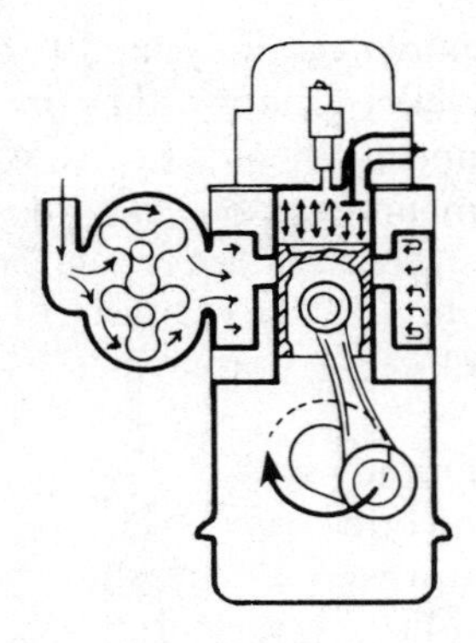

Courtesy General Motors

Fig. 2-5. Cross section of exhaust stroke.

1. One movement compresses air in the cylinder for igniting the fuel oil.
2. The other movement is the power movement—the result of ignition. It is the power stroke; but the strokes are combined, giving the engine the name double-acting. While the piston is being forced downward, in its power stroke, air at the other end of the cylinder is compressed for the reverse stroke; this back-and-forth power movement continues.

List the series of events occurring in a double-acting, two-stroke-cycle diesel engine.

Answer: The illustration (Fig. 2-6) shows the action taking place in one cylinder of a double-acting, two-stroke-cycle diesel engine as follows:

1. Fuel oil is injected at one end of the cylinder and when ignited, sends the piston downward. In the meantime, scavenging air forces out the burned gases of the previous stroke, leaving clean air for compression.
2. The fuel oil ignites and sends the piston downward, creating the power stroke; and, at the same time, the air in the lower part of the cylinder is compressed.
3. As the piston is about to reach the end of the stroke, it uncovers the exhaust port, permitting the gases to escape while compressing air in the lower portion of the cylinder.

4. The piston continues downward to the end of the stroke, and uncovers the air-inlet port and scavenges, or cleanses, the cylinder of burned gases. The air in the lower part of the cylinder, in the meantime, is fully compressed.
5. The fuel valves open and inject the fuel oil into the compressed air in the lower part of the cylinder, where it produces the power stroke in the opposite direction by sending the piston upward.
6. The fuel oil has ignited and the expanding gases have sent the piston upward, producing the other power stroke; at the same time, it compresses air in the upper part of the cylinder for the repeat stroke.
7. The burned gases escape and the piston continues upward to fully compress the air, and the cycle is repeated; thus power sends the piston first in one direction, then in the other.

How does ignition of the fuel differ in diesel engines?

Answer: In the gasoline engine, an outside igniting device, such as a spark, is needed to set the gasoline and air mixture on fire. In the diesel engine, the fuel oil is ignited as it comes in contact with the hot compressed air in the cylinder.

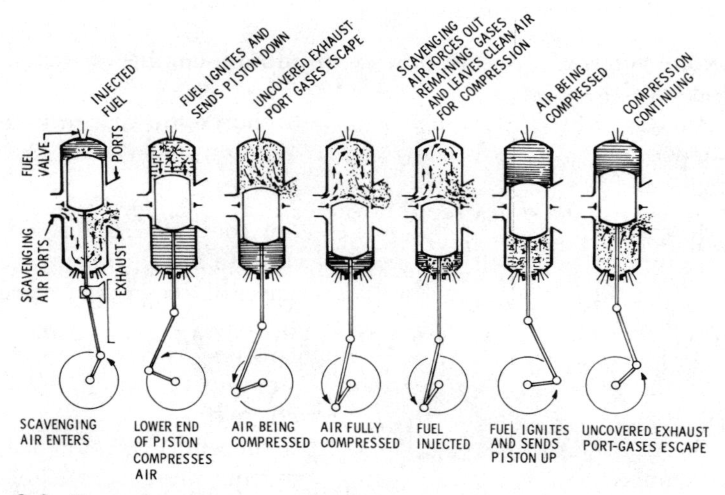

Fig. 2-6. Operating principle of double-acting, two-stroke-cycle diesel engine.

How do diesel engines differ from gasoline engines in the burning of the fuel?

Answer: In the gasoline engine, the spark ignites the mixture causing an explosion and an immediate rise in pressure. In the diesel engine, the lighter bodies of the fuel oil begine burning as soon as they enter the heated air and the particles gradually become fully ignited, thereby creating a gradual increase in pressure in the cylinder of the engine.

Why is the diesel engine known as a constant-pressure engine?

Answer: Because the fuel oil ignites as soon as the first particles enter the cylinder, and it continues to ignite, burn, and expand during the entire admission period of the fuel oil. The created pressure moves the piston, and as more pressure is created, the piston continues moving and giving more space to the expanded gases. The result is an almost constant pressure on the piston.

Why is the gasoline engine known as a constant-volume engine?

Answer: Because a volume of gasoline and air mixture enters the cylinder and is compressed, then ignited by a spark. This creates an explosion and an instantaneous high pressure that decreases quickly as the piston moves downward and gives more space to the burned gases.

How is compression brought about in the cylinder of the four-stroke-cycle diesel engine?

Answer: By the piston drawing air on its downward stroke and compressing it on its upward stroke.

How is compression brought about in the cylinder of the two-stroke-cycle diesel engine?

Answer: By the air being forced into the cylinder under two or three pounds of pressure, while the piston is at the bottom of the stroke, and compressed by the piston on its upward stroke until the heat of compression ignites the injected fuel oil.

What is a four-stroke-cycle engine?

Answer: A four-stroke-cycle engine is one in which four strokes of the piston are required to complete the necessary series of events required to produce one power stroke.

What action occurs after the piston has completed its power stroke in a four-stroke-cycle engine?

Answer: The piston starts on the return or exhaust stroke, forcing out the burned gases.

What action occurs after the piston has completed its power stroke in a two-stroke-cycle engine?

Answer: Air under low pressure is admitted which clears or scavenges the cylinder of burned gases and supplies fresh air for compression in the next cycle.

What is a cycle?

Answer: A cycle is a series of events repeated in regular order in the operation of an engine.

Name the series of events in their order of occurrence in the diesel engine cycle.

Answer: Air intake; air compression; fuel injection; fuel ignition; expansion of gases; exhaust and scavenging.

What is an air-intake, or air-inlet, valve?

Answer: A valve that admits air into the cylinder for compression.

What is meant by exhaust?

Answer: The removal of the burned gases after the fuel oil has been burned and the expanding gases have done their work in producing the power stroke.

What is scavenging?

Answer: Clearing a cylinder of exhaust gases by forcing into it a current of air, which provides clean air for the next compression stroke.

What are the four strokes in a four-stroke-cycle diesel engine?

Answer: The first stroke is the suction or intake stroke; the second—the compression stroke; the third—the expansion or power stroke; the fourth—the exhaust or scavenging stroke.

Describe the operation of a four-stroke-cycle diesel engine.

Answer: The clean air is drawn into the cylinder by the downward movement of the piston on its intake stroke. The air is compressed by the upward movement of the piston or compression stroke. At the height of compression, fuel oil is injected which ignites, creates gas that expands and forces the piston downward—this is the expansion or power stroke. The burned gases are forced out by the next upward or scavenging stroke of the piston.

Describe the operation of a two-stroke-cycle diesel engine.

Answer: Air is forced into the cylinder while the piston is at the bottom of the stroke and it is compressed on the upward stroke. Fuel oil is injected and ignited, the burned gases are exhausted, and the cylinder is scavenged on the downward stroke. All these steps are accomplished in just two strokes of the piston.

Where do the cylinders get air for compression?

Answer: From the atmosphere. Air is drawn in by the suction, or intake, stroke of the piston, in the four-stroke-cycle engine, or it is forced in under slight pressure, in the two-stroke-cycle engine. A supercharger, or a blower, may be used to provide an adequate supply of scavenging air for the cylinders.

How do the cylinders get fuel oil for ignition to create the power stroke?

Answer: By forcing the fuel oil into them either by injection pumps or by high fuel oil pressure. The fuel oil enters the cylinder through fuel injectors which spray the fuel into the cylinder.

What is meant by turbulence as applied to diesel engine operation?

Answer: It means the disturbance or agitation of the sprayed fuel oil and the air within the combustion chamber or cylinder.

What methods are used to provide the air turbulence?

Answer: Designs of spray nozzles, piston heads, and precombustion chambers, which tend to create more violent agitation.

How does turbulence aid combustion?

Answer: By causing a more thorough mixing of the hot compressed air with the injected fuel oil, thereby bringing about more complete combustion and more even cylinder pressure.

What is a precombustion chamber on a diesel engine?

Answer: A small chamber directly connected with the cylinder through a restricted channel into which the fuel oil is injected before it reaches the main power cylinder. Many of the old diesel engines were designed with precombustion chambers. Some of the later designs use an intake-air preheater to aid in starting at cool temperatures. Preheater equipment usually consists of a hand priming pump, to pump fuel oil into the intake manifold, and a glow plug, which is electrically heated by the battery. Fuel oil is burned in the intake manifold. This heats the intake air.

What are the advantages of a precombustion chamber?

Answer: The fuel oil injected into a precombustion chamber is more thoroughly atomized and also partly ignited in advance of reaching the cylinder. This produces a more uniform cylinder pressure and a smoother running engine.

How does a precombustion chamber cause better atomization and more uniform pressures?

Answer: It creates a better turbulence—a better mixing of the fuel oil and air—through the whirling motion given to the air and fuel, and forces the fuel mixture through the narrow neck between the chamber and the cylinder.

Explain the process that air and fuel oil undergo when a cylinder is equipped with a precombustion chamber.

Answer: The piston, compressing the air, forces it through the narrow neck which gives it a whirling motion around the chamber. At the same time the friction produced by the air as it is forced through a narrow neck gives it a higher temperature. This ignites particles of the fuel oil entering the chamber, which in turn creates a pressure and sends the mixture forcibly into the cylinder spreading itself to all parts, thus creating a more uniform combustion and pressure. Design of pistons and cylinder heads has been altered to increase the whirling motion and give improved turbulence.

What is meant by supercharging?

Answer: Increasing the total amount of charging air in the working cylinder of the engine.

What is the purpose of supercharging or increasing the supply of charging air?

Answer: It increases the power which may be required at intervals and gives more power where space does not permit larger engines.

How does supercharging increase the power of the engine?

Answer: By the increase in amount of air which provides the fuel oil with more oxygen for combustion of a larger charge of fuel.

How is an engine supercharged?

Answer: By an air compressor pump or blower, which supplies the additional air.

What is meant by a single-acting engine?

Answer: A single-acting engine is one in which the pressure that produces the power stroke is exerted upon only one side of the piston. Single-acting engines work on both the two- and four-stroke-cycle principle (Fig. 2-7).

What is meant by a double-acting engine?

Answer: A double-acting engine is one that operates similarly to a single-acting engine, except that pressure producing the power strokes is exerted first on one side of the piston, then on the other end, which makes each piston stroke a power stroke (Fig. 2-8).

Describe the operation of a two-stroke-cycle diesel engine that has the charging-air and scavenging-air valves located in the head and the exhaust ports located at the side.

Answer: Charging air, under pressure, enters through valves in the cyclinder head and is compressed by the upward stroke of the piston. Near the top of the stroke, fuel oil is injected, burns, creates gas that expands and forces the piston downward, producing the power stroke. Near the end of the power stroke, the piston un-

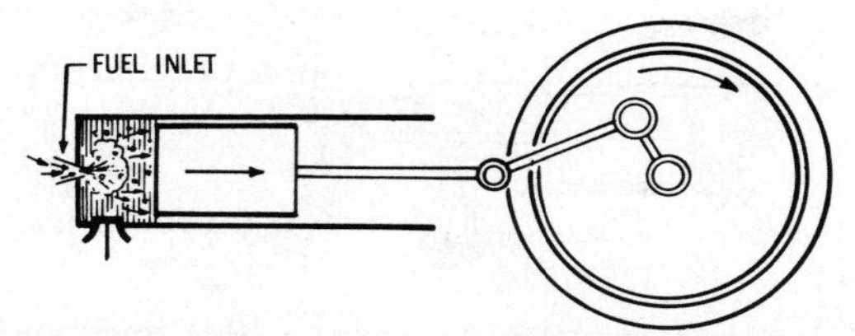

(A) Pressure against the piston head forces the piston outward.

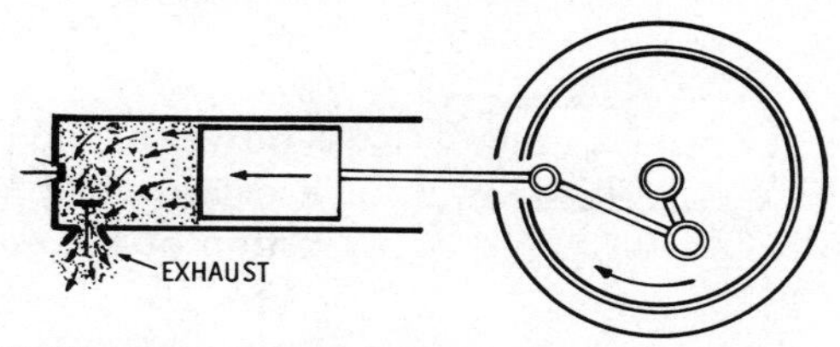

(B) Momentum of the flywheel returns the piston.

Fig. 2-7. Principle of a single-acting engine. These engines depend on the momentum of the flywheel to return the piston for the next power stroke. The force for the power stroke is applied to one end of the piston. Two-stroke-cycle, single-acting engines produce one power stroke per revolution of the crankshaft. Four-stroke-cycle, single-acting engines produce one power stroke per two revolutions of the crankshaft.

covers an exhaust port at the side of the cylinder, which permits the gases to escape. The scavenging- and charging-air valves, in the head, then open and air, under pressure, scavenges the cylinder and refills it for the next cycle (Fig. 2-2).

Describe the principle of operation of a two-stroke-cycle diesel engine with the air-inlet port at the side of the cylinder and the exhaust valve in the head.

Answer: Charging air enters through the side port in the cylinder and is compressed by the upward movement of the piston. Fuel oil is injected, burns, creates gas that expands and forces the piston downward, producing the power stroke. Near the bottom of the stroke, the exhaust valve opens, permitting the exhaust gases to escape. Then the side port is uncovered by the piston, admitting air under pressure, which scavenges the cylinder and refills it for the next cycle (Fig. 2-3).

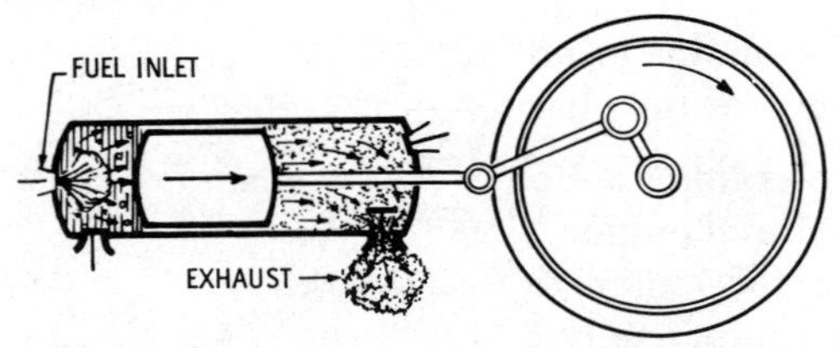

(A) Pressure against the piston head forces the piston outward.

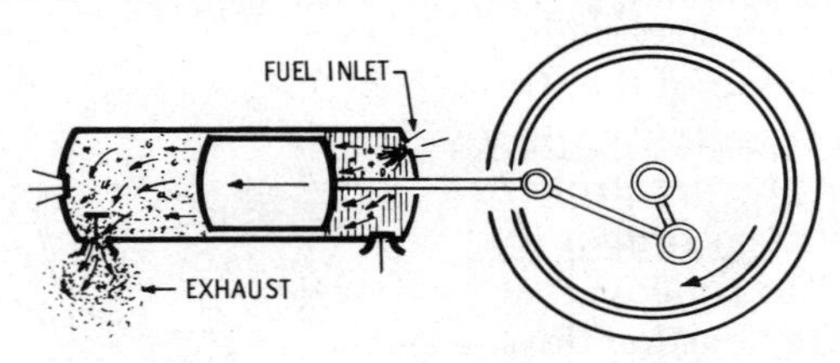

(B) Pressure against the piston base forces the piston inward.

Fig. 2-8. Principle of a double-acting engine. These engines have force applied at each end of the piston alternately, which produces twice the number of power strokes as single-acting engines. Two-stroke-cycle, double-acting engines produce two power strokes per revolution of the crankshaft. Four-stroke-cycle, double-acting engines produce one power stroke per revolution of the crankshaft.

Describe the port-charging and port-exhaust type of two-stroke-cycle diesel engine.

Answer: This type has a scavenging-air and charging-air port opposite that of the exhaust port at the side of the cylinder, at or near the end of the piston stroke. The opening of the exhaust port is just above the opening of the scavenging-air port. The piston, on its downward stroke, uncovers the exhaust port, permitting some of the exhaust gases to escape. The scavenging-air and charging-air port then is uncovered, which clears out the remaining gases and refills the cylinder with clean air (Fig. 2-4).

Describe the double-port scavenging and charging-air type of engine.

Answer: In this engine the scavenging-air, charging-air, and exhaust occurrences are similar to that in the single-port type except that additional charging air is admitted through a series of valves in a port located in the side of the cylinder above the charging-air port. These valves are held closed by pressure in the cylinder and remain closed until the pressure drops below that of the air received or the charging-air pressure.

Describe the crankcase-compression and receiver type of two-stroke-cycle diesel engine.

Answer: In this engine the crankcase is used as an air-compression chamber and receiver. The piston, on the upward stroke, compresses air in the cylinder and at the same time draws air into the crankcase for scavenging and refilling the cylinder. When the piston is at the top of the stroke, fuel oil is injected, which ignites, burns, creates gas that expands and forces the piston downward, producing the power stroke and at the same time compressing the air that had been drawn into the crankcase. On its downward stroke, the piston uncovers an exhaust port, permitting the burned gases to escape. Farther downward it uncovers the scavenging-air port that leads from the crankcase, and compressed air from the crankcase forces out the remaining gases and leaves clean air in the cylinder for the next compression.

Describe the opposed-piston type of two-stroke-cycle engine.

Answer: In the opposed-piston type of engine, two pistons work in each cylinder in opposite directions (Fig. 2-9). When they come together, they compress a charge of air; into this air is injected fuel that, when ignited, expands and forces the pistons apart. In their travel they uncover first an exhaust port and then a scavenging port. The gases are expelled, the cylinder is cleared, new air in the cylinder is compressed, and the cycle is repeated. In order to accomplish this, the pistons are connected to crank journals set at a 180° angle. The upper piston is connected to the crankshaft of the lower piston by side rods (Fig. 2-9).

What is the temperature of the air in the cylinder when the diesel engine is operating at full load?

Answer: Between 800° and 1,100° Fahrenheit.

What are the exhaust gas temperatures at normal operating conditions?

Answer: Between 400° and 700° Fahrenheit. They are usually lower in two-stroke-cycle engines because the scavenging air of two-stroke-cycle engines somewhat reduces the temperature.

How are the temperatures of exhaust gases measured?

Answer: By high-temperature thermometers or pyrometers.

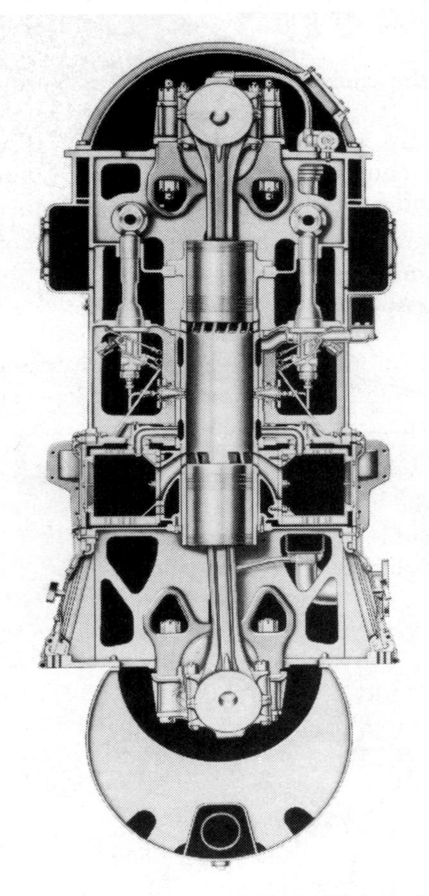

Courtesy Fairbanks, Morse & Co.

Fig. 2-9. Transverse of two-stroke-cycle, opposed-piston Fairbanks-Morse Model 38D8 ⅛ diesel engine.

Where are thermometers or pyrometers installed to indicate the exhaust temperatures?

Answer: In the exhaust pipes or manifolds of the various cylinders.

Why is it important to have the temperatures of the various cylinders as nearly uniform as possible?

Answer: To be sure that each cylinder is getting its share of air and fuel and is delivering its share of power.

What may be the cause of variations in cylinder exhaust temperatures?

Answer: Faulty valves, varied fuel injection pressures or amounts, faulty combustion, faulty operating mechanism of the various parts, and faulty or clogged exhaust piping (Fig. 2-10).

How can faulty or clogged exhaust piping affect the temperature of the exhaust gases?

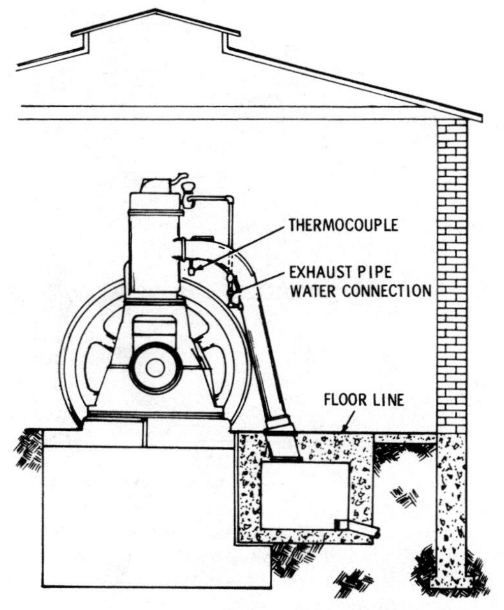

Courtesy Fairbanks, Morse & Co.

Fig. 2-10. The underground conduit method of conducting exhaust gases to the atmosphere. The various cylinder exhaust pipes are connected to a conduit or closed trough arrangement. The trough has a chimney or stack at one end through which the gases escape to the air, and it is drained to a sewer to carry off water , which is often applied to cool and condense the exhaust gases.

Answer: By causing back pressure and friction of the exhaust gases.

What becomes of the exhaust gases after leaving the engine cylinder?

Answer: They are either wasted or utilized for heating water or producing steam for other operating purposes. They may be used for heating in connection with the engine or equipment (Fig. 2-11).

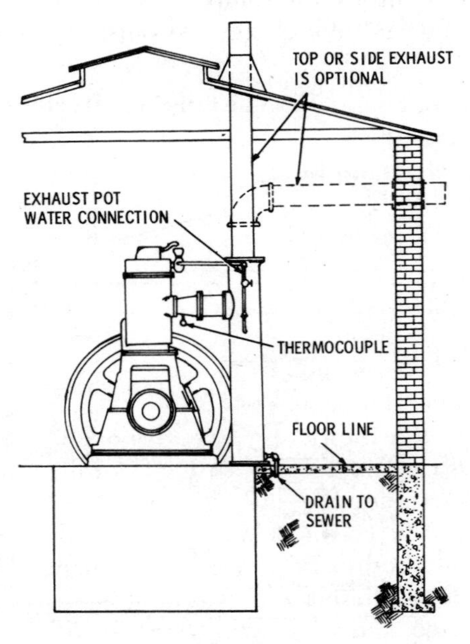

Courtesy Fairbanks, Morse & Co.

Fig. 2-11. The exhaust-pot method of conducting gases to the atmosphere. Two or three cylinders are exhausted through one exhaust stack. Several stacks are required to exhaust a multicylinder engine. Water drains are provided to carry off cooling water.

How is the noise of exhaust gases reduced?

Answer: By mufflers or silencers.

How is muffling or silencing of exhaust gases brought about?

Answer: By reducing their pressure before they reach the air and by having the gases strike a material that muffles their sound.

How are gas pressures reduced?

Answer: By having them enter pipes or chambers that are larger than the original ones, thus giving them space to expand before reaching the air, which reduces the force with which they strike the air.

What is the chief advantage of a diesel engine?

Answer: The lower fuel consumption combined with the lower cost of low-grade fuel oil results in a pronounced saving in operating costs.

What do you understand by the terms semi-diesel and full-diesel engines?

Answer: They are ridiculous misnomers.

State the basic difference between a diesel engine and a gasoline engine.

Answer: In the diesel engine, the heat for igniting the fuel oil is in the air in the cylinder prior to the introduction of the fuel. In the gasoline engine, the fuel is already mixed with the air during admission and is compressed prior to the introduction of heat—the electrical spark to ignite the fuel.

How does the method of breaking up the fuel oil into very fine particles differ in the diesel engine cylinder from that employed in the gasoline engine, and why?

Answer: Since diesel fuel oil is heavier than gasoline and much harder to vaporize because of its natural formation, it must be sprayed or atomized by mechanical means.

How is the fuel broken up in the gasoline engine cylinder?

Answer: Pneumatically. That is, it is forced out of a nozzle due to a difference in pressure at the inlet and the outlet of the nozzle. This pressure difference is created by the action of the engine during the intake stroke—objectionably and erroneously called the "suction" stroke. The fuel is sprayed into the combustion chamber by the injector and not atomized. The term "atomized" in this instance is ridiculous and may be considered a loose term.

How do the essential parts of a diesel engine compare with those of a gasoline engine?

Answer: The diesel engine has many parts in common with the gasoline engine, but they are made heavier to adapt them to the more severe stresses of the diesel cycle. These parts in both engines are classed as the stationary parts and the moving parts.

What parts are different?

Answer: The diesel engine has a fuel-injection system, which replaces the carburetor, and a small combustion chamber for thermal ignition, which replaces the spark plug.

How can smooth operation be obtained in spite of the severe diesel cycle?

Answer: It is necessary to have a stiffer crankshaft and a more rugged crankcase in order to prevent any considerable deflection of the crankshaft.

How can this deflection be prevented?

Answer: By reinforcing the crankcase just above the surface of the oil-pan flange (Fig. 2-12).

Describe the common type of reinforcement.

Answer: Typical practice consists of a special type of through-bolt construction in which the main bearing caps, crankcase, cylinder block, and cylinder heads are all bolted into one structure by means of long alloy-steel bolts. Not all engines have the through-bolt type of construction. These alloy-steel studs or bolts have an eccentric head or center portion which is sealed in a corresponding hole or recess in the top of the crankcase. This eccentric portion prevents the studs from turning. The main bearing end of the studs is of a larger diameter than that of the upper end. This permits the main bearing to be adjusted independently of any adjustment made on the top part of the engine. This construction also results in the stresses being placed on strong alloy-steel parts, the cast members all being under compression.

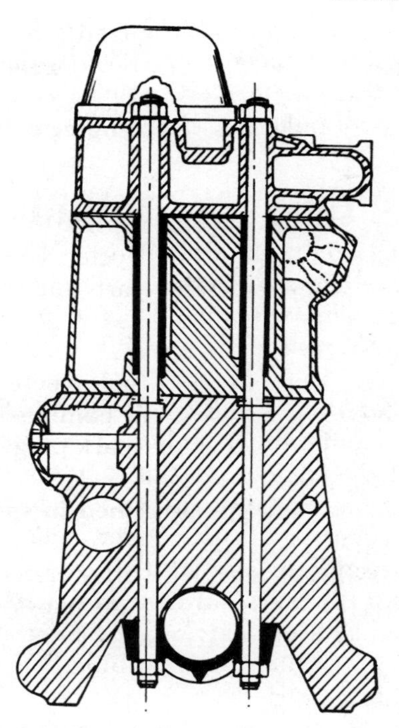

Fig. 2-12. Method of reinforcing the crankcase just above the surface of the oil pan flange.

How are the crankcase and cylinder block designed?

Answer: In some engine designs, they are a single casting with separate removable sleeves or liners fitted into the block (Fig. 2-13). These liners are called wet sleeves, which means that the cooling water comes in contact with the outside of the liners. The sleeves are made watertight in the block to prevent cooling water from reaching either the crankcase or the combustion space.

How are the sleeves sealed near the top?

Answer: By a ring-shaped copper gasket placed between the shoulder on the outside of the sleeve and the counterbore in the block.

What provision is made for watertight joints?

Answer: Near the bottom of each sleeve there are two ring-shaped rubber gaskets which fit into grooves in the bore of the block and fit tightly against the sleeve when it is pressed into place, which makes the joint tight.

Describe the dry sleeves.

Answer: Dry sleeves are removable sleeves or cylinder liners that are pressed into the cylinder bore and do not come in direct contact with the boiling water of the block. Therefore, they do not require special gaskets or sealing rings to prevent water from getting into the crankcase.

How are the cylinder block and crankcase usually constructed?

Answer: The cylinder block and crankcase (which is the main structural part of the engine) are typically box-like in form and may be cast in one piece of alloy cast iron. In mass production, the blocks for three-, four-, and six-cylinder engines are identical in design and dimensions, except for the necessary length for the additional bores of the four- and six-cylinder engines. The upper halves of the main bearing seats are cast integral with the block. Drilled passages in the block carry lubricating oil to all moving parts and eliminate piping.

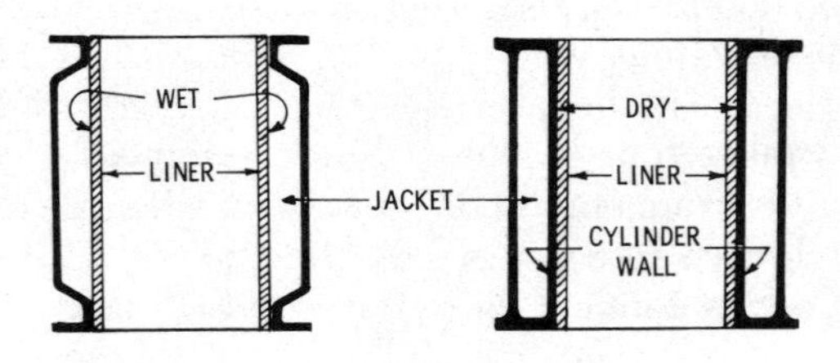

Fig. 2-13. Wet and dry cylinder liners or sleeves.

How are crankshafts usually constructed?

Answer: They are one-piece, high-carbon steel drop-forgings that have been carefully machined and heat treated. Surfaces of main bearing journals and connecting-rod bearing journals are hardened.

How many bearings are provided?

Answer: Two-cylinder engines may have two or three bearings; four-cylinder engines may have two, three, or five bearings.

What are these bearings called to distinguish them from other bearings?

Answer: Main bearings on the crankshaft main bearing journals and connecting-rod bearings on the connecting-rod journals.

What is the usual construction of these bearings?

Answer: They usually consist of a pair of removable Babbitt-lined shells having either a steel or a bronze backing.

What materials are used in piston construction?

Answer: Various materials are used: aluminum alloy, malleable iron, etc.

What is the advantage of the aluminum piston?

Answer: Lightness and high conductivity of heat. The latter property results in the heat from the top of the piston being distributed rapidly into the piston rings and the skirt of the piston.

How many piston rings are usually provided on diesel engines?

Answer: Six rings on each piston. The four top rings are conventional compression rings while the ring immediately above the wrist pin and the one in the piston skirt are oil control rings. Some manufacturers place four compression rings above the wrist pin and two oil control rings below the wrist pin to scrape off excess lubricating oil thrown onto the cylinder liner by the crankshaft and the lower end of the connecting rods.

What is the best location for the top piston ring, and why?

Answer: It should be located a considerable distance below the top of the piston so that it will not become too hot to permit satisfactory lubrication (Fig. 2-14).

How is the connecting rod usually made?

Answer: A connecting rod is made of drop-forged, heat-treated carbon steel and forged to an I-section with a closed hub at the upper end and an integral cap at the lower end.

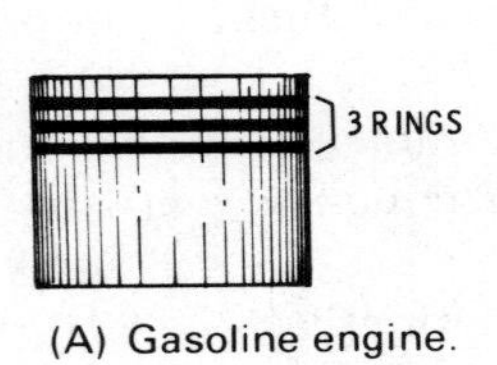

(A) Gasoline engine.

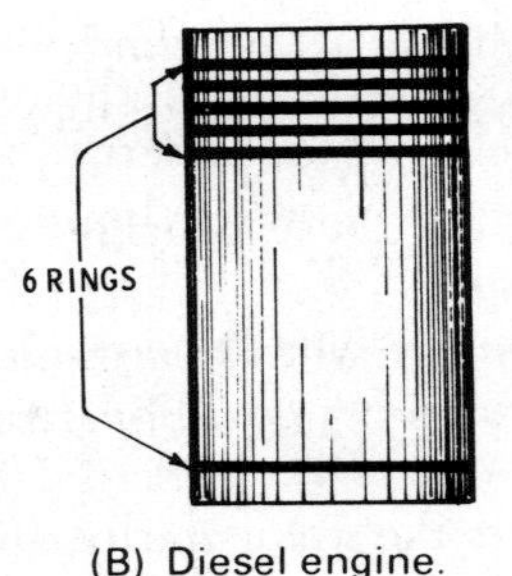

(B) Diesel engine.

Fig. 2-14. Differences in ring location on gasoline and diesel engines.

What provision is made for lubrication?

Answer: A connecting rod is rifle-drilled for lubrication of the upper end and is equipped with an oil spray jet for cooling the piston head. The lower end of the connecting-rod shank is fitted with an orifice which meters oil to the rifle-drilled duct. The crank journal bearings are shells of the precision type without shim adjustments. The upper and lower halves of the connecting-rod bearing shells are different, hence they are not interchangeable, but they are replaceable without machining.

What provision is made in the upper-crank journal bearing shell for lubrication?

Answer: It is grooved midway between the bearing edges a short distance upward from each parting line, with an oil hole through the shell at the termination of each groove.

How is the lower-crank journal bearing shell constructed for lubrication?

Answer: It has an oil groove in line with that of the upper bearing shell and circles the shell from parting line to parting line. These grooves maintain a continuous alignment with the oil holes in the crank journals. They supply lubricating oil to the wrist-pin bearings and the spray jet.

How is the wrist-pin bearing constructed?

Answer: It consists of a helically grooved steel-backed bronze bushing pressed into each side of the upper end of the connecting rod. A cavity of 3/16 inch between the inner ends of these bushings, aligned with the oil passage in the connecting rod, forms a duct around the wrist pin whereby the wrist-pin bearing is lubricated and oil can be forced to the spray jet.

What speeds are considered high speed?

Answer: As diesel engines operate from 100 revolutions per minute (100 rpm) to 2,000 or 3,000 revolutions per minute, it is difficult to draw a dividing line; but any speed over 1,000 rpm may be considered high speed.

How are high-speed diesel engines started?

Answer: By an electric motor generally; also, if a generator system is connected to the diesel engine, the diesel started first may be considered its own power source. The generator that the diesel is turning while running may have terminals connected to it. This will enable other diesels to be jump-started.

What are radial engines?

Answer: Engines in which the cylinders radiate from a central point as in the spokes of a wheel. Types of radial engines in use may have ten or more cylinders.

What kind of diesel fuel oil is used in high-speed engines?

Answer: The manufacturer's recommendation should be followed. Most manufacturers recommend No. 2D diesel fuel for high-speed engines. Diesel fuel oil should be clean, stable, and noncorrosive. Diesels use a lighter fuel oil than do heavy low-speed engines.

What grade of lubricating oil is suitable?

Answer: Again, follow manufacturer's instructions. Temperature is the most important factor. Most diesel engines require an SAE 30 lubricating oil. Extreme temperatures or conditions may require another weight.

How are high-speed engines operated and maintained?

Answer: Similarly to moderate-speed engines, as given in the chapter on engine operation (Chapter 23).

How many pounds of pressure are maintained by the fuel transfer pump, and how is it maintained?

Answer: A low pressure (12 to 60 lbs. depending upon design) is built up by a restricted fitting in the fuel oil return line.

What is the capacity of the fuel transfer pump?

Answer: It pumps more fuel oil than is required by the engine; the excess is returned to the tank (Fig. 2-15).

Describe the sediment sump.

Answer: This is a small chamber located at the bottom of the fuel tank; it has a drain cock to allow sediment and water to collect and be drained.

Why is the water at the bottom of the sump?

Answer: It sinks because water is heavier than fuel oil.

How many filters are usually provided, and what are they called?

Answer: Two, a first-stage and a second-stage filter. They may be called primary and secondary filters. Some manufacturers use a third filter, called the final-stage filter, through which the fuel oil passes just before entering the fuel injection pump (Fig. 2-16).

How do the valve mechanisms of gasoline and diesel engines compare?

Answer: Diesel engines have the same type valves as gasoline engines; also the valve gear has much in common with the gasoline engine gear (Fig. 2-17). As to location of valves, the valve-in-head engine is the prevailing type, although it requires more attention than the L-head engine. The valve-in-head construction requires more mechanism than the L- or T-head types, hence the valve gear is subject to more wear and requires more frequent adjustment.

What condition is necessary to permit the valves to seat properly?

Answer: A specified valve clearance must be provided because of the expansion of the valve stem and the engine parts when they get hot.

What may be the difficulty if a diesel engine will not start or is hard to start?

Answer: **Cause 1.** No fuel in tank.

Remedy. Refill fuel tank and bleed or vent the fuel injection system.

Cause 2. No fuel in fuel pump.

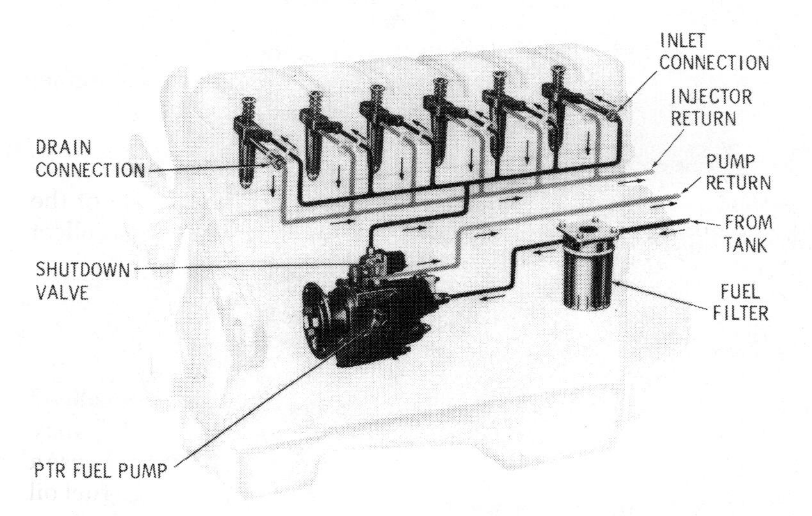

Courtesy Cummins Engine Co., Inc.

Fig. 2-15. PTR fuel-system flow on Models H and NH series Cummins Diesel engines.

What is the effect of too much valve clearance?

Answer: This affects the timing of the valves as well as the smooth operation of the engine, and also produces noise.

What is the valve clearance called in adjusting the valve gear?

Answer: Valve lash.

Remedy. When the engine has a hand priming pump: loosen the check-valve fitting and by using the hand priming pump pull the fuel from the tank and force it through the filter, located between the transfer pump and the injection pump.

It is best to leave vent cocks open until all the air is out of the system up to this point. Then close vent cocks and pump fuel into the injection pump until a solid stream of fuel oil comes through the opening created by loosening the check-valve fitting.

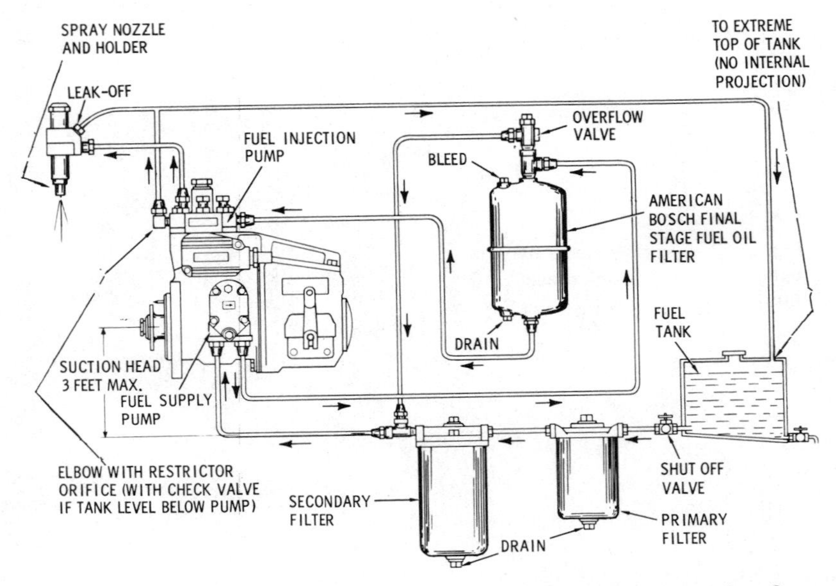

Courtesy American Bosch Arma Corp.

Fig. 2-16 Schematic of a typical through-flow fuel supply system. This filtering system is a progressive one, consisting of three filters in series. The primary unit is a metal-edge filter to remove heavy particles. The secondary filter has a cleanable cloth or felt element capable of removing practically all the fine abrasives that pass through the primary filter. The last, or final, stage-filter consists of a sealed or noncleanable unit located immediately before the injection pump and is capable of removing particles from the oil as small as three to five microns (0.00012 in. to 0. 00020 in.) that otherwise would be harmful to the injection equipment.

Be sure the cam that operates the fuel transfer pump does not hold the transfer-pump plunger in its outermost position, which will cause the hand priming feature to become inoperative.

Cause 3. Weak batteries will not turn the engine over rapidly enough.

Remedy. Recharge batteries.

Cause 4. Fuel oil too heavy to flow through pipes properly.

Remedy. Substitute a more freely flowing fuel oil.

Cause 5. Water in fuel.

Remedy. Drain fuel system and tanks. Change fuel supply.

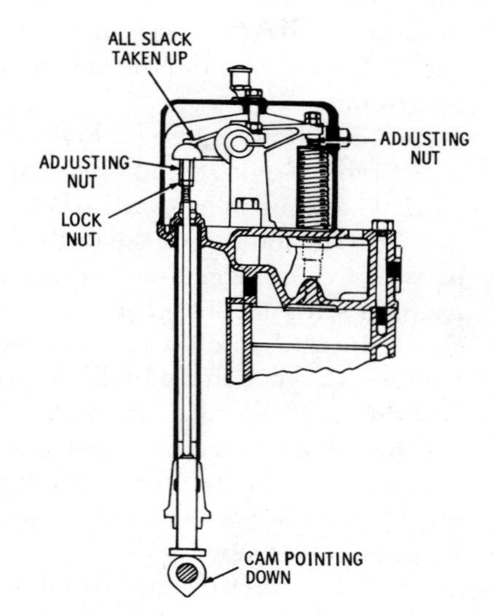

Fig. 2-17. Section at the exhaust valve of a Mack truck diesel engine.

Cause 6. Dirt in fuel between filter and pump.

Remedy. Replace the fuel filters. Clean the fuel tank.

Cause 7. Piston rings or cylinder walls worn badly.

Remedy. Replace with new parts.

Cause 8. Intake or exhaust valve seats pitted or worn.

Remedy. Regrind valves.

Cause 9. Leaking head gasket.

Remedy. Replace with new gasket.

Cause 10. Air cleaner plugged, not allowing sufficient air to pass through.

Remedy. Clean the air cleaner.

Cause 11. Air traps.

Remedy. Bleed the fuel injection system of air. Check for dirty fuel filters, insufficient fuel, and inoperative fuel transfer pump or an air leak in the inlet line (Buda) .

What may be the difficulty if the engine misfires on one or two cylinders?

Answer: First determine which cylinder or cylinders are misfiring. Loosen the nuts connecting the fuel lines to the fuel spray nozzles one at a time. If the engine speed remains the same and the exhaust sound is the same, that is the missing cylinder. If the engine speed slows down and the exhaust loses its rhythm, the cylinder is functioning.

Cause 1. Spray valve is clogged.

Remedy. Remove and clean. Spray nozzles should be cleaned by first soaking them either in kerosene or in clean fuel oil to soften the dirt. The interior of the body can be cleaned with a small strip of wood dipped in the cleaning oil and the spray hole can be cleaned with a pointed piece of wood. The spray nozzle valve should be rubbed with a clean oil-soaked (but not fluffy) soft rag. Neither hard nor sharp tools, emery paper, crocus cloth, grinding powder, nor abrasives of any kind should be used.

Cause 2. Air or gas in either the fuel pump or the fuel lines.

Remedy. Usually when testing to see which cylinder is misfiring, this condition will be cleared up as opening the nut allows the air or gas to escape.

Cause 3. Either exhaust or intake valve stuck.

Remedy. Remove the valve cover and check the stuck valve. Free the valve by pouring kerosene, gasoline, or alcohol down the valve stem. Alcohol is the quickest solvent. If the valve still sticks, remove the head and determine the cause.

Cause 4. Leaky intake or exhaust valve.

Remedy. Regrind valve.

Cause 5. Intake- or exhaust-valve spring or valve spring retainer key broken.

Remedy. Replace with new part.

Cause 6. Improper exhaust- or intake-valve clearance between the valve and the rocker arm.

Remedy. Check clearance and reset to proper clearance.

Cause 7. Fuel pump delivery valve leaking or stuck.

Remedy. Remove and clean with soft cloth and either clean fuel oil or gasoline. Do not use either an abrasive or a sharp tool on this part.

If cleaning does not free the valve, remove both the valve and the valve seat and install new parts. These valves and seats must be used as an assembly, as parts are not interchangeable, that is, one seat with another valve.

Cause 8. Fuel pump delivery valve spring broken.

Remedy. Replace with new part.

Cause 9. Either piston rings or cylinder walls badly worn.

Remedy. Replace with new parts.

What may be the difficulty if an engine misfires erratically or intermittently on all cylinders?

Answer: **Cause 1.** Improper fuel or fuel oil with poor burning qualities.

Remedy. Drain the fuel system including the fuel tank and refill with suitable fuel oil.

Cause 2. Water in the fuel.

Remedy. Drain the fuel system including the tank of all water and sediment. Refill with clean fuel oil.

Cause 3. Sticking injection nozzle valve stems, or fuel pump delivery valves, or both.

Remedy. Remove the stuck parts and clean. Usual cause is dirty fuel. Clean the entire system after draining and refill with clean fuel oil.

Cause 4. Worn piston rings, or cylinder sleeves, or both.

Remedy. Replace with new parts.

Cause 5. Leaky intake valves, exhaust valves, or both.

Remedy. Regrind the valves.

Cause 6. Plugged air cleaner reducing the volume of air admitted into the cylinders.

Remedy. Clean the air cleaner.

Cause 7. Piston seizure due to lack of lubrication.

Remedy. Remove the piston and replace with a new piston if badly scored. Also inspect cylinder sleeve for scores—replace if necessary. Change lubricating oil after thoroughly cleaning oil pan, oil liners, and oil filter.

Cause 8. Bearing seizure due to lack of lubrication.

Remedy. If not too badly warped, scrape clean, and reinstall. If badly warped, replace with new parts.

Cause 9. Broken fuel pump driving chain.

Remedy. Replace with a new driving chain and time the engine.

Cause 10. Fuel pump adjustable coupling slipped due to being improperly tightened.

Remedy. Retime the fuel pump.

CHAPTER 3

Fuel Oil

What factors have made diesel engine operating costs lower than those of the gasoline engine?

Answer: The use of cheaper fuel oil, the elimination of ignition devices, and the elimination of fuel and air mixing devices.

How is the fuel oil ignited in the diesel engine?

Answer: By direct heat produced by compressing air in the working cylinders. The fuel oil, upon coming in contact with the hot compressed air, becomes ignited.

How is the fuel oil mixed with the air for burning or combustion?

Answer: By being sprayed into the charge of hot air already compressed in the working cylinder.

What kind of fuel is used in diesel engines?

Answer: A low-grade diesel oil, similar to domestic furnace oil, but with the impurities removed. Domestic furnace oils are not considered satisfactory for most diesel engines.

What is crude oil?

Answer: An oil practically as it comes from the ground—unrefined. It is the raw product or the basic oil from which the higher oils, such as gasoline, are made.

What is petroleum?

Answer: Another name, or the technical name, for crude oil.

What is diesel fuel oil?

Answer: Diesel fuel oil is any oil that fits the requirements for use in diesel engines. Different types use different oils. Slow-speed engines can use a heavy fuel oil, while high-speed engines require a fuel oil that flows more easily. The heavy fuel oils remain after the high-grade oils have been removed or distilled from the crude petroleum.

Where is petroleum obtained?

Answer: From many countries. It is usually found in sandy and rocky formations. It has a heavy dark-brown color with a greenish tint.

What makes fuel oil burn?

Answer: The hydrocarbons in the oil—the carbon and hydrogen elements.

What is carbon and what is hydrogen?

Answer: Carbon is a chemical substance found in nature that burns and is the basis of lampblack, charcoal, other coals, and liquid fuels.

Hydrogen is a gaseous element that burns readily in its pure state but forms water when combined with oxygen. Two volumes of hydrogen and one volume of oxygen form water (H_2O).

What is the chemical composition of diesel fuel oil?

Answer: Fuel oil is composed of practically 85 percent carbon and 15 percent hydrogen. Fuel oil is not pure oil because it also contains other ingredients, such as water, sulfur, and ash.

Does fuel oil burn because of its own ingredients?

Answer: No. It requires the presence of oxygen in order to burn.

Where, in the diesel cylinder, does fuel oil get its oxygen for burning or combustion?

Answer: From the charge of hot air that is compressed in the cylinder to ignite the fuel oil.

Air is chiefly composed of what elements?

Answer: Of a mixture of oxygen and nitrogen—in the proportion of about one volume of oxygen to four volumes of nitrogen.

How many pounds of air are needed to burn one pound of diesel fuel oil?

Answer: Theoretically, about 14 pounds of air are needed to burn one pound of fuel oil. Actually, it requires about twice that amount of air to bring about complete combustion because of the impurities in the oil, the speed with which the oil is sent into the cylinder, and the incomplete burning of the oil of the previous charge.

What are the general qualities of a good diesel fuel oil?

Answer: Diesel fuel oil should be free from dirt, gum, water, and other foreign substances—especially water, which will work its way into the fuel pumps and hinder ignition and combustion.

What other qualities are required of a good diesel fuel oil?

Answer: It should be free from mechanical impurities held in suspension; such impurities will clog fuel strainers and score fuel-pump barrels and plungers (Fig. 3-1). It should flow freely at ordinary temperatures, should be free of highly volatile elements that produce inflammable gases, and should have a heating value of 18,000 to 19,000 Btu per pound.

Courtesy Caterpillar Tractor Co.

Fig. 3-1. Fuel filter element. Fuel oil entering the combustion chamber must be clean or unburned particles will remain in the cylinder. Dirt clogs the spray valves and prevents a full charge of fuel from entering the cylinder or combustion chamber.

How can an asphaltum-base fuel oil be distinguished from a paraffin-base fuel oil?

Answer: By evaporating a few drops on a piece of white paper. If it contains paraffin, it leaves no color; if it contains asphaltum, it leaves a brownish color.

What is meant by the flash point of an oil?

Answer: The flash point is the minimum temperature at which the vapors, given off by an oil when heated, will flash or catch fire when a flame is held just above the surface of the oil; but the oil itself will not catch fire. The vapor that rises, when mixed with the air, will burn or flare up and then the flame will die out. More vapor must rise, mix with the air, and be relighted to obtain another flash. Therefore, the flash point is the minimum temperature at which a fuel oil will give off vapors that will ignite but will not continue to burn.

What temperature is the flash point of diesel fuel oil?

Answer: According to ASTM (American Society of Testing Materials), No. 1 diesel fuel oil should have a flash point of 100° Fahrenheit; No. 2 diesel fuel oil should have a flash point of 125° Fahrenheit.

What is meant by the fire point or the burning point of a fuel oil?

Answer: The fire point is the minimum temperature at which a fuel oil will continue to burn when lighted. The continuous burning of an oil is dependent upon the vapors continuing to rise from the oil. Therefore, when an oil is heated to the flash point and heating is continued until the vapors, when ignited, continue to burn, the fire point has been reached. This point should be approximately 50° to 125° above the flash point of the oil.

How are the flash point and the fire point of a fuel oil determined?

Answer: Heat a small portion of the fuel oil in an open vessel. Hold either a lighted match or a candle above the surface and note the temperature at which the vapor ignites. A thermometer placed in the oil will give the temperature. The temperature at which the vapors will flare up and then die out is the flash point of the oil. Continue to heat the oil, and the temperature at which the vapors retain the flame and continue to burn is the fire point of the oil.

What are the specifications for a good diesel fuel oil for average use?

Answer: In addition to the ASTM classification, diesel fuel oil must be free from grit, fibrous substances, and other impurities or dirt that may destroy or clog the pumps, valves, strainers, pipes, injection nozzles, and other parts of the engine and equipment.

Diesel fuel oil should have a paraffin base—hydrocarbon oil. It should be heavy enough to be smooth and oily, yet it must flow freely at moderate temperatures. Hot climates require a heavier fuel oil as fuel oil thins with an increase in temperature. A lighter fuel oil that flows more freely is required in colder climates so that the fuel oil may be used without preheating.

The ignition quality of a diesel fuel oil should be high enough so that it ignites readily on coming in contact with the hot compressed air in the cylinder in order to avoid a delay that causes unburned fuel oil to remain in the cylinder as a residue, which may cause valves and piston rings to stick.

Atmospheric temperature, engine speed, load, and application of the engine influence selection of the fuel oil. Sulfur content, cetane number, and distillation range are three important properties for optimum service. For satisfactory operation, the engine manufacturer should be consulted for the grade of fuel oil suitable for a particular engine and operating conditions.

What is meant by viscosity?

Answer: The fluidity of an oil—the thickness or stickiness of it—the rate at which it will flow at a given temperature. Viscosity is determined by the number of seconds required for a measured quantity of oil to flow out through a measured opening.

What is the Saybolt method of determining viscosity?

Answer: A Saybolt viscosimeter is used to determine viscosity.

Describe the Saybolt viscosimeter.

Answer: The Saybolt viscosimeter is an instrument that consists, in its essential parts, of a cup with a small opening at the bottom through which the oil is to flow out. The cup is filled to a standard level which is 60 milliliters of oil, at the temperature desired, and it is then permitted to flow out. The number of seconds required for the oil to flow out is the measure of its viscosity.

Example: A good average fuel oil for a diesel engine is one that will flow out in from 32 to 70 seconds at a temperature of 100° Fahrenheit.

What is the Baumé scale of liquid measurement?

Answer: The Baumé scale is an arbitrary measurement that indicates the specific gravity, or weight, of a liquid by inserting into the liquid a weighted tube (called a hydrometer), which is marked off into divisions and then noting the depth to which it sinks. The scale is read at the level of the surface of the liquid. The lighter the liquid, the lower the tube sinks.

Baumé scale hydrometers for oil are known as for "liquids lighter than water" and marked accordingly. There are also Baumé scales for "liquids heavier than water." The comparison is made with distilled water at the temperature of 60° Fahrenheit—the oil is also tested at 60° Fahrenheit.

How is the hydrometer used for testing an oil?

Answer: The hydrometer method is a convenient form used for testing the specific gravity of an oil—fuel or lubricating. The oil is poured into a jar and a Baumé or API (American Petroleum Institute) hydrometer floated in it. The degrees are read at the surface of the oil.

The meaning of the degrees in specific gravity is ascertained from a table as given in the test. The basic principle of the hydrometer is that the lighter the oil, the lower the weighted tube will sink. The heavier oil will offer a more resistive force and keep the hydrometer from sinking so low.

Temperature must be taken into account. Therefore, a thermometer is inserted or a hydrometer is used which has a thermometer attached. Temperature is extremely important in determining specific gravity.

Specific gravity can be converted to a Baumé reading by the formula:

$$\text{degrees Baumé} = \frac{140}{\text{sp. gr.}} - 130$$

Baumé reading can be converted to specific gravity by the formula:

$$\text{specific gravity} = \frac{140}{130 + \text{degrees Baumé}}$$

Example: If the Baumé reading of an oil is 26°, the specific gravity of the oil would be calculated by the formula:

$$\text{sp. gr.} = \frac{140}{130 + 26} = .897$$

How are temperature corrections made for oil tests?

Answer: A rule for roughly converting the specific gravity at any temperature to that at the standard 60° Fahrenheit is: *For every 10°F. above 60°F., subtract one degree from the Baumé reading; and for every 10°F. below 60°F., add one degree.*

$$\text{degrees Baumé at } 60°\text{F.} = \text{degrees Baumé at indicated temperature} - \frac{(\text{indicated } °\text{F.} - 60°\text{F.})}{10}$$

Example: If a hydrometer indicates a specific gravity of 27.5° Baumé at an oil temperature of 75°F., what is the Baumé reading at 60°F.?

$$°\text{Baumé at } 60°\text{F.} = 27.5° - \frac{(75°\text{F.} - 60°\text{F.})}{10}$$
$$= 27.5° - \frac{15}{10}$$
$$= 27.5° - 1.5 = 26° \text{ Baumé}$$

Example: If the hydrometer indicates 24° Baumé at an oil temperature of 40°F., what is the Baumé reading at 60°F.?

$$°\text{Baumé at } 60°\text{F.} = 24° - \frac{(40°\text{F.} - 60°\text{F.})}{10}$$
$$= 24° - \frac{(-20)}{10}$$
$$= 24° - (-2°)$$
$$= 24^{0} + 2° = 26° \text{ Baumé}$$

Table 3-1. Specific Gravities, Degrees Baumé, and Degrees API
(at 60° Fahrenheit)

Degrees Baumé	Specific Gravity	Degrees API	Specific Gravity
10	1.0000	10	1.0000
15	.9655	15	.9659
20	.9333	20	.9340
25	.9032	25	.9042
30	.8750	30	.8762
35	.8485	35	.8498
40	.8235	40	.8251
45	.8000	45	.8017
50	.7777	50	.7796
55	.7568	55	.7587
60	.7368	60	.7389
65	.7179	65	.7201
70	.7000	70	.7022
75	.6829	75	.6852
80	.6666	80	.6690
85	.6511	85	.6536
90	.6363	90	.6388

What is the API (American Petroleum Institute) scale of measurement of petroleum products?

Answer: The API specific gravity measurement is a hydrometer measurement similar to the Baumé except in scale divisions.

Specific gravity may be converted to degrees API by the formula:

$$\text{degrees API} = \frac{141.5}{\text{sp. gr.}} - 131.5$$

Degrees API may be converted to specific gravity by the formula:

$$\text{specific gravity} = \frac{141.5}{131.5 + \text{degrees API}}$$

How is ignition quality of a fuel oil determined?

Answer: By the Critical Compression Ratio at which the fuel oil will fire within three seconds after being injected into the cylinder of an engine externally driven at 600 revolutions per minute, with the intake air maintained at 100°F. and the cylinder jacket at 212°F. The fuel feed is 9 milliliters per minute, the advance angle is 12 degrees, and the injection pressure is 1,500 pounds per square inch.

By the Measurement of Actual Ignition Delay, which is determined by means of a pressure-time indicator in a large, slow-speed, solid-injection engine, at a compression pressure of 450 pounds per square inch. This delay is compared with the delay obtained on mixtures of cetane and alpha-methyl-naphthalene. The percentage of cetane in the mixture that gives the same delay as the fuel under test is the *cetane number*.

Why is it important to know the flash point and the fire point of a fuel oil?

Answer: They are of importance in determining the fire hazard—the temperature at which the fuel oil will give off inflammable gases or vapors, the ease with which an oil will catch fire from an open flame or spark, and its suitability for use in the diesel engine.

What are the safe qualities of diesel fuel oil?

Answer: Diesel fuel oil is inert or lifeless, nonexplosive, difficult to ignite in bulk, and not subject to spontaneous ignition or combustion—that is, it will not catch fire of its own accord.

What are the dangerous qualities of diesel fuel oil?

Answer: The vapor is explosive when mixed with a proper amount of air. The vapor is heavier than air and tends to settle or collect in low places, such as bottoms of tanks, bilges, oil containers. The vapor is often unnoticed until it becomes ignited by an exposed flame or spark. The vapor is always present in partly filled tanks and in tanks which formerly contained fuel oil but have been emptied. Vapor is also present in vent openings when tanks are being filled. If a leak in an oil pipe is unnoticed and oil has accumulated in closed or out-of-the-way places where air does not circulate freely, a lighted match or spark may cause the vapors to ignite, resulting in a serious fire or explosion (Fig. 3-2).

How may carelessness cause ignition of the fuel vapors?

Answer: By a thoughtless manner in handling lighted matches,

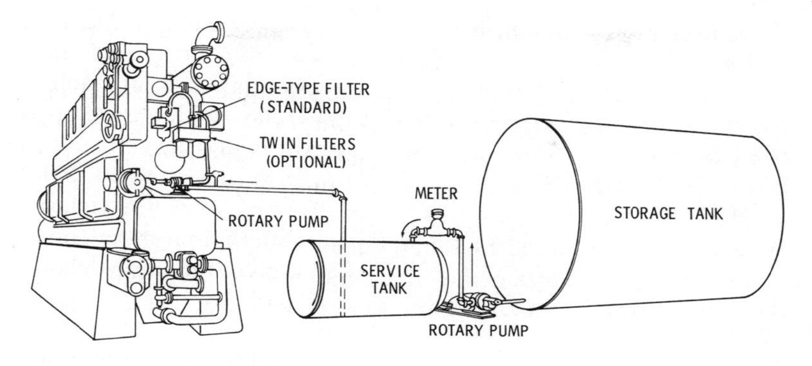

Fig. 3-2 Diagram of injection system and handling of fuel oil in a typical four-stoke-cycle diesel engine. Fuel oil is injected by means of high-pressure injection pumps and multiholed spray nozzles, operated on the differential pressure system, through the cylinder head directly into the combustion chamber. Fuel pumps are housed in an isolated, recessed section cast into the side of the engine frame. This section is sealed off from the crankcase. Overflow from the by-pass of the fuel nozzle drains into this section, where other excess oil drains off. To keep a constant pressure on the fuel injection-pump intake, an attached fuel oil service booster pump is supplied with each engine. The booster pump can take the fuel supply from a service tank located below the base line of the engine or from an elevated tank, whichever is more convenient.

cigarette butts, open lamps, lanterns, sparks from welding or cutting tools, broken electrical contacts, sparks from a fire in a stove, or from a chimney or galley pipe. There are innumerable ways in which thoughtlessness may cause a spark or fire to come near a vapor-laden tank or pocket.

What precautions should be taken to prevent accidental explosions and oil fires?

Answer: No one should be permitted to carry an open flame, a lantern, or an electrical device that may spark, or to smoke or to strike a match within 25 feet of an oil hose, tank compartment, fuel tank, or fuel tank vent.

Unprotected electrical lights, fuses, switches, or apparatus liable to produce sparks should not be permitted either in a fuel oil tank or in a pump compartment at any time.

What precautions should be taken before entering an empty fuel tank?

Answer: No one should enter an empty fuel tank until after it has been cleared of all gases either by purging with compressed air or by filling with water and then emptying. There should always be another person present in event of mishap.

What precaution should be taken when filling fuel tanks?

Answer: Make certain that the vents are not exposed to any flame or spark that might ignite vapors while the tank is being filled.

How can an oil fire be extinguished?

Answer: By the use of a foam or carbon dioxide fire extinguisher. Sand and steam are also effective in extinguishing oil fires. Water should not be used.

What care shall be given fuel oil tank vents?

Answer: The vents should be kept in good serviceable order and the strainers always be kept open and clear.

Should the valves of the oil tank gauge glass be kept open or shut?

Answer: The gauge glass valves should be kept shut except when testing for the height of the oil.

Is it a desirable practice to run fuel oil through a centrifugal purifier?

Answer: Yes. It removes impurities that often cause undue wear in the cylinders and it eliminates water, which interferes with the operation of the fuel pumps. The purifier also prevents ignition troubles.

CHAPTER 4

Diesel Injection Pumps

How does a diesel injection pump work?

Answer: The injection pump on a diesel engine may well be compared to the human heart in that it supplies the life blood that makes functioning of the engine possible. The diesel engine depends on the pulsations, timing, and operating pressure of the injection pump to furnish its life blood, or fuel, for successful operation.

What is the injection pump's major job?

Answer: The injection pump's major job is to create pressure to feed to injectors at such a rate that compression is accomplished in a split second.

What is the purpose of the type APF fuel injection pump?

Answer: The purpose of the APF fuel injection pump (Fig. 4-1) is to meter and deliver fuel accurately and under high pressure at a precise moment to the spray nozzle, which injects it into the engine cylinder.

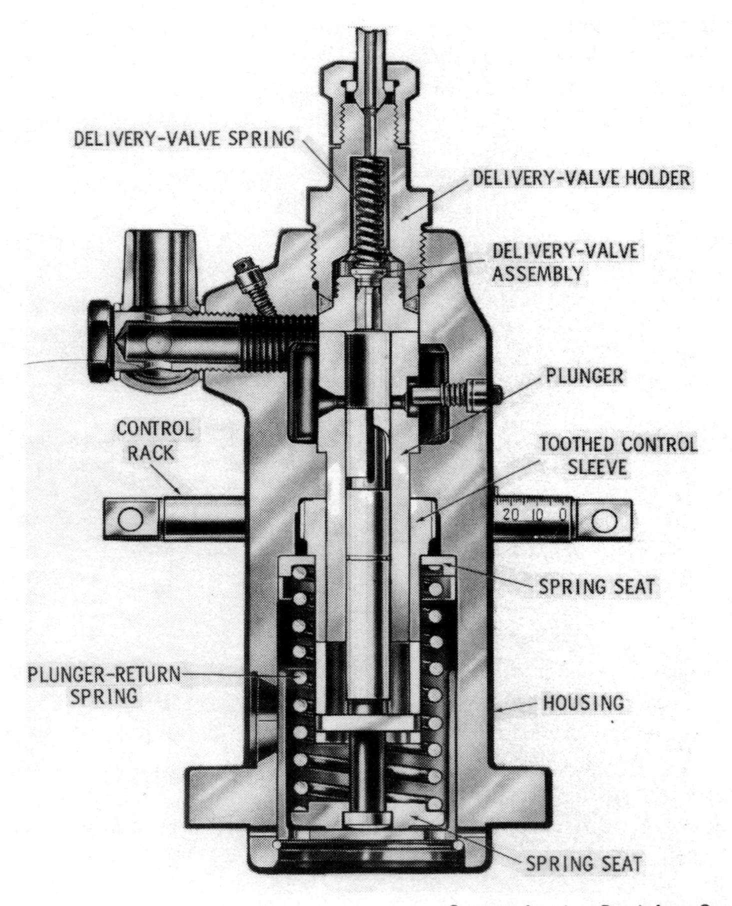

Courtesy American Bosch Arma Corp.

Fig. 4-1. Cross section of an American Bosch APF fuel injection pump. These pumps are of single-cylinder construction. One APF pump serves only one engine cylinder. A six-cylinder engine requires six of these pumps.

An engine must have one APF fuel injection pump for each cylinder; thus a six-cylinder engine requires six of these pumps. Pumps are made for engines having only a few horsepower to several hundred horsepower per cylinder.

What is the action of the nozzle holder and the spray nozzle?

Answer: The nozzle holder contains a spring and a means of pressure adjustment to provide for proper action of the spray-nozzle valve (Fig. 4-2). It also provides a means of conducting fuel oil to the nozzle and holds the spray nozzle in its correct position.

The fuel injection pump meters the quantity of fuel oil entering the nozzle holder through an inlet connection. The fuel passes through connection ducts to the pressure chamber just above the

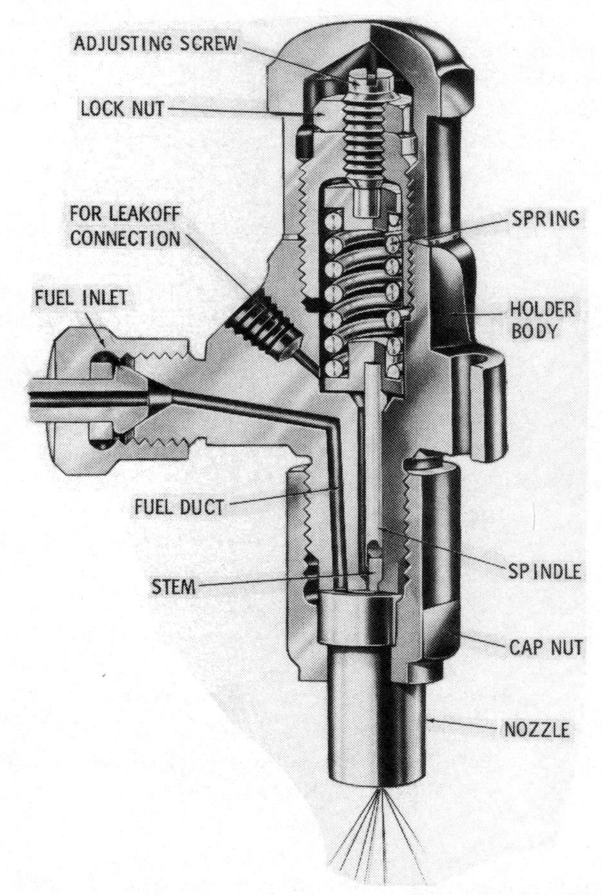

Courtesy American Bosch Arma Corp.

Fig. 4-2. Cross section of an indirect-cooled nozzle holder and spray nozzle.

spray-nozzle valve. At the instant fuel pressure acting on the spray valve differential area exceeds the predetermined spring load, the valve lifts from its seat and fuel oil flows until delivery from the injection pump ceases. The spring force snaps the valve to its seat instantaneously to cut off fuel and avoid any possibility of dripping or dribbling of fuel.

Misfiring of a single cylinder indicates improper functioning of a spray nozzle. Spray nozzles must have periodic cleaning, depending upon operating conditions, fuel used, and other factors. Loosening the high-pressure line fitting on each spray nozzle in turn, which prevents fuel entering the cylinder, will aid in locating the faulty spray nozzle. The cylinder that least affects engine performance usually has the sticking or improperly functioning spray nozzle.

What is the purpose of the APE fuel injection pump?

Answer: The APE fuel injection pump (Fig. 4-3) accurately meters the fuel and delivers it under high pressure at a precise moment to the spray nozzles by which the fuel oil is injected into the respective engine cylinders. A single type APE pump is needed for a multicylinder engine. Only one type APE six-cylinder fuel injection pump is required for a six-cylinder engine.

What is the function of a continuous-pressure fuel injection pump?

Answer: To maintain a continuous pressure of fuel oil in fuel accumulators or in fuel lines. In straight-rail systems, it supplies fuel at constant pressure to the spray nozzles.

What is meant by a constant-stroke fuel injection pump?

Answer: A constant-stroke fuel injection pump (Fig. 4-4) is one that delivers a definite quantity of fuel on each stroke; the quantity needed to maintain the speed of the engine is controlled by a governor mechanism that controls the suction valve, by-passing part of the fuel back to the storage tank; or the control may be by hand.

How is the length of an injection pump stroke controlled?

Answer: By means of an eccentric or quadrant, which is a hand adjusting device that lengthens or shortens the pump stroke, and

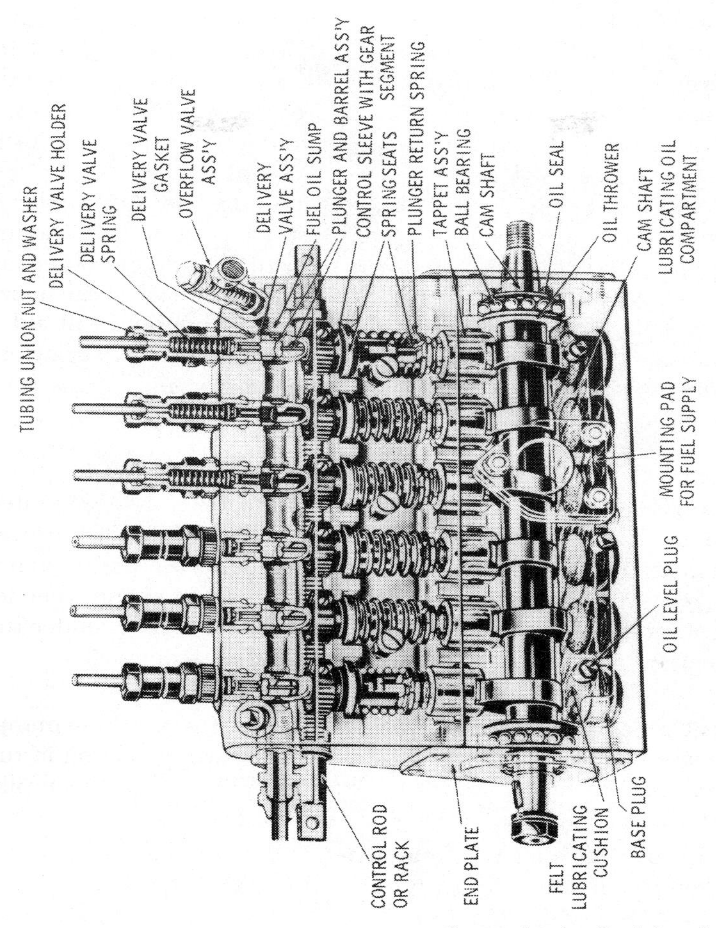

Courtesy American Bosch Arma Corp.

Fig. 4-3. American Bosch APE six-cylinder injection pump. These pumps are of multicylinder construction. One APE six-cylinder pump serves a six-cylinder engine.

by the engine governor, which moves a wedge and automatically either lengthens or shortens the stroke (Fig. 4-5).

How is fuel oil, under pressure, provided for starting an engine when the main fuel pump is directly connected to the engine?

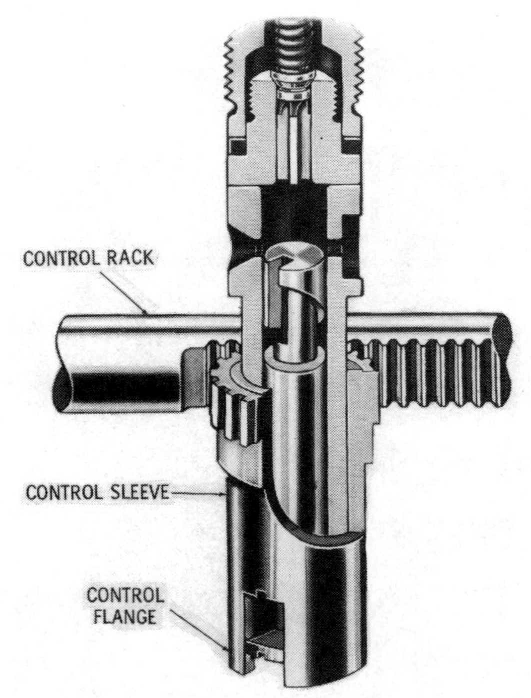

Courtesy American Bosch Arma Corp.

Fig. 4-4. Sectional view of fuel injection pump element. The plunger is rotated by its flanges, which engage slots in the control sleeve. The control-sleeve gear is actuated by the control rack.

Answer: By a hand gear attached to the pump or by a separate pump provided for the purpose.

How can one tell whether all the cylinders in an engine are getting fuel and are firing?

Answer: By the temperature reading of the exhaust, by training the ear to detect differences in sound of the various cylinders, and by feeling the water jackets around the cylinders. A cold water jacket indicates that the cylinder is misfiring.

What methods are used to control the fuel supply from the injection pump to the spray nozzles?

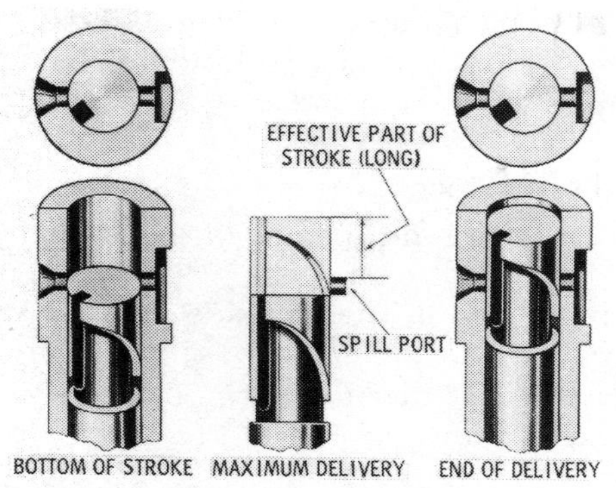

(A) Plunger rotated in position of maximum fuel delivery, where effective part of the stroke is relatively long.

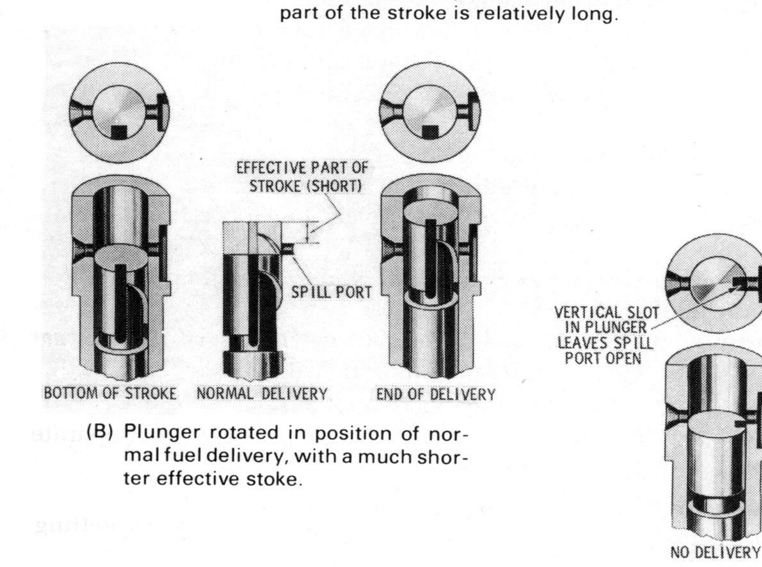

(B) Plunger rotated in position of normal fuel delivery, with a much shorter effective stoke.

(C) Plunger rotated so that the vertical slot is in line with the spill port and the entire stroke is noneffective, resulting in no fuel delivery.

Courtesy American Bosch Arma Corp.

Fig. 4-5. **Metering principle. The amount of fuel delivered is controlled by rotating the plunger, thus changing the length of its effective pumping stroke.**

Answer: Throttling of the suction inlet valve, releasing fuel by a spill valve, and by-passing fuel from the discharge end of the injection pump to the suction end.

Which method of fuel control is most generally used?

Answer: Controlling the opening of the suction valve.

Why is close regulation of fuel possible with the suction-control type of pump?

Answer: By the fact that the suction valve is not subject to severe mechanical strains.

What effect has a slow or delayed action of the fuel suction valve on the operation of the engine?

Answer: Slowed action of the valve delays injection of the fuel oil into the cylinders at the proper time and in the proper quantity; this is quite noticeable when engine loads vary.

How is fuel injection maintained evenly in all cylinders?

Answer: By means of adjustment of the adjusting screws that control the time and flow of fuel into each cylinder.

If one or more fuel spray-nozzle tips were plugged, how would it be indicated?

Answer: By the lowered temperature of the cylinder cooling water and the exhaust gases of that particular cylinder. Also by the unbalanced operation of the engine, which may be detected by the sound of the engine.

What keeps the fuel pressure constant in an accumulator?

Answer: A spring-loaded escape valve.

What is the principal difficulty encountered in fuel accumulators?

Answer: Keeping the escape valve tight.

What method is a good practice for refitting escape valves?

Answer: Face off the disk in a lathe, then caseharden or temper it. Make it tight by hammering gently while turning the stem.

How is the fuel oil supplied to the cylinders in the individual pump fuel injection system?

Answer: In the individual pump fuel injection system (Figs. 4-6 and 4-7), fuel oil is supplied to the cylinders by individual pumps

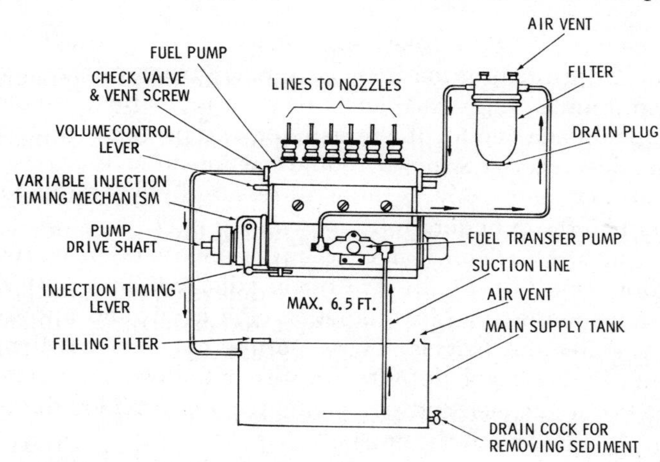

Fig. 4-6. **Individual pump fuel injection system.**

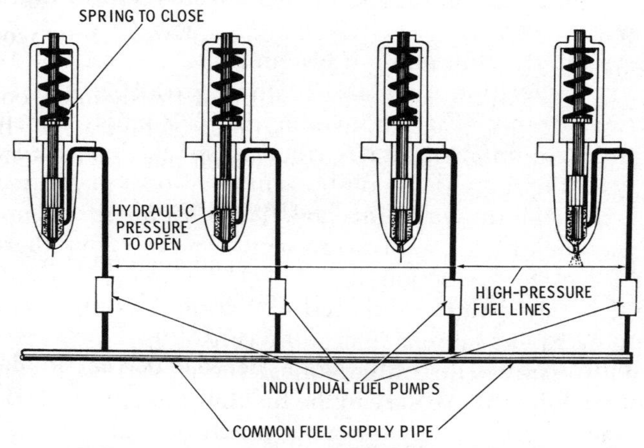

Fig. 4-7. **Hydraulic spray nozzles. A separate pump is used for each cylinder. The pumps work at pressures from 1,000 lbs. to 10,000 lbs. per sq. in. The hydraulic valves are opened by air pressure and closed by a spring.**

operated from cams either on a camshaft or on an auxiliary drive, and the pumps are independent of each other in their operation.

Fuel oil is drawn from the main tank by a fuel supply or operating tank, or directly to the injection pump supply pipe. From there it is taken up by the pump plungers and injected into the cylinders in proper quantity and at a prearranged time.

The pump plunger strokes are usually of the constant-stroke type and adjustment is made by by-passing being different with different pumps. Individual injection pumps can be taken out of service without affecting the remainder; and pressures in the system are low except when a plunger is injecting fuel into the cylinder; and, then only, is there a pressure between the pump plunger and the cylinder.

Filtering the fuel oil is a very important function in engine operation. Injection pumps are finely adjusted, have very little play between parts, and would soon clog if dirt and impurities were to enter the barrels. These pumps must send a definite amount of fuel oil, at a definite pressure, into the engine to maintain it at proper speed and power and for uniform cylinder pressure in all cylinders of the engine.

How does the common-rail fuel injection system supply fuel to the cylinders?

Answer: In the common-rail system (Figs. 4-8 and 4-9), fuel is supplied to a fuel pipe or manifold common to all the cylinders. A high-pressure pump (Fig. 4-8) maintains a constant pressure in this manifold or common rail, as it is called, from which taps are taken off to feed the individual combustion chambers of the various cylinders. Fuel is drawn from a tank through a filter to an operating or day tank by a low-pressure suction pump; from there the fuel oil travels to the suction side of the high-pressure pump from where it is raised to the required operating pressure—3,000 to 5,000 pounds per square inch or more (Fig. 4-9).

A high-pressure relief valve, adjustable to various pressures, maintains a constant pressure in the fuel rail and spray nozzles. If the pressure increases beyond a set point, the valve opens and permits some of the fuel to flow back, or by-pass, into the tank. This system is also called the "constant-pressure system." A low-pressure relief valve by-passes the fuel on the low-pressure or

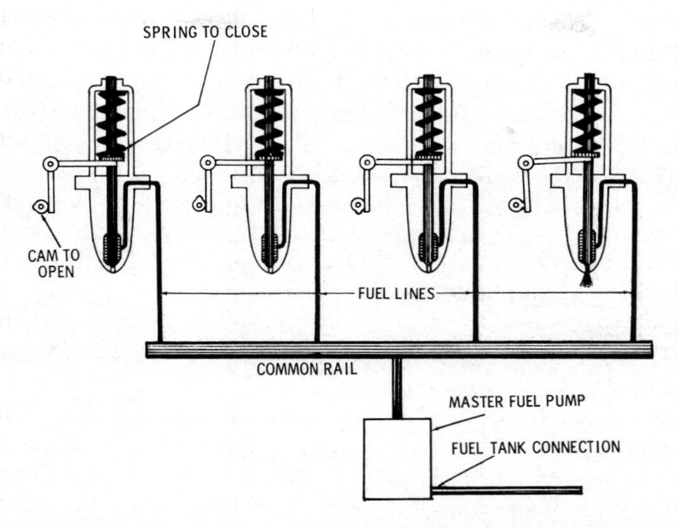

Fig. 4-8. Mechanical spray nozzles. A single high-pressure pump supplies fuel for all cylinders in the common-rail fuel injection system. The mechanical spray valves are opened by a cam gear and closed by a spring.

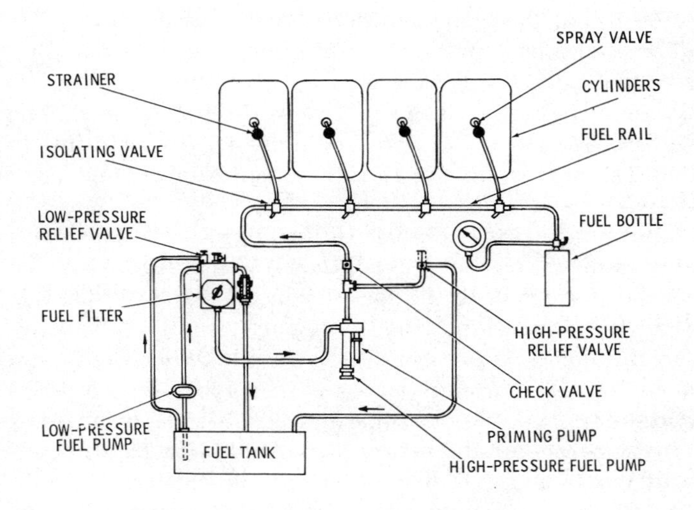

Fig. 4-9. Common-rail fuel injection system.

transfer pump side. Check valves prevent the return of the fuel oil by cylinder or back pressure. Cutoff valves cut off the fuel feed to a cylinder if it becomes necessary to do so for repairs or replacement of parts or nozzles. An accumulator, or fuel bottle, serves as a cushion to smooth out the pulsations produced by the pump. Pressure gauges indicate the pressures in the line. The fuel injector spray nozzles are opened mechanically (Fig. 4-8) and permit the fuel under pressure to pass through.

What is the operating principle of the Roosa Master type DB fuel injection pump?

Answer: A vane-type fuel transfer pump draws the fuel oil through the inlet strainer from the supply tank. The regulating valve by-passes a high percentage of the fuel back to the inlet side. The fuel provided by the transfer pump greatly exceeds the injection requirements.

Transfer pump pressure forces the fuel oil through the drilled fuel passage in the hydraulic head into the annulus. Fuel then flows around the annulus to the top of the sleeve and through a connecting passage to the metering valve. The governor controls the flow of fuel into the charging ring, which contains the charging ports, by controlling the rotary position of the metering valve (Fig. 4-10).

The single charging hole registers with one of the charging ports in the hydraulic head as the rotor revolves and fuel flows through the angled passage to the pumping cylinder at transfer pump pressure. The plungers are forced outward, by the in-flowing fuel oil, a distance proportional to the quantity to be injected on the following stroke.

The plungers move out very little if only a little fuel is admitted to the pumping cylinder, as at idling speed. As more fuel is admitted, the plunger stroke increases. The maximum stroke is limited by the leaf-spring adjustment.

At this charging point the rollers are between the lobes of the cam. Fuel oil is trapped in the cylinder for a slight interval after charging because the rotor charging port has passed out of registry with the head port and the rotor discharge port has not come into registry with an outlet port in the hydraulic head.

As the rotor rotates further, its discharge port is brought into registry with an outlet port of the head, at which point the rollers

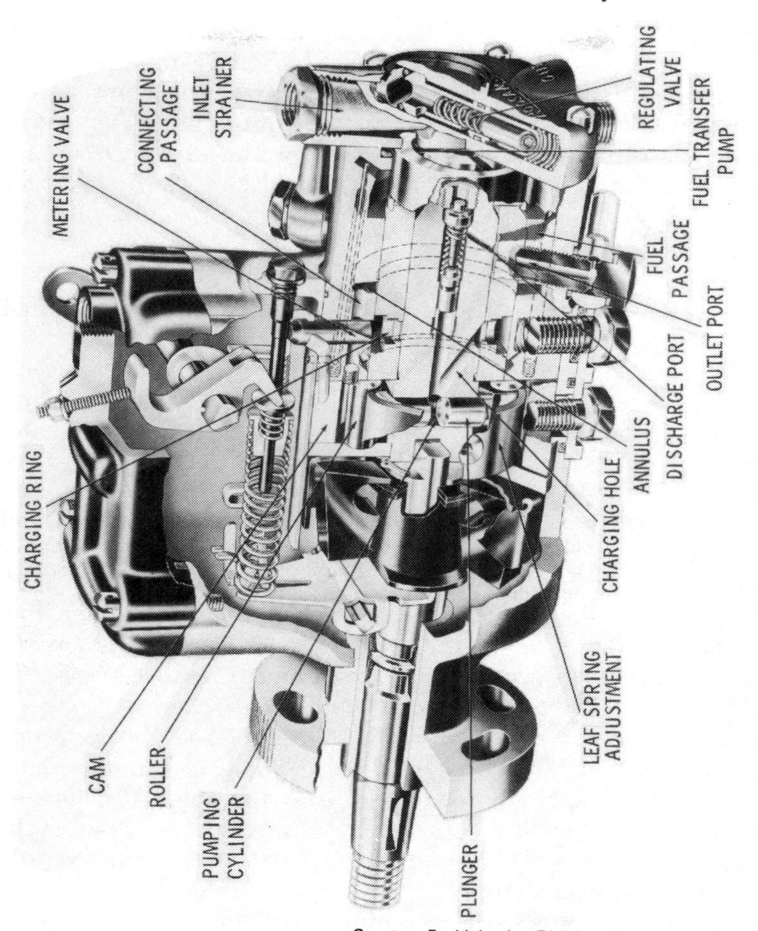

Courtesy Fuel-injection Division, Standard Screw Co.

Fig. 4-10. Roosa Master DB fuel injection pump. This is a single-cylinder, opposed-plunger, inlet metering, distributor-type pump.

contact the opposing cam lobes simultaneously and force the plungers toward each other. The trapped fuel between the plungers is forced from the pump through one of the outlets (Fig. 4-11).

What is the principle of operation of mechanical spray nozzles in the distributor system?

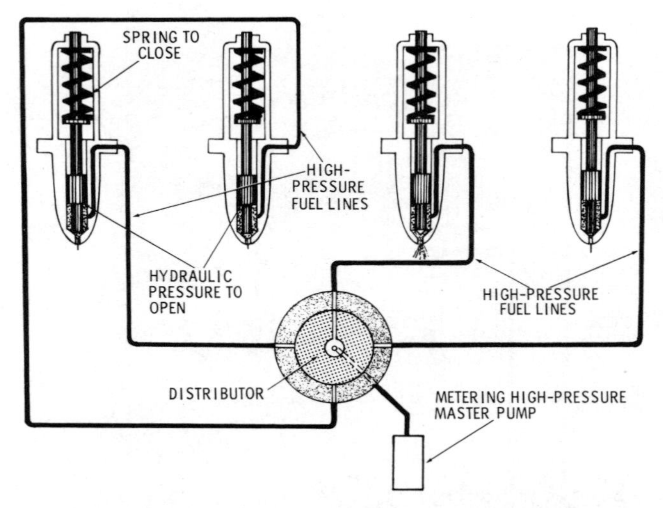

Fig. 4-11. Hydraulic spray nozzles in a distributor-type fuel system. The distributor is a rotary multiport valve and switches the fuel charge delivery by the metering master pump to the several cylinders via hydraulic spray nozzles in proper sequence.

Answer: In the operation of the Graham injector, as the valve closes, the master pump forces (via the distributor) a metered charge of fuel into the receiving chamber at low pressure. During the opening of the valve, this charge is transferred to the discharge chamber where it is preheated and discharged through the multinozzle passages during the closing of the valve. This injector requires no valve adjustment and no high-pressure fuel line. Note in the injector (B in Fig. 4-12) that the reversed cone valve moves downward to open.

According to Research Laboratories Section of General Motors, the pressure in fuel injector pumps may reach 20,000 pounds per square inch. Under the usual conditions of operation the pressure is 15,000 pounds per square inch. The pump forces the fuel oil through the small holes in the spray nozzle at 13 miles per minute—780 miles per hour. This high velocity breaks the oil up into a fine penetrating fog.

Describe a variable-stroke fuel injection pump.

Answer: The pump stroke can be changed to a longer or shorter

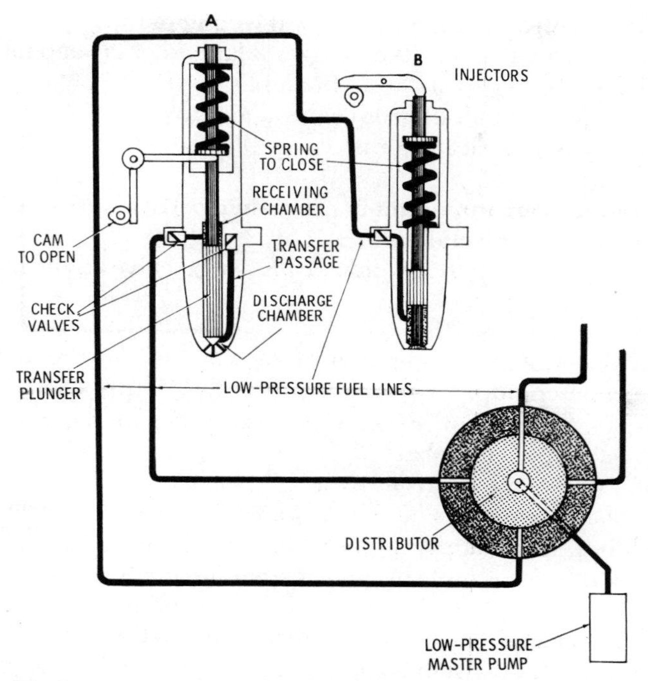

Fig. 4-12. Distributor system having unit injectors. These mechanical injectors are opened by a cam and closed by a spring.

stroke, thereby increasing or decreasing the amount of fuel delivered at each stroke (Figs. 4-5A and 5B).

What principal troubles are encountered in the variable-stroke fuel injection pump system?

Answer: The same as those in other pumps: leaky suction or discharge nozzles, clogged nozzles, sticking nozzles, worn plunger packing, and leak-backs into the fuel line. These are remedied as prescribed.

What is the plunger type of injection pump?

Answer: This consists of a packed or lapped plunger working in a barrel or cylinder. It may be either a single- or a double-acting pump.

What is meant by a lapped-plunger injection pump?

Answer: One in which the plunger is lapped or ground into the cylinder or barrel to make it fit tightly without packing. This plunger has a number of grooves around it to catch the waste oil, which acts as a lubricant for it.

What cautions are important in maintaining efficient operation of packed-plunger injection pumps?

Answer: Keeping the suction or discharge valves tight and the plunger well packed.

What is the principle of operation of the type PSB American Bosch fuel injection pump?

Answer: The pump (Fig. 4-13) is cast in an aluminum housing (13) containing a camshaft (3) supported at the rear by a sleeve bearing (1) and at the front by a ball bearing (4). Four-cylinder pumps have two lobes on the cam and six-cylinder pumps have three lobes on the cam. A spline at the forward end of the camshaft receives the drive hub (5), and a spiral gear (2) is cut into the camshaft adjacent to the cams to drive the lower gear of the quill-shaft assembly (25) of the distributing system.

The quill-shaft assembly transmits the rotary motion of the camshaft to the plunger at one-half camshaft speed. The assembly consists of a quill shaft (25) with an integral spur gear (24) at the top, and a spur gear located with a Woodruff key on the lower end of the shaft.

The gears are marked for left- or right-hand mounting and for proper timing of the plunger.

Between the cam and the lower end of the plunger is the tappet assembly, consisting of a shell (6), guided through a slot by a pin in the housing, and a roller follower turning on a floating bushing on a hardened shaft.

A marked timing window is located in the upper part of the housing for plunger gear lineup during assembly and disassembly. A timing pointer is provided at the drive end of the pump to align with a mark on the gear hub for more accurate setting of the port closing position.

The control lever is located directly above the timing window. It is attached to the housing through a flanged bushing. The lever

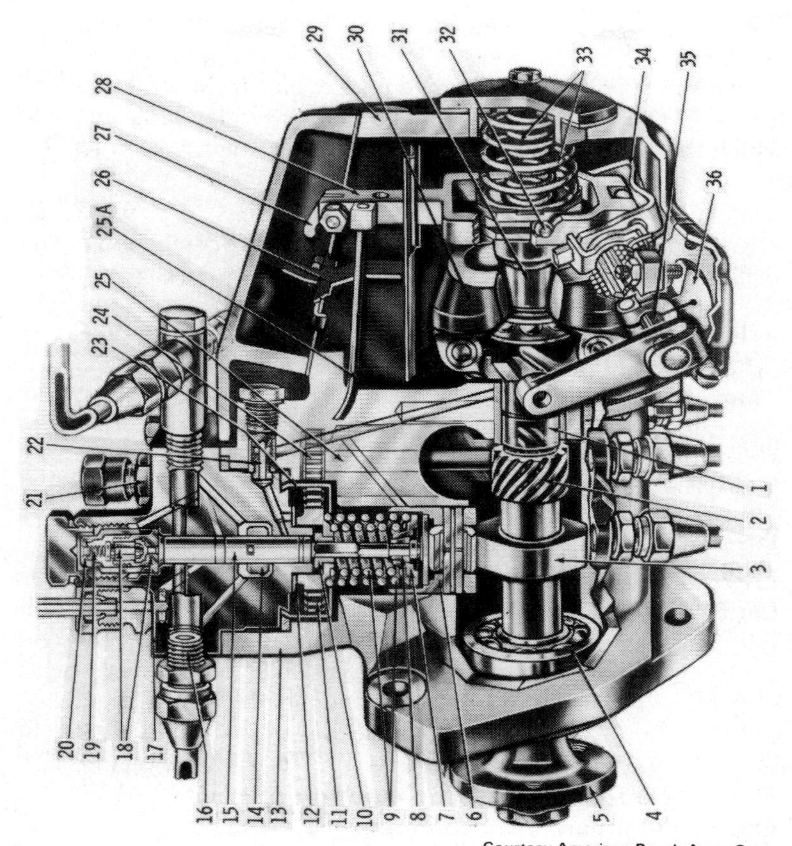

Courtesy American Bosch Arma Corp.

Fig. 4-13. PSB American Bosch fuel injection pump with hydraulic head and internal-spring governor. The hydraulic head is easily removed for service or replacement. Separate heads are required for four-cylinder and six-cylinder engines.

extends into the hydraulic head and has an eccentric pivot pin that engages the control sleeve (14) in the head.

A cover assembly encloses the timing window and the control lever. This screws onto the housing and carries a spring-loaded shutoff rod that the operator can use for shutting off the engine manually.

The hydraulic head assembly is fastened to the housing by four studs. It is easily removed for service or replacement. Separate heads are required for four-cylinder and six-cylinder engines. The head block (17) contains a lapped bore, which is counterbored and threaded at the upper end for the delivery valve (18), and into which the plunger (15) fits. Four discharge ducts (six ducts for six-cylinder units) extend symmetrically from the plunger bore. These ducts end in vertical discharge fittings (21), which are screwed into the top of the block. A downwardly inclined duct is drilled from the delivery valve bore to a horizontal duct entering the plunger bore at the distributing annulus nearly halfway down its length. Two fuel inlet fittings (16) are drilled and tapped in the head. Drilled holes slant through the head into the fuel supply sump at the middle of the plunger bore. The sump also accommodates the control sleeve (14). The head is also drilled and tapped for a lube oil filter (22) in the passage below the sump. This lubricates and seals the lower portion of the plunger's lapped fit.

The plunger (15) is also lapped and fitted into the head and into the control sleeve (14) so that an inseparable assembly is comprised of these three pieces. The lower portion of the plunger is flatted and is locked into the plunger drive gear (11) by a guide (10) that fixes the angular position of the plunger in relation to the drive gear. The thrust of the drive gear is taken by a bronze washer (12) placed between the gear and the head. The plunger has a lower spring seat (7) fastened on by means of split locks (8). The entire assembly is fastened to the hydraulic head by a retainer (23) crimped onto the head, which is cut away to permit meshing of the gears, and for purposes of timing.

At the top of the plunger bore the delivery valve (18) is seated. A retainer (20) sealed by a copper gasket holds the valve assembly and spring (19) into the head. The delivery valve is threaded on the outside for ease of removal.

How is the internal-spring governor action accomplished with the type PSB fuel injection pump?

Answer: The governor assembly (Fig. 4-13) consists of an aluminum housing (29), containing a shaft and weight assembly (30), a sliding sleeve (31), control springs (33), fulcrum lever (28), operating shaft assembly (35), and stop plate (36). The housing is attached to the rear of the pump housing with cap screws. A spider,

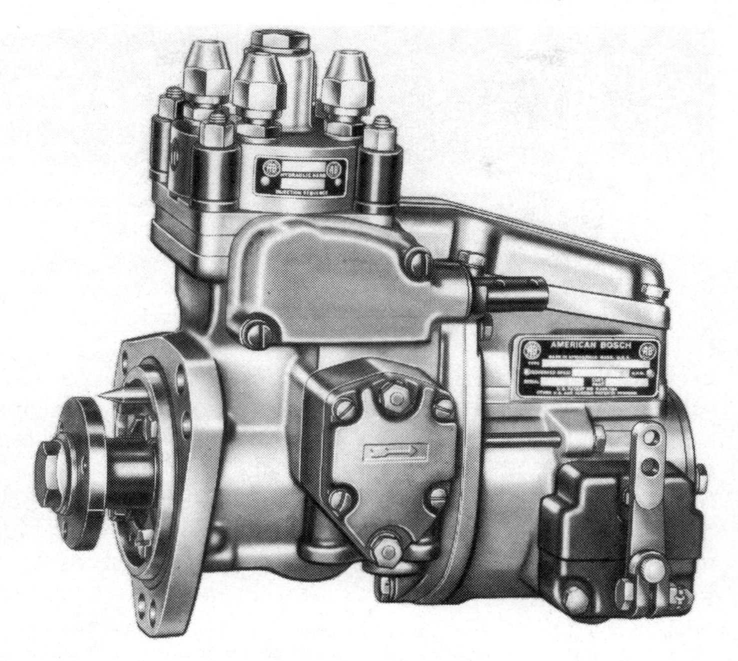

Courtesy American Bosch Arma Corp.

Fig. 4-14. American Bosch PSB pump with internal-spring governor.

comprised of the shaft and weight assembly and two movable weights (30), is pressed onto an extension of the camshaft (3).

The sliding sleeve (31) moves freely on the governor shaft and has a slot on each side to receive the pivot pins (32) of the fulcrum lever (28) which slide in the slots on the sleeve. The control rod extension (25A) connects to the top of the fulcrum lever and transmits the governor movement to the control sleeve in the hydraulic head.

The operating shaft assembly (35) is connected to a yoke (34) through a torsion spring acting on an ear on the shaft and an ear on the fulcrum lever. The fulcrum lever has a smoke-limit cam which limits the maximum fuel delivery attached to its upper end. Maximum fuel delivery, regardless of operating lever position is limited by contacting an adjustable stop plate (26), thus avoiding over-

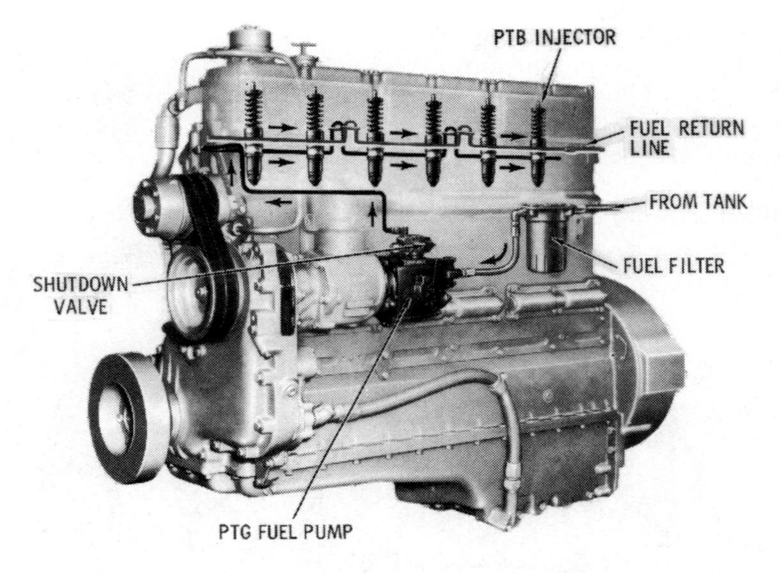

Courtesy Cummins Engine Co.

Fig. 4-15. PTG fuel-system flow in Cummins diesel engine. The PTG fuel pump assembly consists of three main units. The gear pump drains fuel from the fuel tank and forces it through the pump filter screen to the governor. The governor controls the flow of fuel from the gear pump, as well as the maximum engine speed. The throttle provides a manual control of fuel flow to the injectors under all conditions.

loading the engine. The stop plate and limiting screws are easily accessible on the outside for adjusting idling and full-load speeds.

What is a fuel transfer or fuel supply pump?

Answer: It is a pump that transfers fuel oil from the storage tanks to the service or day tanks; this may be driven by power or it may be hand operated.

What types of pumps are used for transferring fuel from the storage tank to the service or day tanks?

Answer: Plunger, piston, rotary gear, and centrifugal types.

Why is fuel transferred to a day or service tank?

Answer: To permit impurities in the fuel oil to settle to the bottom, to allow the air that may be trapped in the transfer pump to escape, and to enable the operators to make certain that a supply of fuel is on hand for use.

What is meant by the piston type of fuel transfer pump?

Answer: In this type, the piston works in a cylinder. It may be

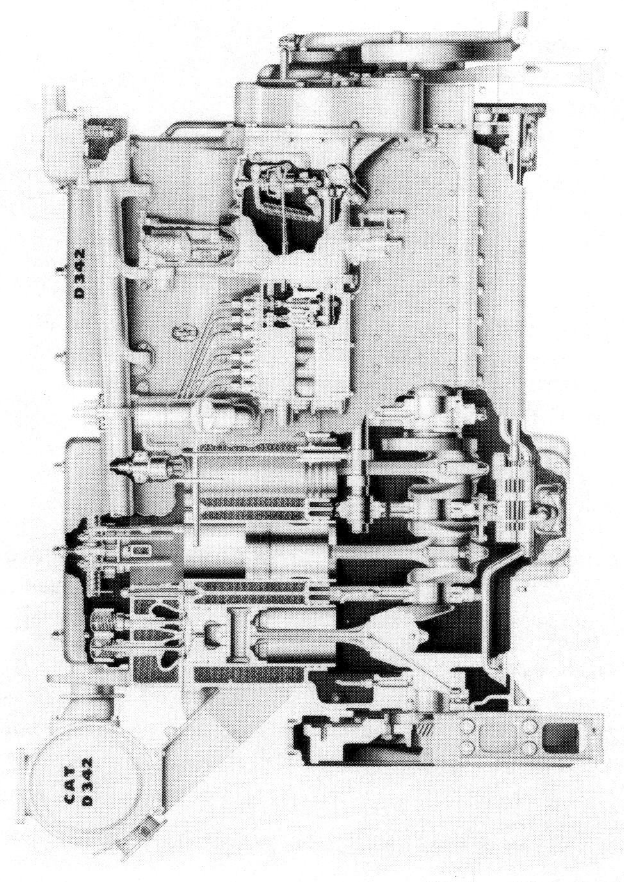

Courtesy of Caterpillar Tractor Co.

Fig. 4-16. Cutaway of Caterpillar D342 diesel engine.

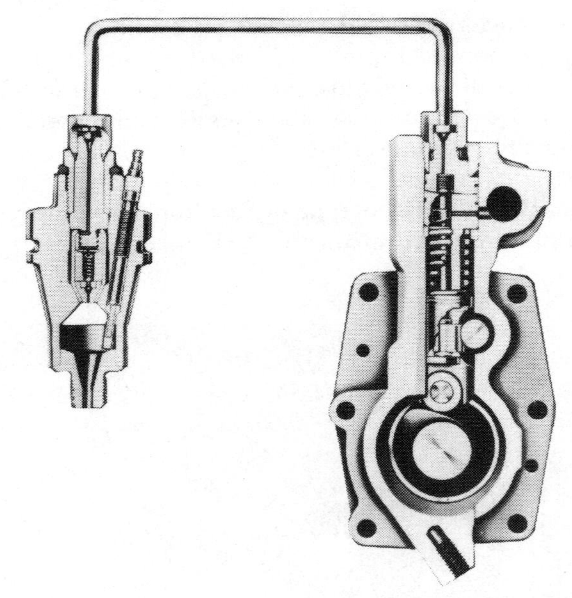

Courtesy Caterpillar Tractor Co.

Fig. 4-17. Fuel pump and fuel injector cutaways with glowplug used on Caterpillar D330-D333 diesel engines.

either single-acting or double-acting—usually a double-acting pump.

Describe the rotary gear type of fuel transfer pump.

Answer: It is a fuel pump, in which, by means of gears which mesh and fit the sides of the housing, the liquid is forced, by the moving teeth, from the suction side to the discharge side. Because of this, the pump is able to create pressures as desired.

What is the centrifugal type of fuel transfer pump?

Answer: This is a type in which a rotor containing a number of vanes or blades revolves in a casing, thus moving the liquid from the inlet side to the discharge side. This type is not generally used where high pressure is required, but centrifugal pumps are made for pressure service.

What is the purpose of the fuel supply pump?

Answer: The fuel supply pump draws fuel oil from the main

supply tank and forces it through a filter or filters to the fuel injection pump (Fig. 4-18).

What is the principle of operation of the American Bosch fuel supply pump?

Answer: The plunger is forced by its spring toward the camshaft (Fig 4-19 and Fig. 4-20). The fuel oil enters the plunger-spring chamber when the intake valve opens as a result of a suction effect created as the plunger moves toward the camshaft (Fig. 4-20A). The fuel is forced out of the chamber through the outlet valve around into the chamber created behind the spring chamber by the forward movement caused by the cam lobe driving the plunger against its spring (Fig. 4-20B). As the fuel injection pump camshaft rotates, the compressed plunger spring forces the plunger toward the camshaft again, forcing the fuel oil behind the plunger out into the fuel line leading to the filters and injection

Courtesy Cummins Engine Co.

Fig. 4-18. Changing the fuel filter. This is a throwaway element.

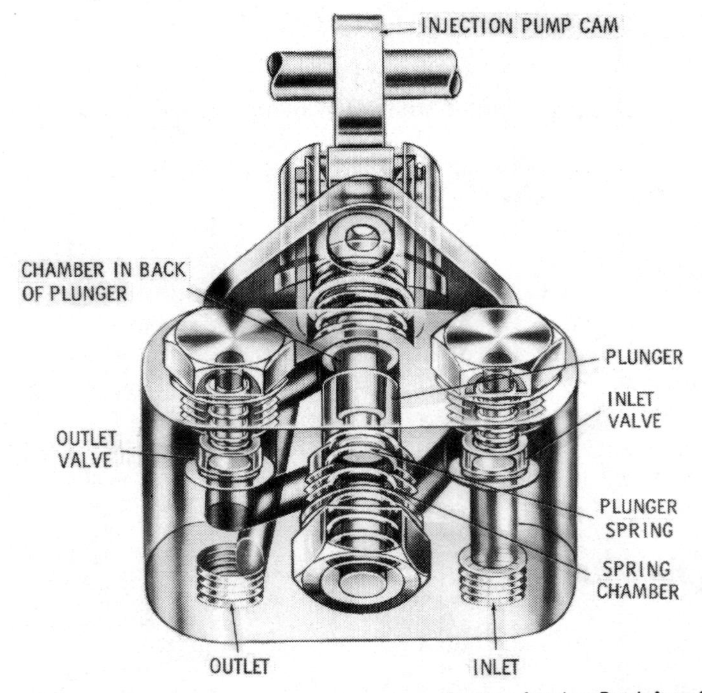

Courtesy American Bosch Arma Corp.

Fig. 4-19. American Bosch fuel supply pump.

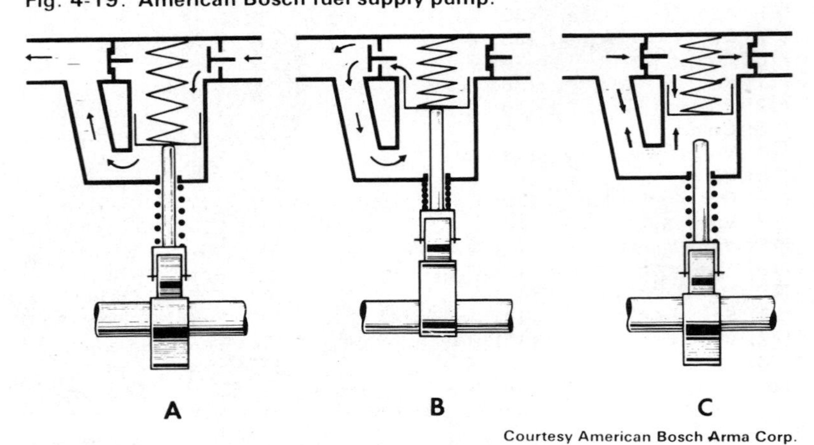

Courtesy American Bosch Arma Corp.

Fig. 4-20. Operation of American Bosch fuel supply pump.

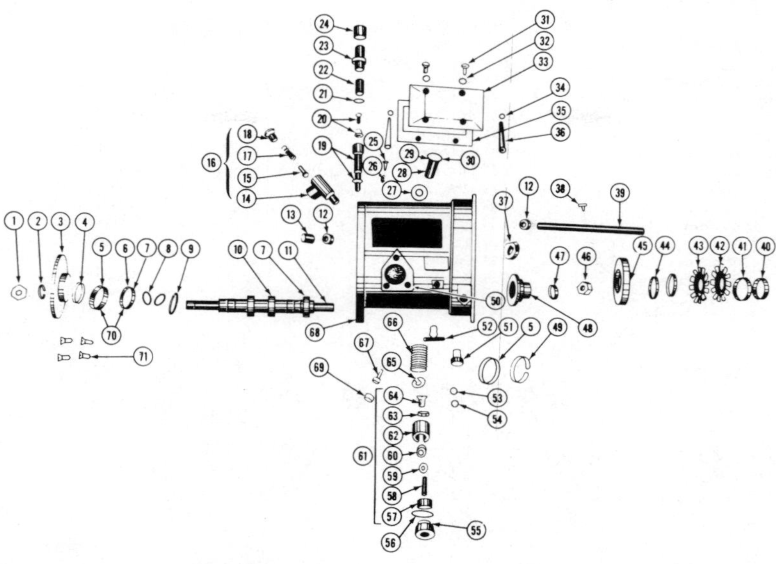

(1) Nut
(2) Lock Washer
(3) End Plate
(4) Oil Seal
(5) Outer Race
(6) Inner Race
(7) Retainer and Balls
(8) Adjusting Washer
(9) Oil Thrower
(10) Camshaft
(11) Key
(12) Bushing
(13) Cover
(14) Valve Body
(15) Valve
(16) Overflow Valve
(17) Valve Spring
(18) Valve Screw
(19) Plunger and Barrel
(20) Valve and Seat
(21) Seat Gasket
(22) Valve Spring
(23) Valve Holder
(24) Cap
(25) Barrel Set Screw
(26) Set Screw Gasket
(27) Spring Seat
(28) Control Sleeve
(29) Screw
(30) Toothed Segment
(31) Screw
(32) Lock Washer
(33) Cover
(34) Cover Gasket
(35) Screw Gasket
(36) Screw
(37) Baffle
(38) Rod Retaining Screw
(39) Control Rod
(40) Hub Nut
(41) Nut Lock
(42) Disc Outer
(43) Disc Inner
(44) Spacer
(45) Governor Drive Gear
(46) Nut
(47) Lock Washer
(48) Drive Gear Assembly
(49) Gearing Retainer
(50) Stud
(51) Drain Plug
(52) Oil Lever Cock
(52) Lock Washer
(54) Nut
(55) Closing Plug
(56) Gasket
(57) Cushion
(58) Tappet Roller Pin
(59) Tappet Roller Bushing
(60) Tappet Roller
(61) Tappet Assembly
(62) Tappet Shell
(63) Lock Nut
(64) Adjusting Screw
(65) Spring Seat
(66) Plunger Spring
(67) Tube
(68) Pump Housing
(69) Overflow Plug
(70) Camshaft Gear
(71) Screw

Courtesy American Bosch Arma Corp.

Fig. 4-21. Exploded view of Bosch "K" pump.

(1) Nut
(2) Washer
(3) Gear
(4) Bearing Plate
(5) Bearing
(6) Oiler
(7) Spring
(8) Bearing
(9) Bushing
(10) Shaft Assembly
(11) Spider
(12) Weights
(13) Weight Pin
(14) Washer
(15) Cotten Pin
(16) Cap Screw
(17) Lock Washer
(18) Shaft and Weight Assembly
(19) Sleeve
(20) Fulcrum Lever
(21) Nut
(22) Screw
(23) Smoke Limit Cam
(24) Washer
(25) Cotten Pin
(26) Link Pin
(27) Link
(28) Nut
(29) Pin
(30) Spring, Inner
(31) Spring, Outer
(32) Spring Seat
(33) Bearing
(34) Washer
(35) Nut
(36) Housing
(37) Dowel Pin
(38) Cap Screw
(39) Cap Screw
(40) Washer
(41) Breather
(42) Cap Screw
(43) Cap Screw
(44) Cap Screw
(45) Inspection Cover
(46) Gasket
(47) Bridge
(48) Bridge Assembly
(49) Stop Plate
(50) Bumper Spring
(51) Lock Washer

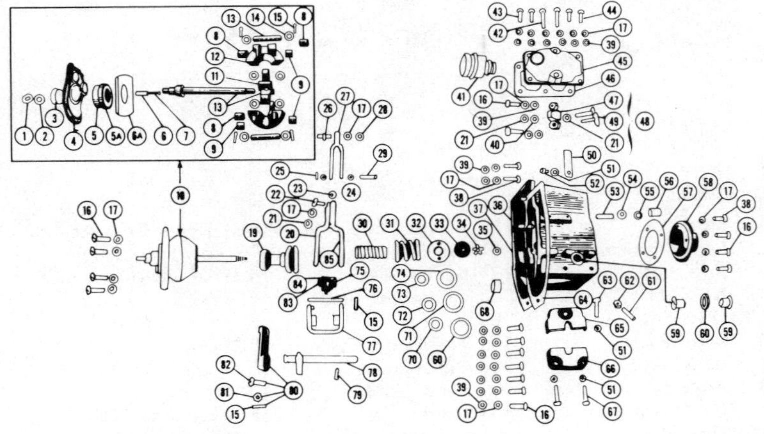

(52) Screw
(52) Adjusting Screw
(54) Lock Nut
(55) Gasket
(56) Cap
(57) Gasket
(58) End Cap
(59) Bushing
(60) Seal
(61) Stop Screw
(62) Nut
(63) Cap Screw
(64) Gasket
(65) Cover-Upper
(66) Cover-lower
(67) Cap Screw
(68) Closing Plug
(69) Spacer, Outer Spring
(70) Spacer, Inner Spring
(71) Spacer, Outer Spring
(72) Spacer, Inner Spring
(73) Spacer, Inner Spring
(74) Spacer, Outer Spring
(75) Operating Shaft Spring and Hub Assembly
(76) Bracket Pin
(77) Bracket
(78) Operating Shaft
(79) Set Screw
(80) Lever
(81) Nut
(82) Cap Screw
(83) Spring Hub
(84) Spring
(85) Pivot Pin

Courtesy American Bosch Arma Corp.

Fig. 4-22. Exploded view of Bosch "K" pump governor.

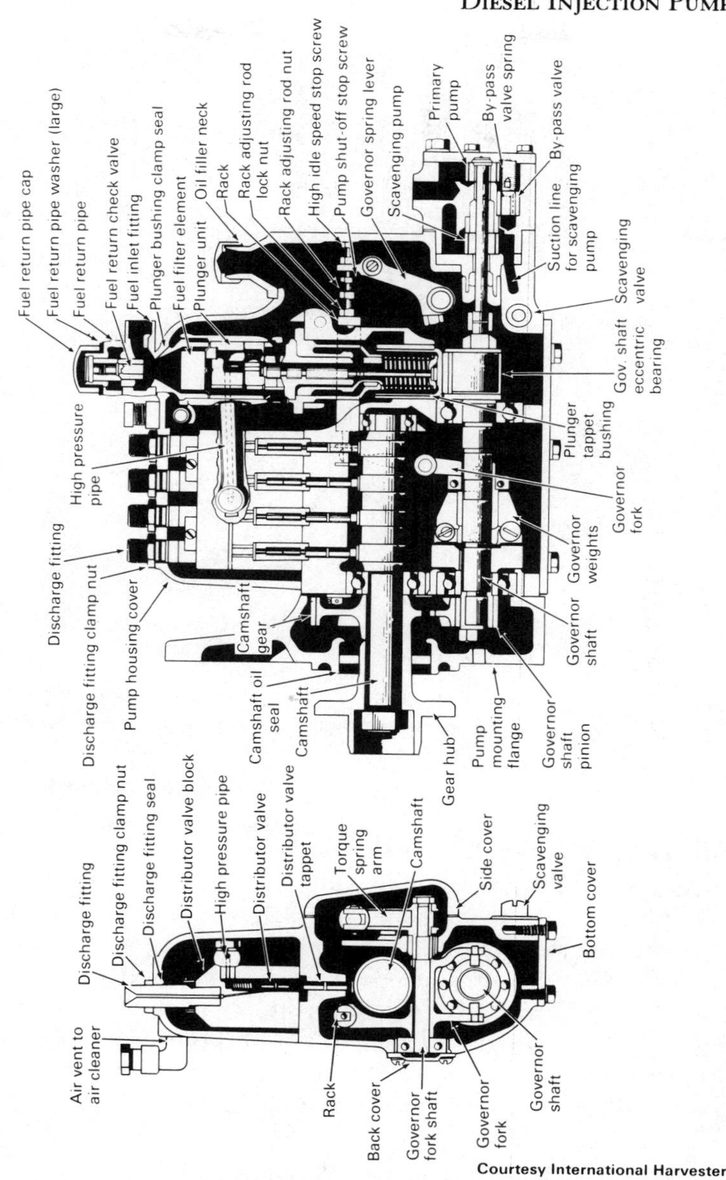

Courtesy International Harvester

Fig. 4-23. Cross section of International Harvester single plunger fuel injection pump.

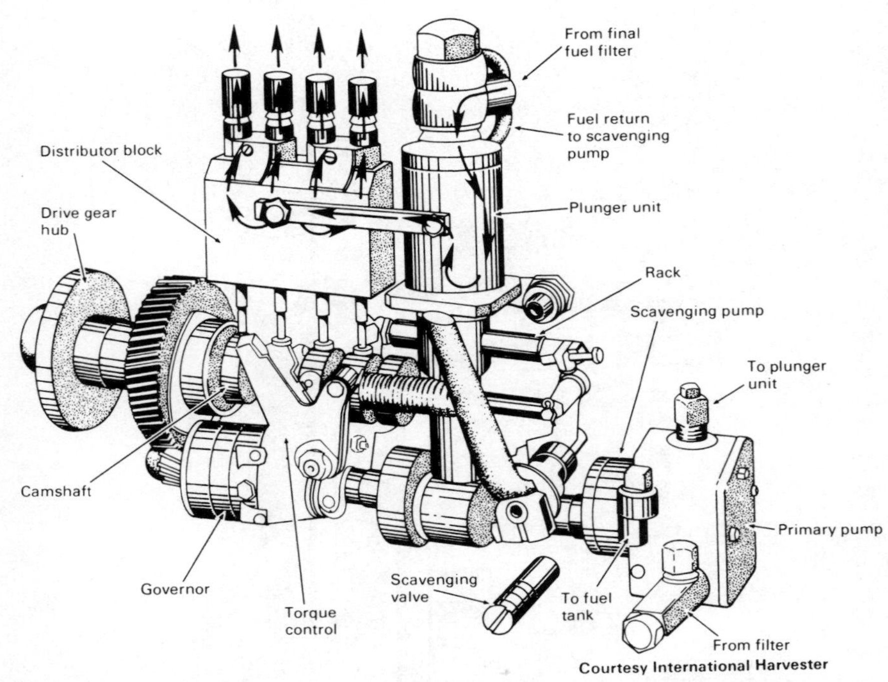

Fig. 4-24. Skeleton view of International Harvester single-plunger mechanism.

pump (Fig. 4-20B). In the same action a suction effect is again created by the plunger, which permits additional fuel to flow into the spring chamber through the inlet valve (Fig 4-20A).

As long as fuel oil is used fast enough by the injection pump, the pumping action will continue. When the supply pressure equals the force exerted by the spring on the plunger, pumping action stops because the pressure between the supply pump and the injection pump holds the plunger stationary against the spring and away from the cam (Fig. 4-20C). When the pressure drops enough to permit the plunger to resume operation, pumping action will start again. The entire cycle is automatic and continues as long as the engine is running.

How is a Bosch "K" pump assembled?

Answer: Each part should be blown dry with compressed air.

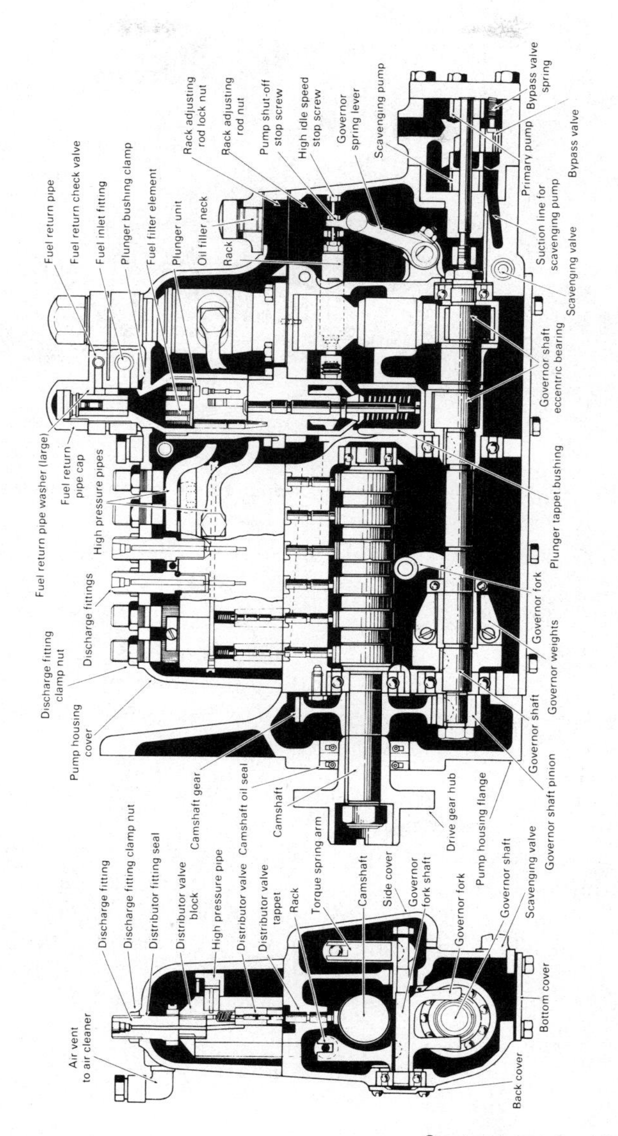

Courtesy International Harvester

Fig. 4-25. Cross section of International Harvester twin-plunger fuel injection pump.

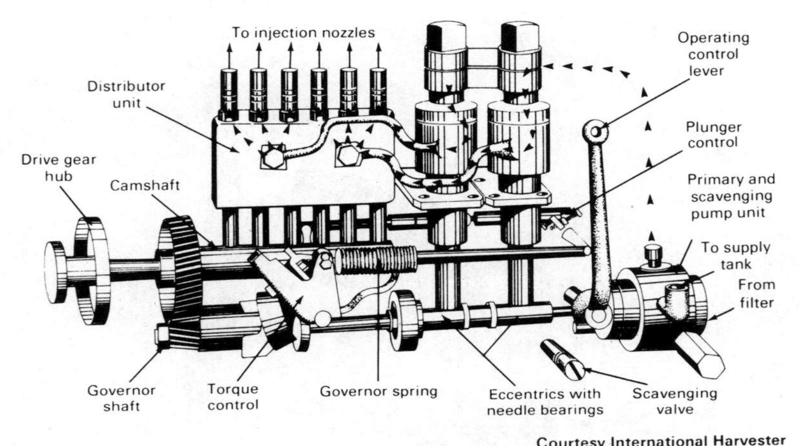

Fig. 4-26. Skeleton view of International Harvester twin-plunger pump mechanism.

This will prevent water and dirt from collecting. Gummy and sticky parts, especially plungers and barrels, should be soaked in acetone (Fig. 4-21).

How is a Bosch "K" pump governor assembled?

Answer: Again, each part of the governor should be free from dirt and moisture before assembly. Use compressed air to prevent moisture (Fig. 4-22).

What are the single-plunger and twin-plunger fuel pumps?

Answer: The International Harvester diesel single-plunger and twin-plunger pumps are of the flange-mounted type; that is, they are mounted to the crankcase front plate. An exception is the 18-A Series engines, which are base mounted to the side of the engine crankcase. The single- and twin-plunger pumps are distributor-type pumps. A single eccentric cam for the single plunger and two eccentric cams for the twin plungers operate the lapped plungers at a constant stroke. The removable plunger units are mounted on top of the pump housing and to the rear of the distributor block. The governor shaft is located below the camshaft in the lower half of the housing; the governor weights are on the front end of the shaft; and the eccentrics that operate the plungers are on the rear end of the shaft. See Figs.4-23 through 4-26.

CHAPTER 5

Fuel Spray Nozzles

What is a fuel spray nozzle?

Answer: It is a nozzle or device used for atomizing or cracking fuel oil and through which the fuel oil is injected into the working cylinders of diesel engines.

What is the function of a fuel spray nozzle?

Answer: The function of a fuel spray nozzle (Fig. 5-1) is to admit fuel to the cylinders at the proper time, in the proper quantity, and in properly atomized form to make ignition easy.

Why is the fuel spray nozzle important?

Answer: It must operate under high pressures, break up or atomize the fuel for ignition, act quickly, and withstand high temperatures. It is a very complicated valve and the successful operation of the engine is dependent upon it.

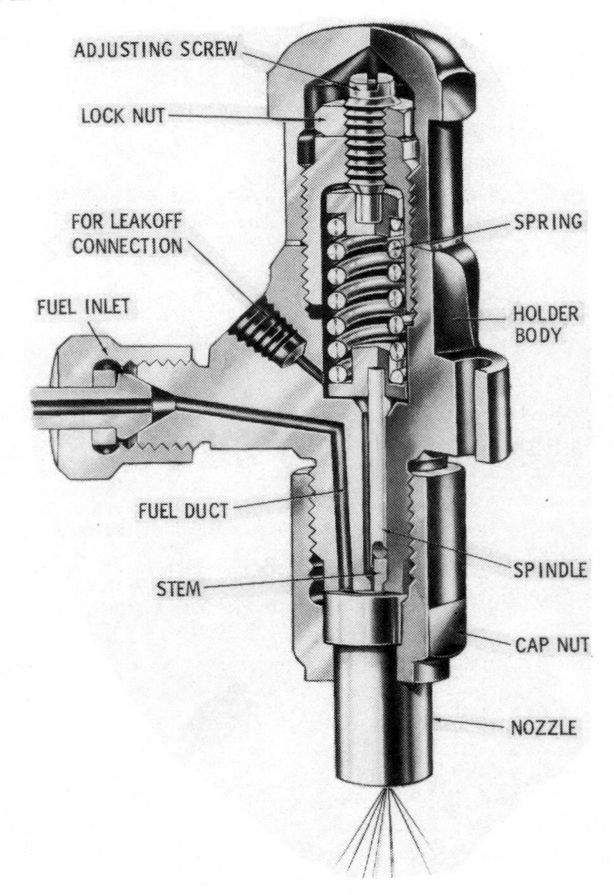

Courtesy American Bosch Arma Corp.

Fig. 5-1. Fuel spray-nozzle holder and nozzle.

What are the normal working air pressures in the power cylinders?

Answer: 350 to 500 pounds per square inch.

What are the normal temperatures in the working cylinders?

Answer: From 800 to 1,000 degrees Fahrenheit.

In what length of time must a fuel spray nozzle act?

Answer: In a fraction of a second—depending upon the density of the fuel oil and the size and the speed of the engine.

What is meant by an air-injection fuel system?

Answer: The use of compressed air that forces a definite amount of fuel, mixed with the air, into the cylinder. Usually a measured amount of fuel is delivered by a fuel pump to the spray valve and this amount is injected. The amount delivered for each power stroke is dependent upon the power required. The fuel burns more readily because it is mixed with air. The more recently developed diesel engines do not use the air-injection system.

What quantity of fuel must a fuel spray nozzle admit into the cylinder?

Answer: A quantity sufficient to produce complete combustion for the power required. The quantity of fuel used is small.

What methods are used for fuel injection into the cylinders?

Answer: Mechanical and hydraulic methods.

What names are usually given to the methods of fuel injection?

Answer: Air injection, mechanical injection, and solid fuel injection are all methods of fuel injection.

What governs the timing of the spray nozzle?

Answer: Timing of the spray valve is governed by the speed of the engine, by the resistance which the pulverizer offers to the passage of fuel oil, and by the size of the holes in the flame plate.

At what part of the cycle should the spray nozzle open?

Answer: At 3 to 20 degrees before the piston reaches dead center preceding the power stroke—depending upon the speed of the engine.

On what factors do timing variations depend?

Answer: Mainly upon the speed of the engine. At high speed, the timing may be in advance of piston dead center from 10 to 20 degrees.

Why is fuel admission retarded?

Answer: To permit the fuel oil time to become ignited—to give the working, or power, stroke full benefit of the expansive force of the entire charge of fuel.

What are the dangers either of too much retarding or of too early fuel injection?

Answer: Either too much retarding or too early fuel injection may cause pressure in the cylinder before the piston reaches its dead center preceding the power stroke and set up stresses on the crankshaft and frame caused by the tendency of the pressure to reverse the rotation.

How is too early fuel injection recognized?

Answer: By a cracking sound and by a pounding noise in the cylinder.

What is the importance of the spray-nozzle tip?

Answer: The spray-nozzle tip is an extremely important part of the spray nozzle. It is located directly in the combustion space of the cylinder, subject to the intense heat of combustion and the pressure exerted by the compressed air and the burning fuel. It must spread the fuel charge over the required area of the combustion chamber, properly atomized, and it must be easily replaceable. Two types, the pintle type and the open-hole type, are extensively used. They are shown in Fig. 5-2.

The pintle-type nozzle tip projects through the nozzle body, into the combustion chamber. Fuel oil is fed to it by means of small tunnels that terminate in a little reservoir behind the face of the nozzle seat. As the valve is lifted by the fuel pressure, the fuel is ejected through the opening.

The open-hole-type tip is cone shaped, and when opened, permits the fuel to be ejected in a stream and at an angle depending upon the drilling of the holes, which may be at any angle and in any number.

What effect has late fuel injection on the working of the engine?

Answer: Late fuel injection has the effect of increasing fuel consumption; cutting down the mean effective pressure; losing power; causing excessive temperatures in the cylinder, piston, and exhaust; destroying the lubrication of the cylinder; burning and

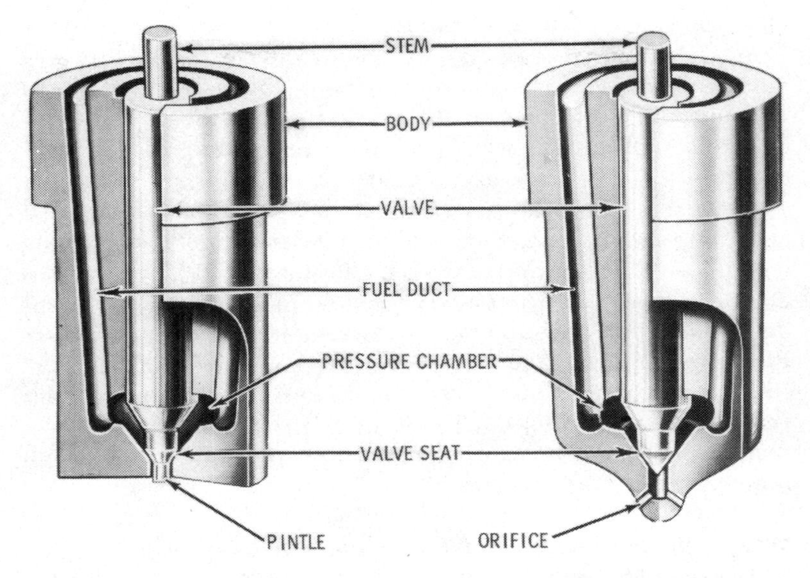

Courtesy American Bosch Arma Corp.

(A) Pintle-type nozzle. (B) Open-hole-type nozzle.

Fig. 5-2. Cross section of spray nozzles.

cracking the pistons; and doing other damage. Late fuel admission in a diesel engine is similar to a late spark ignition in a gasoline engine.

How is the timing of a fuel valve regulated?

Answer: By adjustment of the fuel cam toe on the camshaft.

What effect has a worn cam nose on the time action of a spray valve?

Answer: It delays its action and decreases its lift. It should, therefore, be set ahead sufficiently to compensate for the wear.

What are the most common troubles encountered in spray valves?

Answer: Leaking of either air or fuel into the cylinders and sticking of stems and needles. These are often caused by improper handling—resulting in bent stems and ruined needles.

How are valve troubles remedied?

Answer: By first removing the stem from the valve body and determining whether it is straight or bent (Fig. 5-3).

Remove the stem, take out the old packing and thoroughly clean both the guide and the stem; apply a thin coat of Prussian blue to the stem, place it back in the guide and move it up and down several times—do not turn it. If the stem is straight, both the stem and guide will show an even coating; if it is not straight, an uneven coating will show. If the stem is bent, it should be replaced, for no amount of grinding will remedy a bent stem. If straight, clean and dry again and rub together dry with the valve seat to note whether the valve is tight. If the valve is not tight, grind it tight with the finest grade of grinding compound. Remove the compound and finish grinding in fuel oil. Do the smallest amount of grinding necessary to make the valve tight. Do not remove more metal than is actually necessary for a tight fit.

What is the result of excessive grinding of a spray valve?

Answer: The stem becomes thin and enters the valve seat too

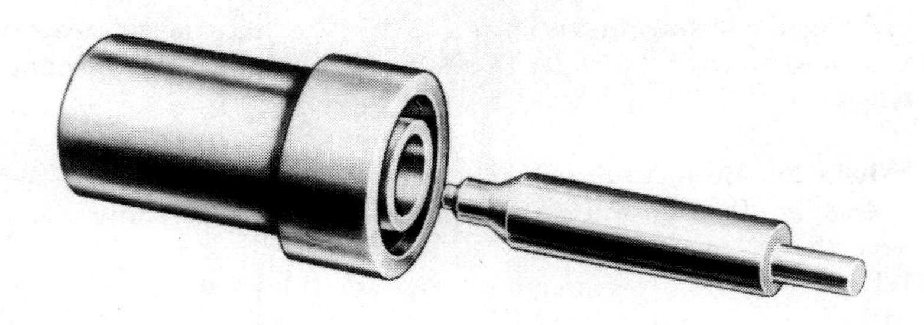

Courtesy American Bosch Arma Corp.

Fig. 5-3. Spray-nozzle body and nozzle valve. The two parts are shown separately. The body and the valve of both nozzle types are lapped to form a mated assembly. Therefore, the body and the valve cannot be exchanged singly, but must be kept together at all times.

deeply, causing a lagging of the fuel admission that results in late combustion and loss of power.

How should a spray-valve stem be packed and tested for friction tightness?

Answer: By using a graphite metallic packing and stuffing each turn into its place separately with a piece of brass or copper pipe—saturating each turn with lubricating oil and moving the stem up and down as each turn of packing is inserted until the stuffing box is full. If properly packed, the stem should drop into its place on the seat by its own weight.

What will cause spray valve chattering when seating?

Answer: Insufficient spring pressure.

If the spring pressure of a spray valve is too great, what will be the effect?

Answer: The needle will hammer when seating. The proper pressure is between the two—a medium pressure.

How can one tell while the engine is operating whether the needle of a spray valve is seating properly?

Answer: By placing one's finger on the exposed end of the valve stem and noting its feel. If the needle seats properly, the feeling will be that of a metallic contact (Fig. 5-5).

What is the proper seat width of a spray valve?

Answer: About 1/16 inch. A valve will remain tight longer and seat more promptly at that width.

What is a good practice to follow in keeping valve stems smooth and in proper working order?

Answer: Giving the valve stem a few turns every four hours; this cleanses the stem, loosens any foreign matter that may cling to the stem and seat, and keeps the valve properly lubricated all around.

What would be the effect on the engine if a fuel valve were leaking?

Answer: A leaking fuel valve causes misfiring and irregular speed, particularly noticeable on light loads.

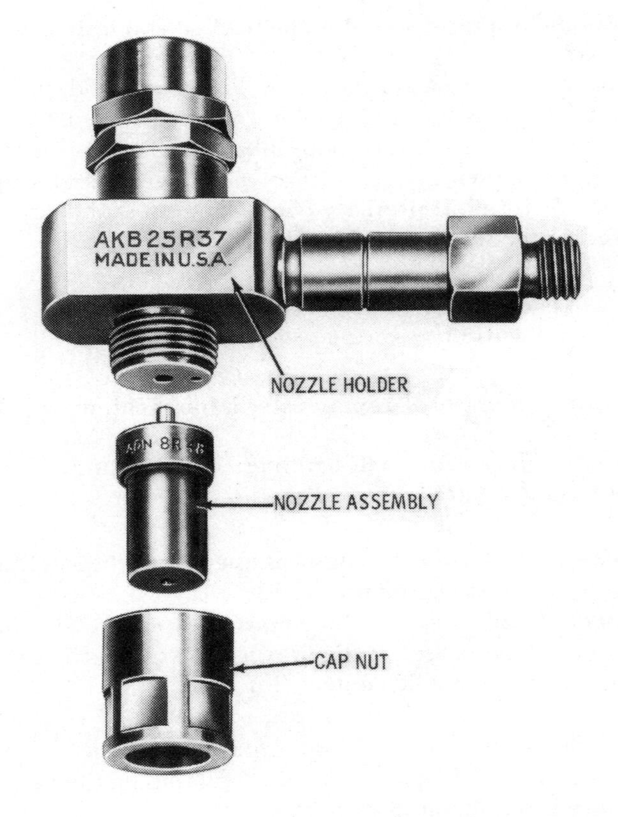

Courtesy American Bosch Arma Corp.

Fig. 5-4. American Bosch AKB spray-nozzle holder and spray-nozzle assembly.

If a fuel valve either remained open or failed to close at the proper time, what might be the result?

Answer: A fuel valve remaining open could cause explosions in the cylinder and subject the engine and engine parts to high working stresses and strains.

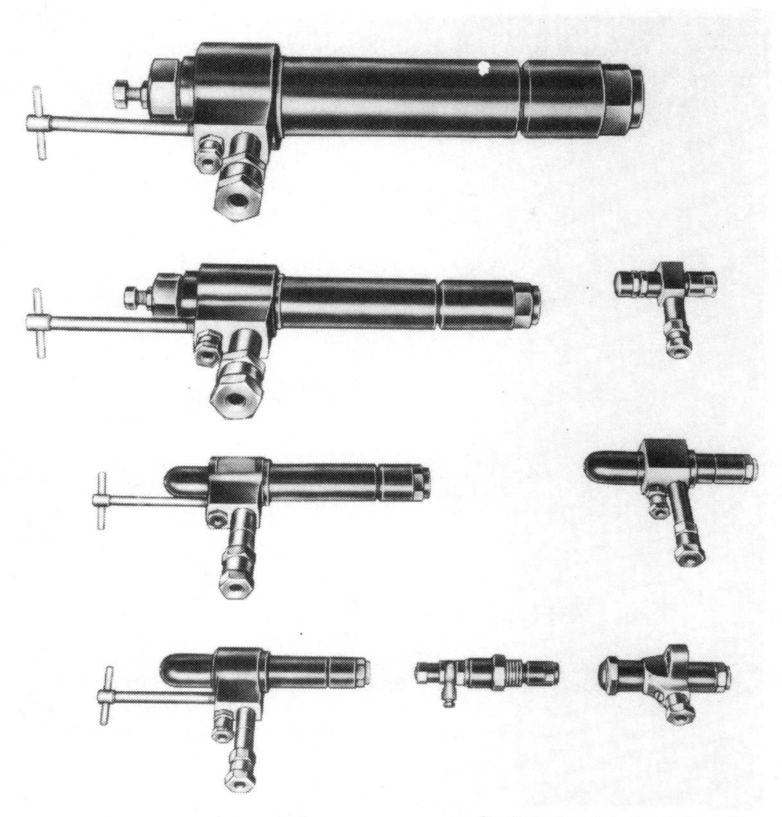

Courtesy American Bosch Arma Corp.

Fig. 5-5. Eight typical spray-nozzle holders. The nozzle holder is used to hold the nozzle in its correct position in the engine cylinder and to provide a means of conducting fuel oil to the nozzle. The holder also contains the necessary spring and means of pressure adjustment to provide proper action of the nozzle valve.

What is the most common cause of explosions in the fuel spray valves and fuel lines?

Answer: The needle in the spray valve not closing promptly, thus setting fire to the fuel in the valve and backfiring into the line.

What is meant by an accumulator or common-rail pump injection system?

Answer: This system is one in which the fuel oil is pumped to pressures of 3,000 to 4,000 pounds per square inch and is held in bottles or accumulators ready for use when the spray valve is opened by the fuel cams (Fig. 5-6).

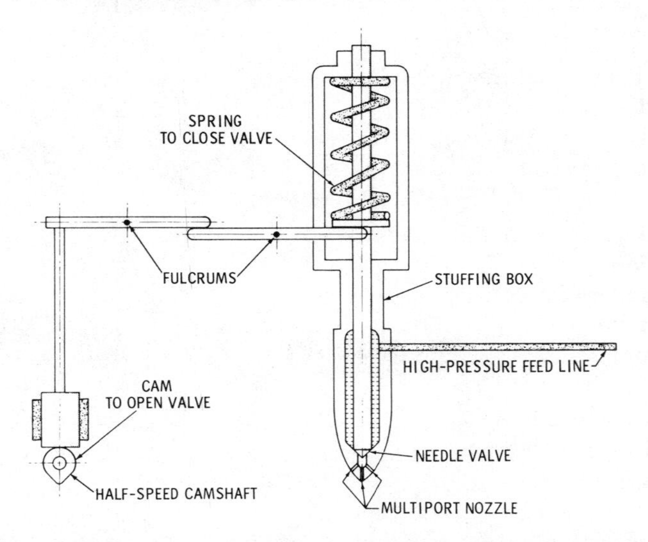

Fig. 5-6. Operating principle of a mechanical injection spray nozzle. The valve is opened by cam action and closed by spring tension. This type of nozzle is used in the common-rail fuel injection system.

What is an accumulator?

Answer: An accumulator is a cylindrical vessel connected in the fuel system to absorb the shock of the injection pumps and to provide a reservoir at high pressure for supplying an even flow of fuel oil to the cylinders.

How is the pressure in an accumulator regulated?

Answer: By spring adjustment of the escape valve—excess fuel oil is returned to the storage tank.

What troubles may be encountered in an accumulator-type fuel injection system?

Answer: Leaks in the escape valve, clogging of holes in the spray tip by accumulation of foreign matter, and carbon deposits.

How is excess pressure in an accumulator prevented?

Answer: By a spring-loaded escape valve that rises and opens when the pressure exceeds the predetermined amount and allows the excess fuel oil to flow back into the storage tank through a by-pass.

How is the quantity of fuel needed for proper operation determined?

Answer: By the temperature readings of the exhaust gases and by the color of the exhaust gases.

How is pressure maintained in the accumulator or plain-rail pump injection system?

Answer: By constant-pressure pumps, usually two or more, which keep the pressure at a predetermined height in the accumulator.

In the accumulator system, what is the lowest pressure at which diesel fuel oil will be atomized sufficiently for ignition when injected into the cylinder?

Answer: About 1,500 pounds per square inch—depending upon the density of the fuel, and the number and size of holes in the spray tips.

How is fuel oil injected in proper quantity?

Answer: By the governor regulating the time and lift of the spray valve. If too much fuel is being injected for the predetermined speed of the engine, the pressure in the accumulator should be reduced so that the proper amount is injected into each cylinder.

Name another system of solid fuel injection.

Answer: Direct-pump injection wherein the fuel is measured and pumped directly into the power cylinders.

What is the essential difference between the accumulator injection system and the direct-pump injection system?

Answer: The direct-pump injection system requires neither a pressure regulator relief valve nor an accumulator for maintaining a supply of fuel oil for injection (Fig. 5-7).

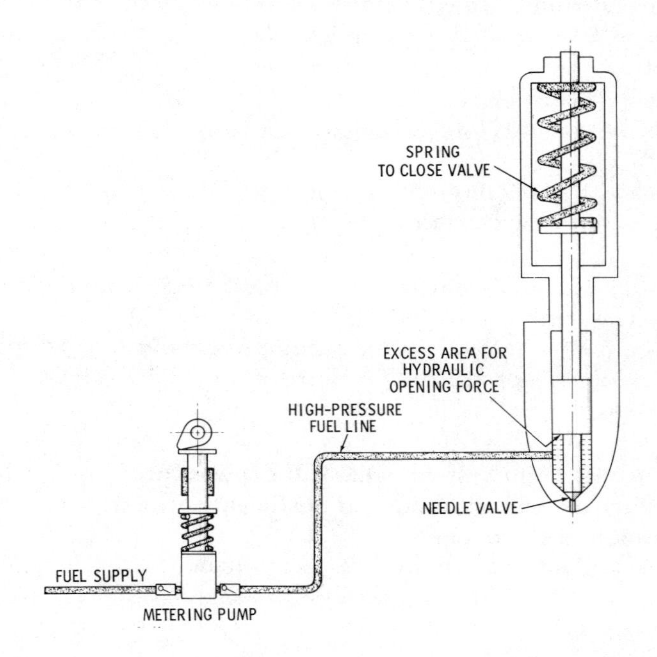

Fig. 5-7. Operation of a hydraulic spray nozzle. The needle valve is opened by fuel pressure and closed by spring tension. This type of nozzle is used in the metering-pump system of fuel injection.

What are the usual working pressures in direct-pump injection systems under full load?

Answer: From 3,000 to 6,000 pounds, depending upon the type of spray-nozzle tip, the density of the fuel, and the size of the engine; some engines may require pressures of 10,000 pounds.

What principal advantages has the solid fuel injection system over the air-injection system?

Answer: It does not require an air compressor and eliminates possible trouble resulting from air getting into the fuel lines and causing flash-backs and explosions.

What is the function of the nozzle holder and spray-nozzle assembly?

Answer: These units receive a metered quantity of fuel from the injection pump and direct it into the combustion chamber of the cylinder in such a manner as to produce the most efficient engine performance.

CHAPTER 6

Inlet and Exhaust Valves

What are the purposes of air-intake and exhaust valves?

Answer: To admit air for compression and to allow burned gases to escape.

Can air-intake and exhaust valves be interchanged?

Answer: Yes, if exhaust valves are used as air-intake valves, they do not require water cooling.

Why should exhaust valves be water cooled while air-intake valves are not?

Answer: Because of the high temperature of the exhaust gases that pass through them. Air-intake valves admit air at the temperature of the atmosphere.

How hot are the exhaust gases?

Answer: Average temperature is from 400 to 700 degrees Fahrenheit, under ordinary operating conditions.

How are exhaust valves cooled?

Answer: By a water jacket supplied with water through direct connected piping or rubber hose connection.

What troubles may be encountered in exhaust valves?

Answer: Pitting of the valves and valve seats, and sticking of valve stems. This may be due either to improper lubrication or to carbonization of the fuel or lubricating oil, or both.

What may be the result of permitting an exhaust valve to leak?

Answer: It will be burned in the region of the leak and may become useless.

How can leaks in exhaust valves be detected.?

Answer: By loss of compression, by undue heating of the valve and valve cage, and by the sound of either escaping gas or air.

What will be the effect of excessive heating on either an inlet or an exhaust valve.?

Answer: It may warp the valve cage and cause leakage.

If an exhaust valve shows excessive heating, how can it be determined whether the heating is due either to a leak in the valve or to a leak in the joint between the valve cage and the cylinder head?

Answer: By determining whether the sound of escaping gas comes from the exhaust outlet or from around the valve cage flange.

How should a valve cage be reset on a cylinder head?

Answer: By first making sure that the contact parts of the valve cage and the cylinder head are perfectly clean. Then place the copper–asbestos gasket so that it fits snugly and set the valve cage in place while feeling all around to make sure that the flange of the cage is at the same distance from the cylinder head at four equally distant points of its diameter.

What is the value of the turbocharger on the diesel engine?

Answer: The turbocharger provides air for an optimum air–fuel ratio. Turbo-blower aftercooling lowers initial cycle temperatures for greatest efficiency. Air must be drawn into the cylinder of the diesel engine to produce the heat of compression for ignition of the fuel. The turbocharger permits a smaller aftercooler and reduces cooling water requirements (Fig. 6-1).

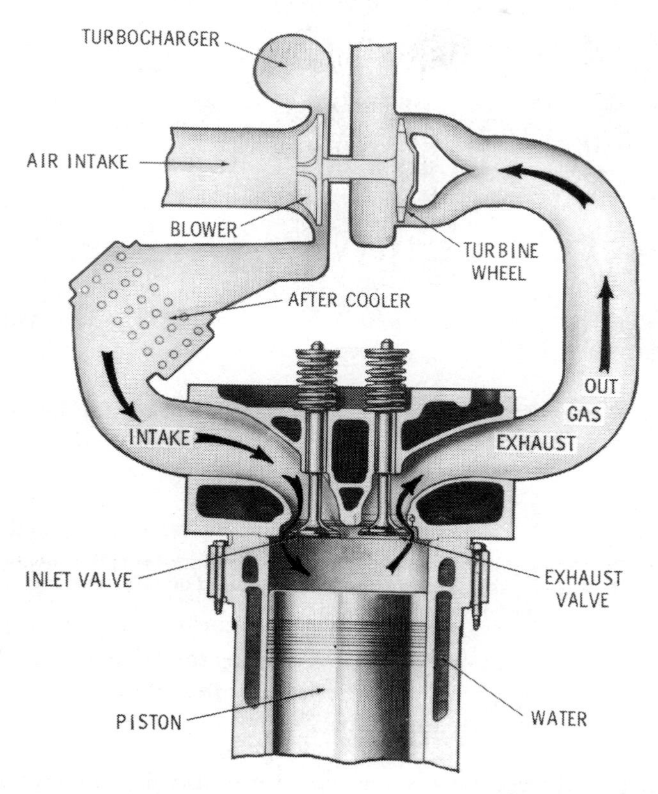

Courtesy Worthington Corp.

Fig. 6-1. Turbocharger for four-stroke-cycle Worthington SW engines. The turbocharger provides optimum air-fuel ratio and reduces cooling water requirements of the engine.

What precaution should be taken in setting a valve cage onto a cylinder?

Answer: First equalize the temperatures of the valve cage and the cylinder head before fastening down. If set cold on a hot

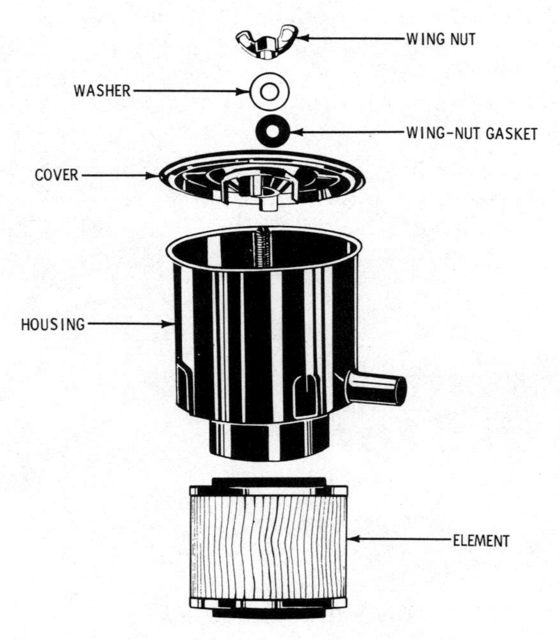

Courtesy Cummins Engine Co.

Fig. 6-2. Crankcase-breather paper element. It is important that the air be clean because dust destroys the lubricating quality of oil and forms a grinding compound that wears parts.

cylinder head, the valve cage may either become warped and leak or break the hold-down bolts. Also make sure that the valve cage is drawn down evenly by the bolts.

Why are copper–asbestos gaskets preferable to plain copper gaskets?

Answer: Because copper–asbestos material lends itself more fully to expansion and contraction, and to uneven surfaces.

CHAPTER 7

Air-Starting Valves

What is an air-starting valve?

Answer: An air-starting valve is a mechanically or pneumatically operated valve for opening and closing the compressed-air supply used in starting diesel engines.

For what other purpose is an air-starting valve used?

Answer: For rotating the engine while testing and adjusting it.

Name two main types of mechanically operated air-starting valves.

Answer: The poppet type and the disk type.

What are the essential differences between poppet-type and disk-type valves?

Answer: The poppet-type valve is a conically shaped valve actuated either directly or through levers from cams on the camshaft by the up-and-down action of the cams and admits air to the various cylinders at the proper time for starting.

The disk-type valve consists of two disks, one of which is stationary and has openings that lead by pipes to the working cylinders. The other disk has one opening and revolves over the stationary disk admitting air through that opening to the various cylinders in accordance with its speed.

What is a pilot valve or pneumatically operated air-starting valve?

Answer: It is a small piston valve actuated by a cam on the camshaft. The pilot valve operates the main valve that admits starting air to the working cylinders.

What condition is necessary to ensure proper operation of a pneumatic air-starting valve?

Answer: The maintenance of uniform air pressure to the valve cylinders that actuate the air-starting valves.

How is the air maintained at uniform pressure?

Answer: By an automatic pressure regulator in the air pipe line between the tanks and the air distributor or pilot valve.

Why is uniform air pressure so essential?

Answer: To ensure the opening of the air-starting valves at the proper time and speed.

What is a master air-starting valve?

Answer: The master or pressure regulating valve controls the pressure of the air supplied to the air-starting valves that lead to the various cylinders.

At what position of the piston should the air-starting valve open?

Answer: When the piston is at the top or dead center, and while the exhaust, inlet, and fuel valves are closed.

How long should the air-starting valve remain open?

Answer: Until air has been admitted during half of the working stroke.

What is an air-starting check valve?

Answer: It is a valve that checks or prevents the working air and the gas pressures in the cylinders from backing into the air-starting lines (Fig. 7-1).

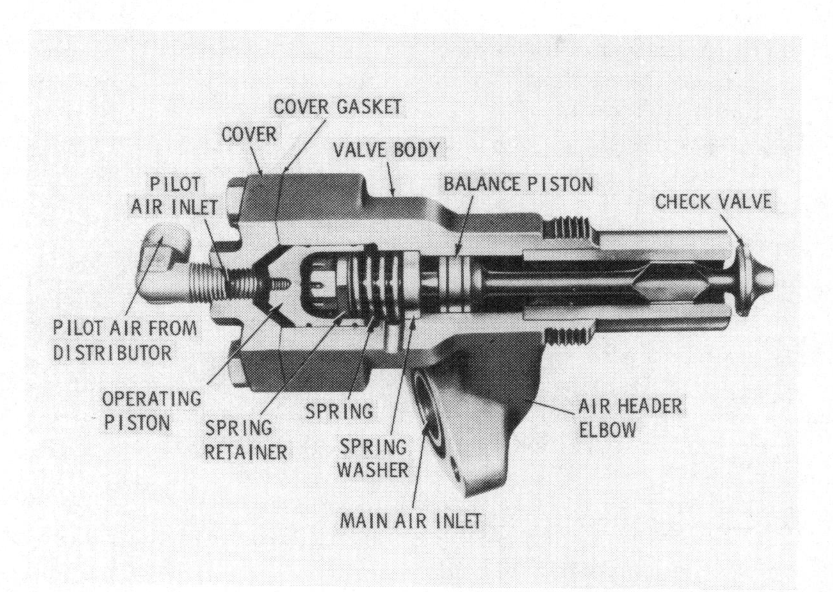

Courtesy Fairbanks, Morse & Co.

Fig. 7-1 Air-starting check valve. These valves receive lubricating oil with the air from the air-start distributor. The valves are cooled by water from the cylinder-liner water jacket.

Where is an air-starting check valve located?

Answer: In the head or upper part of the cylinder. It is used when the air-starting valves are located at a distance from the cylinders.

What difficulties are often encountered in the operation of air-starting systems?

Answer: Leaking air-starting check valves, sticking valves because of carbon deposits, and broken valve springs.

How can a leaking air-starting valve or check valve be detected?

Answer: By the air line heating near the valve connection, and by the uneven or sluggish movement of the engine while operating on starting air.

What effect may a leaking air-starting valve or check valve have on the operation of the engine?

Answer: The cylinders may become air-bound and keep the engine from rotating. The high working pressure in the cylinders may back into the air-starting lines which are at lower pressures, and damage or clog them. It may also produce difficulties in ignition of the fuel oil.

How are air-starting valves or check valves tested for tightness?

Answer: One method of testing for valve tightness is to thoroughly clean both the valve disk and valve seat and make several pencil marks across the seating part of the disk. Then, making sure that the stem and disk are in their true working position, by means of the valve-stem guide, rub the disk and seat together. If the disk is true, all the marks will be rubbed off. If the disk is not true, the remaining marks will show where it is out of true. The same procedures may be used to test the seat of the valve.

If the disk is out of true, it may be faced off on a lathe or other suitable machine. It is difficult to grind a valve to a tight fit if the disk is not true. However, it is comparatively easy to grind a valve to fit tightly if the disk is true. Another method of testing a valve for tightness is to assemble the valve, cover the disk with kerosene, and watch for leaking kerosene drops.

What care should valves be given?

Answer: Valves, along with their stems and guides, should be kept clean and should be ground when necessary.

What is a good method for grinding valves?

Answer: Pour a little kerosene on the valve seat, rub the disk and seat together, separate them, and dry both parts. The amount of grinding required will then be shown by the bright spots or circles on the seat and disk. File smooth. Either grind or file off any shoulders resulting from the grinding. Do not grind or file any more than is necessary to make the valve tight.

What compounds should be used in valve grinding?

Answer: Use any of the well-known grinding compounds on the market. Use a coarse-grade compound for the main grinding, when there is a large amount of material to be removed, and a fine grade of grinding compound for removing small amounts and for finishing.

Describe the mechanical process of grinding valves.

Answer: Provide a spring or device for holding the valve off the seat. It should not be so strong that it prevents the disk from coming in contact with the valve seat. Clean the disk and valve seat with fuel oil. Apply the grinding compound to the disk evenly and rotate the disk back and forth on the valve seat—permitting the spring to raise the valve off the seat from time to time.

CHAPTER 8

Basic Mercedes Diesels

What are the three basic Mercedes-Benz diesels?

Answer: The 615 and 616 Mercedes diesels are four-cylinder engines; the 617 is a five-cylinder engine.

What is the functional description?

Answer: The 615, 616, and 617 diesels have a closed, service-free crankcase breathing system. Engine blow-by gases and cylinder crankcase vapors will flow through connection in cylinder head cover to water separator with by-pass bore at rear pair of intake pipe (Fig 8-1).

What are the cylinder heads made of?

Answer: The three diesels have aluminum heads. The reason for this is better heat distribution. The head of the engine gets hot very rapidly. Aluminum withstands heat much better than cast steels do. Aluminum also reduces the overall weight of the engine.

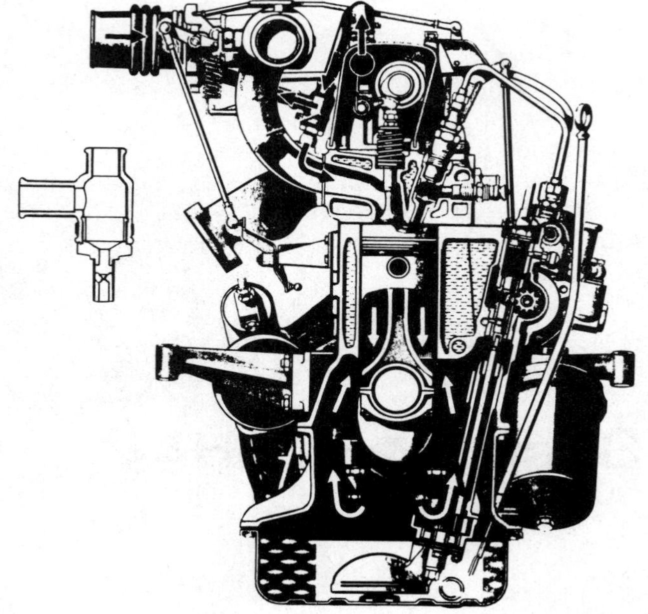

Courtesy Mercedes-Benz

Fig. 8-1. Cross section of 615 and 616 diesel engine.

What type of valve is used?

Answer: The valve used in these diesels is hollow at the top and contains sodium. This enhances the working time of the engine because heat from the engine will be controlled. Compared to the solid steel valve used in most Cummins, Mack, and Detroit diesels, the sodium valve is far superior.

What type of valve seal is used?

Answer: Valve seals for these diesels are made of a thermoset nylon. This plastic material withstands the expansion and contraction produced by heat from the engine. Compared to the rubber valve seals that most American diesels use, the thermoset nylon valve seal lasts longer. Rubber tends to dry rot and crack due to heat, shortening engine work time.

What type of camshaft is used?

Answer: An overhead camshaft is used in all three diesels.

What roles do the timing chain and tension rail play?

Answer: The timing chain plays an important role, for without it the engine will not run. Most problems encountered with the timing chain result from wearing and stretching. The Mercedes 615, 616, and 617 diesels come equipped with a tension rail that enables slack to be removed in the timing chain. Longer engine time is thus accomplished. Most American diesels have a problem with their timing chain stretching, which makes engine work time shorter (Fig. 8-2) .

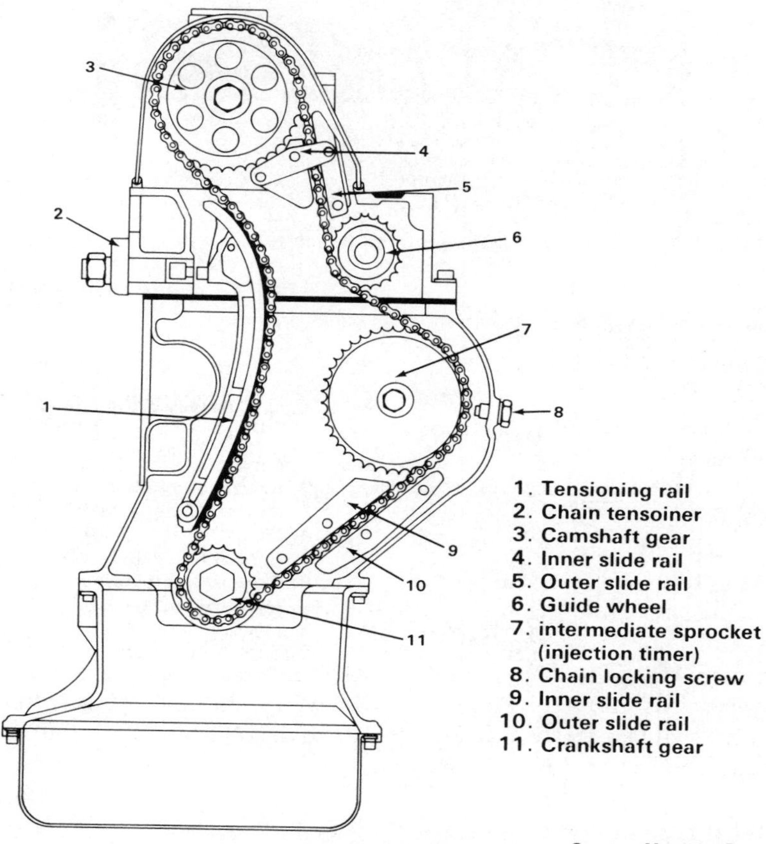

Fig. 8-2 Front view of timing chain and tension rail.

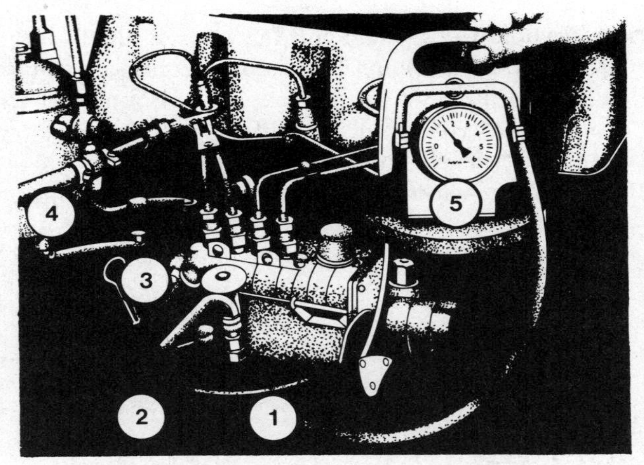

1. Fuel pump — fuel inlet (suction end)
2. Fuel pump — fuel outlet (suction end)
3. Injection pump — fuel inlet
4. Fuel main filter — fuel outlet
5. Tester for fuel pump

Courtesy Mercedes-Benz

Fig. 8-3. Pictoral view of side section of Bosch fuel pump.

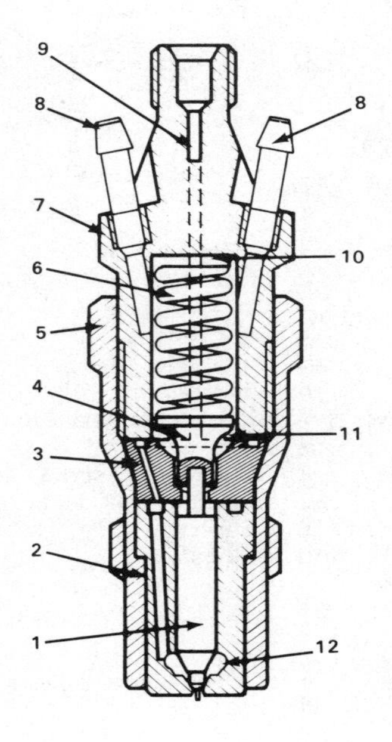

Fig. 8-4. Cutaway of fuel injector.

1. Nozzle needle
2. Nozzle body
3. Nozzle holder insert
4. Thrust bolt
5. Injection nozzle bottom
6. Compression spring
7. Injection nozzle top
8. Leak oil connection
9. Fuel connection
10. Steel washer
11. Ring groove and feed holes
12. Pressure chamber in nozzle body

Courtesy Mercedes-Benz

What type of injection pump is used?

Answer: Again, the 615, 616, and 617 diesel engines contain a Bosch fuel pump (Fig. 8-4).

Why are the injectors important?

Answer: The injectors play an important part because they spray the fine, vaporous fuel into the cylinder without which, combustion will not take place, stopping the working of the diesel.

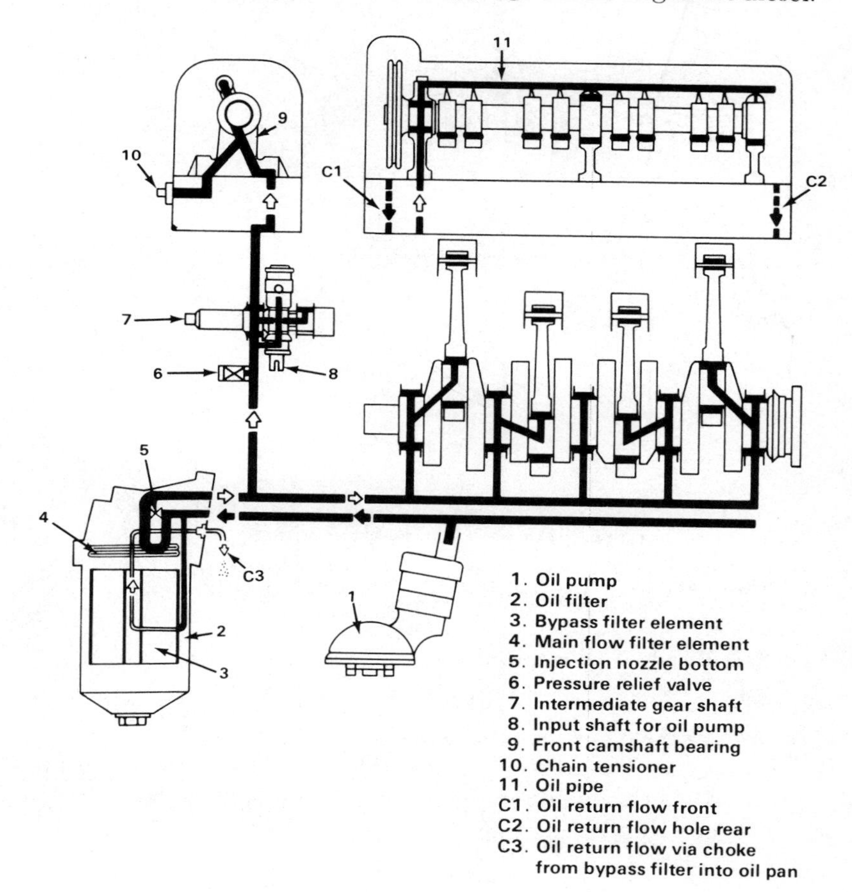

Courtesy Mercedes-Benz

Fig. 8-5 Pictorial view of oil circuit.

How does the lubrication oil circuit work?

Answer: Lubricating a diesel engine and understanding the path the oil takes is a very important process. Much more heat is produced by a diesel engine than by a gasoline engine. Understanding oil lubrication is vital with these engines (Fig. 8-5).

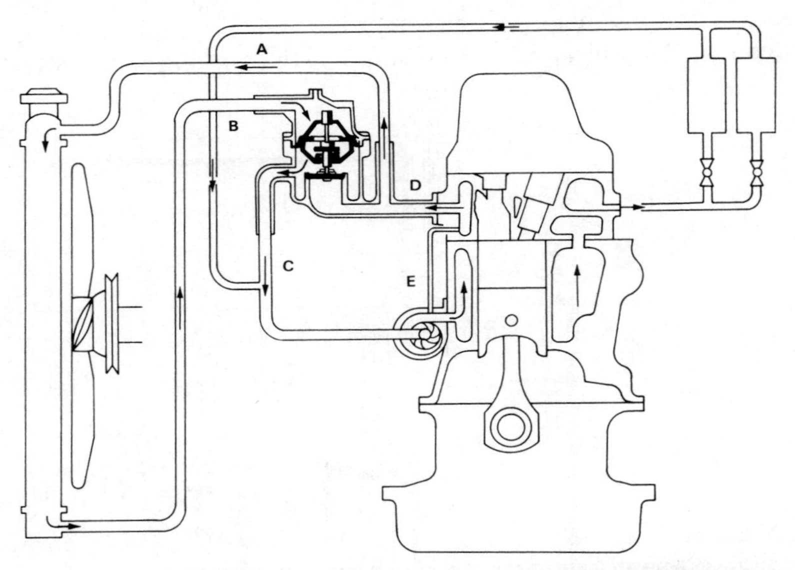

Courtesy Mercedes-Benz

Coolant circuit under full load operation — at high ambient temperatures

A. Toward radiator
B. From radiator
C. Bypass line from thermostat housing to water pump
D. From engine
E. Venting line

Fig. 8-6 Pictorial view of cooling circuit.

How does the coolant circuit work?

Answer: Cooling these engines and understanding the coolant circuit at high ambient temperatures is also vital (Fig. 8-6).

CHAPTER 9

Scavenging-Air Compressors

What is the function of a scavenging-air compressor in diesel engine operation?

Answer: To supply air for clearing the cylinders of the burned gases after each power stroke, and to supply charging air for compression.

How do scavenging-air compressors differ from starting-air compressors?

Answer: They supply air at much lower pressures but in larger quantities (Fig. 9-1).

What air pressures are used for providing scavenging air for the cylinders?

Answer: Two to six pounds, under ordinary operating conditions.

How is scavenging air supplied to the cylinder?

Answer: Through air receivers supplied by the compressors.

Courtesy Worthington Corp.

Fig. 9-1. Starting-air compressors and storage tanks. Air-cooled compressors can be powered with an electric motor or a gasoline engine.

What types of air compressors are usually used for supplying scavenging air?

Answer: The reciprocating and the rotary types.

What type of valve is usually used with reciprocating air compressors?

Answer: A plate or feather valve because it requires minimum attention.

How is air pressure regulated in scavenging-air equipment?

Answer: By regulating the quantity of air taken in by the compressor. This is done by means of a hand-regulating valve located in the suction line.

How is air pressure controlled in scavenging-air receivers?

1 Water pump coupler half
2. End Cover
3. Gasket
4. Gasket
5. Bolt for bearing cages
6. Press bearings
7. Bearing cages
8. Bearing cage seal rings
9. Oil collector rings
10. Housing dowels
11. End plate
12. Housing
13. Bushings
14. O rings
15. Oil pressure ferrules
16. Shaft ring seals
17. Rotor shafts
18. Dowels
19. End plate
20. Thrust washers

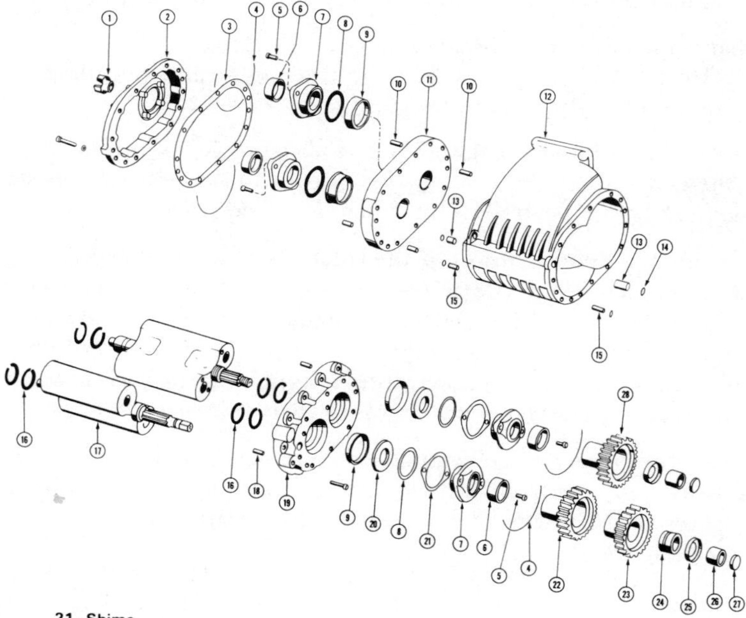

21. Shims
22. Timing gear
23. Drive gear
24. Out Board journal
25. Lockwasher
26. Shaft nut
27. Lock ring
28. Timing gear

Courtesy Cummins Engine Co.

Fig. 9-2. Exploded view of Cummins C-J series supercharger.

Answer: By means of a spring-loaded plate that acts as an excessive pressure relief valve.

What are the objections to the reciprocating type of air compressor for scavenging-air purposes?

Answer: It is large and cumbersome when used in large installations, consists of many moving parts, and is noisy in operation.

What is a supercharger?

Answer: A supercharger is a mechanism mounted to the exhaust manifold. Exhaust gases pass through the supercharger changing excess exhaust fumes to added horsepower (Fig. 9-2).

What is meant by turbo-scavenging?

Answer: Scavenging in which the scavenging air is supplied by turbo- or rotary-type air compressors.

How is the turbo-scavenging type of air compressor operated?

Answer: Either by a chain drive from the crankshaft of the engine or by an electric motor when operated separately.

What are the good features of the rotary-type supercharger?

Answer: The rotary-type supercharger converts exhaust into workable horsepower and if kept clean of carbon from exhaust, usually requires less space, requires a smaller amount of lubrication, is almost noiseless in operation if kept clean of carbon from exhaust, and gives an even flow of air (Fig. 9-2).

How is the supply of air regulated in a rotary-type compressor?

Answer: By controlling the quantity of air passing through the compressor, by means of a hand valve in the suction line, and by controlling the speed.

CHAPTER 10

Diesel Engine Cylinders

What are engine cylinders?

Answer: The cylinders are the chambers located inside the engine in which the air is compressed, the fuel is ignited, and the power is produced (Fig. 10-1). In addition to withstanding both high pressure and intense heat, they must withstand the friction of the pistons working back and forth, as well as the mechanical strains and stresses of the engine. Therefore, the cylinders must be accurately and strongly constructed.

How is the temperature in a cylinder kept within the working range to permit proper lubrication?

Answer: By a water jacket surrounding the cylinder, in which cooling water circulates.

How is a diesel engine cylinder constructed?

Answer: Usually of a strong close-grained cast iron, accurately bored and ground, in which a piston can travel freely.

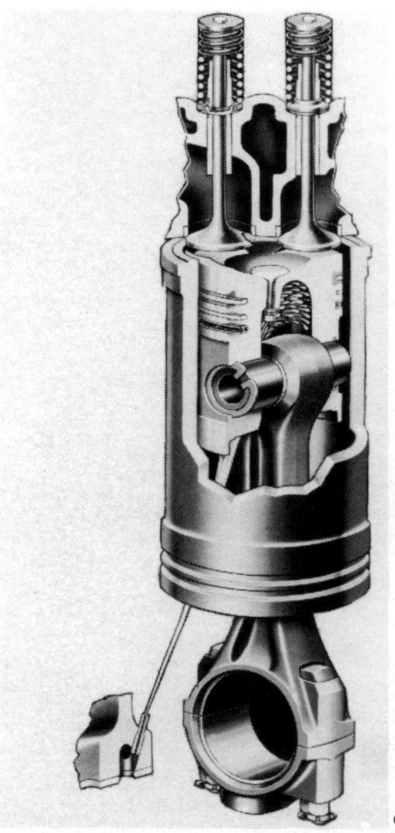

Courtesy Caterpillar Tractor Co.

Fig. 10-1. Cylinder cutaway with valves.

What two methods of construction are generally adopted for diesel cylinders?

Answer: The one-piece and the sectional construction. In addition, they are cast either single or multiple and with removable liners (Fig. 10-2).

Describe one-piece construction and sectional construction of diesel engine cylinders.

Answer: In one-piece construction the cylinder is cast so that it includes the cylinder, the cylinder head (Fig. 10-3), and the water jacket.

In sectional construction, each cylinder section is cast separately and fastened together.

What is a disadvantage of one-piece construction?

Answer: When it is necessary to renew the cylinder liner, the entire casting including the cylinder head and the water jacket must be discarded.

What are other disadvantages of this construction?

Answer: It is difficult to tell whether the cylinder is of uniform thickness throughout, whether the cooling water space is uniform, and whether the cylinder head is flawless.

Why is this method of construction in use?

Answer: It lowers construction and assembly costs because it

Courtesy Worthington Corp.

Fig. 10-2. Cylinder head for four-stroke-cycle Worthington SW engine. This cylinder head has two air-inlet and two exhaust valves. Gaskets are not required between the cylinder head and liner.

Courtesy Worthington Corp.

Fig. 10-3. Cylinder head for four-stroke-cycle Worthington SEH engine. This cylinder head seats on the liner. A pressure-tight seal is provided with the use of cylinder-head gaskets.

does not require fastening and connecting as is required in sectional types of construction.

What is meant by single-block or multiple-block construction?

Answer: Construction in units of one or more cylinders in which both the cooling water jacket and the cylinder liner are cast in one piece (Fig. 10-4).

What are the objections to combined water-jacket and cylinder liner construction?

Answer: If it becomes necessary to renew the cylinder wall, the entire casting must be scrapped. Also, there is more possibility of an uneven thickness of the cylinder wall which causes overheating where the wall is thickest. In addition, there is a possibility of

Courtesy Worthington Corp.

Fig. 10-4. Multiple-cylinder block for Worthington SW diesel engine.

stoppages of the cooling water channels in the water jacket and unevenness of expansion between the cylinder liner and the water-jacket castings.

What is the effect of lack of uniformity in the thickness of the cylinder wall?

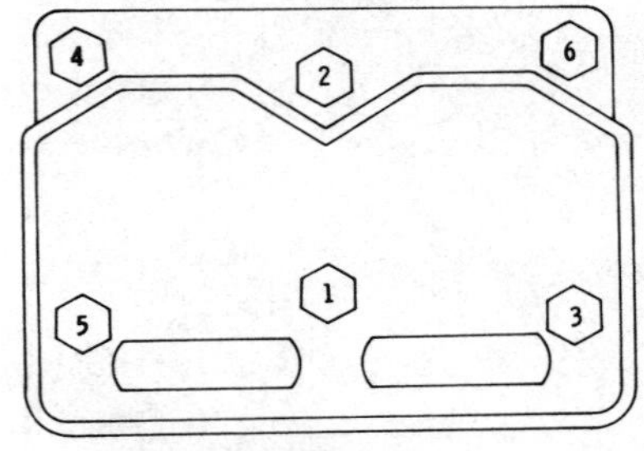

Courtesy Cummins Engine Co.

Fig. 10-5. Sequence for tightening cylinder head on Cummins 4⅞-inch and 5⅛-inch bore engines. Continue tightening in rotation and in 100 foot-pound steps with torque wrench until all cap screws have been tightened to 460/480 foot-pounds.

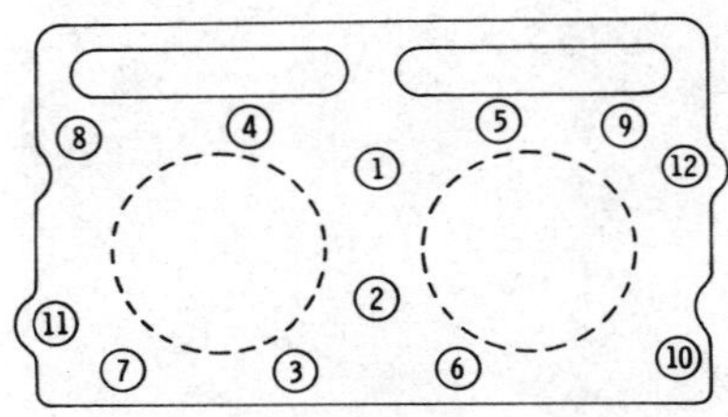

Courtesy Cummins Engine Co.

Fig. 10-6. Sequence for tightening cylinder head on Cummins 5½-inch bore engine. Continue tightening with torque wrench in rotation in 100 foot-pound steps to 315/335 foot-pounds.

Answer: An uneven quantity of water circulates and leaves hot spots, which affect lubrication of the cylinder.

What is meant by separable sleeves or liners?

Answer: They are the removable sleeves that form the working cylinders in the cylinder block.

What are the advantages of removable sleeves in the cylinder block?

Answer: They provide an even thickness of the cylinder wall

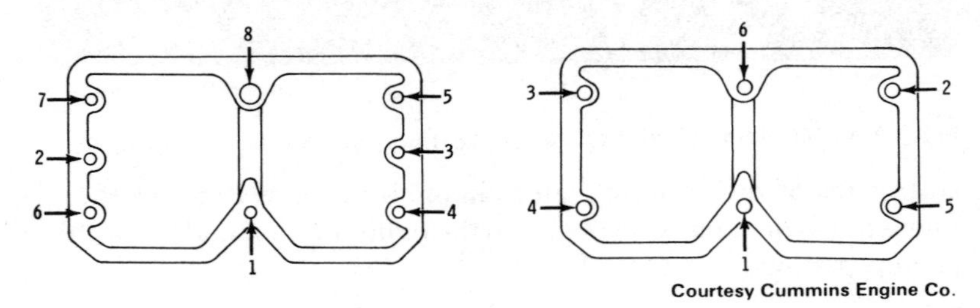

Courtesy Cummins Engine Co.

Fig. 10-7. Sequence for tightening rocker housing on Cummins diesel engine. Tighten in sequence to 55/75 foot-pounds.

which lends itself to more uniform heating and lubrication. Removable sleeves take care of differences in expansion between the sleeve and the water-jacket casting. They also lower the cost of cylinder wall renewal.

How are cylinder sleeves made watertight to prevent water entering the working cylinder and the crankcase?

Answer: A copper gasket is inserted underneath the flange at the top of the cylinder sleeve (Fig. 10-8) where it is set in a groove

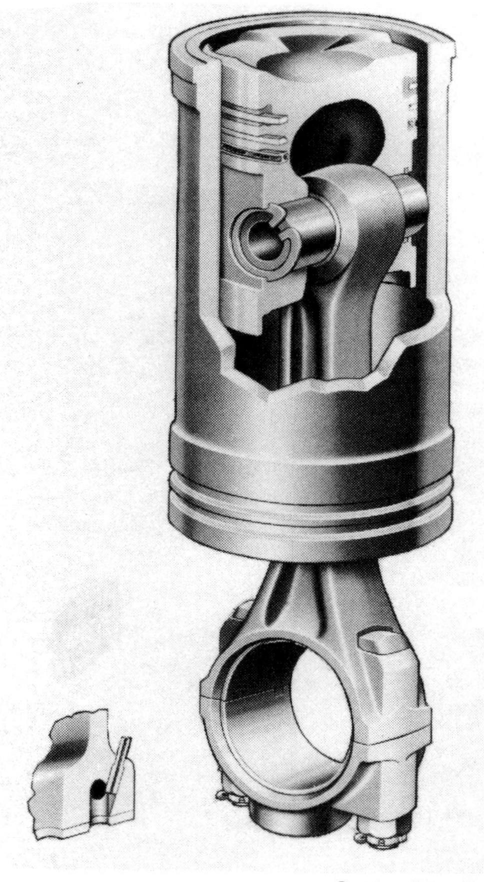

Courtesy Caterpillar Tractor Co.

Fig. 10-8. Cutaway of cylinder and piston.

provided for it. A gasket set in a groove in the lower part of the water-jacket casting also prevents entrance of water.

In addition to watertightness, what other purposes do the gaskets serve?

Answer: They serve as cushions to provide for differences in expansion and contraction of parts.

CHAPTER 11

Pistons and Rings

What material is used for piston construction?

Answer: Close-grained cast iron, although alloys to make pistons strong yet light are frequently used. Steel rings for reinforcement are embodied in many of them because pistons must retain their shape and be airtight under the intense heat and pressure to which they are subjected almost continually.

What types of pistons are used in diesel engines?

Answer: The plain, or uncooled, and the cooled types (Fig. 11-1) .

Why is it advantageous to cool pistons?

Answer: Because the heat, 800 to 1,000 degrees Fahrenheit, reduces the effect of the lubricant and affects the metal itself.

How are pistons cooled?

Answer: By circulating lubricating oil or cooling water through

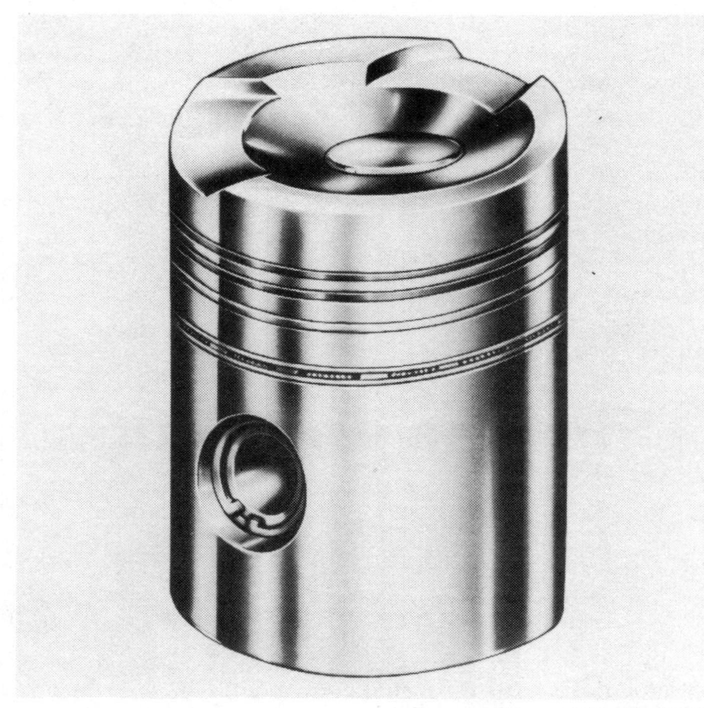

Courtesy Caterpillar Tractor Co.

Fig. 11-1. Piston with piston rings from a Caterpillar diesel engine.

the cavities or spaces in the piston—the method varying with different designs. The lubricating oil or cooling water then passes through coolers where its temperature is reduced and from where it is recirculated.

Why is water not generally used for cooling pistons?

Answer: Because of the possibility of leaks in the piping and water, which may do considerable damage, getting into the lubricating system.

What other consideration must be given to the design and operation of pistons?

Answer: The effect of temperature on metals—expansion.

Greater allowance must be made for expansion of the metal at the top of the piston where it comes in contact with the heat, than at the lower side or bottom. If allowance for expansion were not made, the intense heat would cause the piston to wedge.

Why are pistons dished or somewhat hollowed at the top?

Answer: To provide a greater combustion space and more logical surface for the fuel gases to act upon. Also, the stresses are more centrally applied. The top of the piston is designed to compress the air closer to the fuel oil spray (Fig. 11-2).

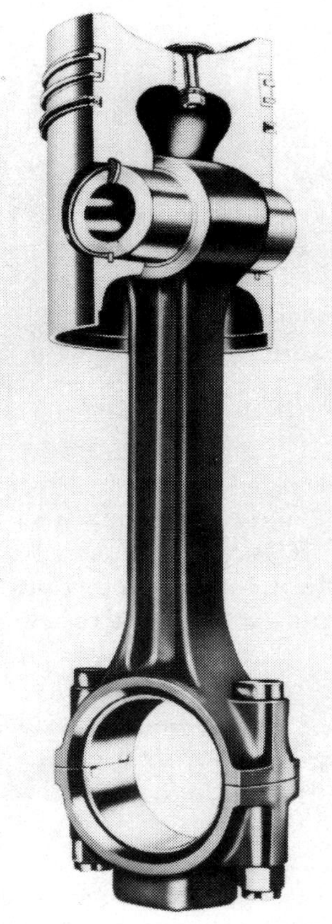

Courtesy Caterpillar Tractor Co.

Fig. 11-2. Partial cutaway view of Caterpillar diesel engine 6.25-bore piston and connecting rod.

What other construction features are found in pistons?

Answer: Many of them have replaceable tops or heads for easy replacement in event of damage. These replaceable heads are securely bolted to the piston body.

What does pounding in a cylinder indicate?

Answer: That the piston may be excessively heated or needs lubrication.

What should be done when a cylinder pounds?

Answer: The fuel supply to that cylinder should be cut off, the engine slowed down, and the lubrication increased until the pounding stops.

How is a cylinder taken out of power service?

Answer: In emergency, by cutting off the fuel to the affected cylinder and by opening and removing the indicator cock or spray nozzle to relieve compression.

What is meant by piston seizure?

Answer: Binding of the piston and the cylinder wall as a result of the lubrication having been destroyed by excessive temperature and friction.

How is piston seizure relieved?

Answer: By taking the cylinder out of power service, as given above, by increasing the lubrication, and by temporarily decreasing the quantity of cooling water circulating around the cylinder wall which thereby permits the cylinder wall to heat and expand—relieving the friction.

How can approaching piston seizure be recognized?

Answer: By increased temperature of the cylinder water jacket or the cooling water of that particular cylinder and by a slowing down of the engine.

How does expansion of the cylinder wall relieve piston friction?

Answer: By increasing the clearance space between the cylinder wall and the piston. The increased heat of the cylinder expands the metal in the cylinder wall, but does not expand the metal in the piston to the same degree.

What action should be taken after the temporary treatment of piston seizure?

Answer: An effort should be made to avoid a recurrence by determining the cause and eliminating it.

What serious damage can be the result of piston seizure?

Answer: The sudden seizure or stoppage of a piston may break the connecting-rod bolts, spring or twist the crankshaft, score the cylinder wall and piston, and result in extensive and costly repair bills for the engine.

Why can such damage be caused by piston seizure?

Answer: Because the momentum of the flywheel and the power of the other pistons tend to continue the movement; and something must give.

How can a sprung or twisted crankshaft be detected?

Answer: By excessively high temperature, although the engine is properly lubricated.

What is meant by cylinder or piston scoring?

Answer: Grooves in the cylinder wall or piston, or in both. It is caused by the piston scraping the cylinder wall in its movement without proper lubrication.

What usually causes scoring?

Answer: Lack of proper lubrication, overheating of the piston, and lowering of the temperature of the cylinder wall by an increased quantity of cold circulating water.

How can a cylinder wall be smoothed after being scored?

Answer: By rubbing down the surface, first with a coarse carborundum shaped to the contour, or curve, of the cylinder and then finishing off with a smooth or finer stone. The piston can also be smoothed in the same manner.

Will smoothing a cylinder wall after scoring result in good operation?

Answer: No, because removing metal from the wall of a cylinder increases the space between the cylinder wall and the piston, which results in improper fit.

What can be done to prevent scoring?

Answer: Prevent the pistons scraping against the cylinder walls by proper lubrication and by maintaining correct temperatures of pistons and cylinders.

When a cylinder wall has become badly scored, what action should be taken?

Answer: Regrind the cylinder wall and fit a new piston.

What is a piston skirt?

Answer: That portion of the piston which extends below the piston pin and serves as a guide for the piston and connecting rod. It also serves to distribute lubricating oil to the piston and cylinder wall.

What are piston rings?

Answer: Rings located in the grooves of the piston, usually near the top and sometimes near the bottom. The top set of rings makes the piston and the cylinder walls air tight and these rings are called compression rings, while the lower set is a factor in lubrication and these rings are called oil rings (Fig. 11-3).

Why are piston rings needed on diesel engine pistons?

Answer: To make the pistons pressure tight. They adjust themselves to the contour of the walls of the cylinders and thereby cause a tight fit (Fig. 11-4).

How many rings does a diesel engine piston usually have?

Answer: Six or more compression rings at the top—depending upon the size and type of piston. Some have one or two additional rings near the bottom of the skirt which serve as scrapers or fins to keep excess lubricating oil from the crankcase getting up into the

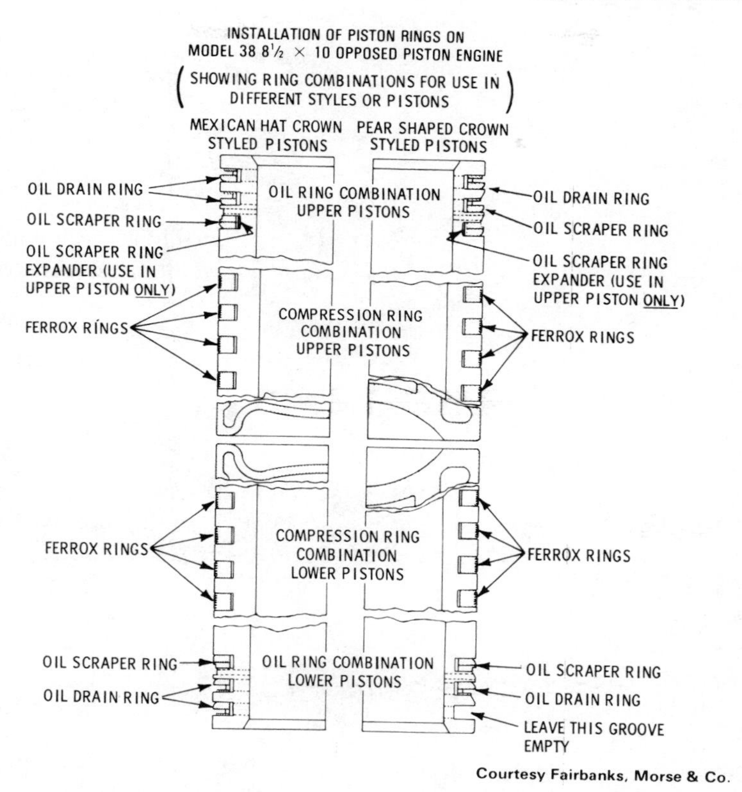

Fig. 11-3. Piston-ring combinations for use in different styles of pistons on Fairbanks-Morse Model 38, 8½ × 10, opposed piston engine.

combustion chamber of the cylinder. Oil rings near the bottom are necessary where the splash system of lubrication is used.

What are the qualities of a good piston ring?

Answer: Flexibility—to be able to conform to irregularities of the cylinder wall; elasticity—to keep an even tension; and durability—to withstand a reasonable amount of service wear.

What is the result if more than the required amount of lubricating oil gets into the working cylinder or compression chamber?

Answer: The oil fouls the cylinder wall and piston rings because

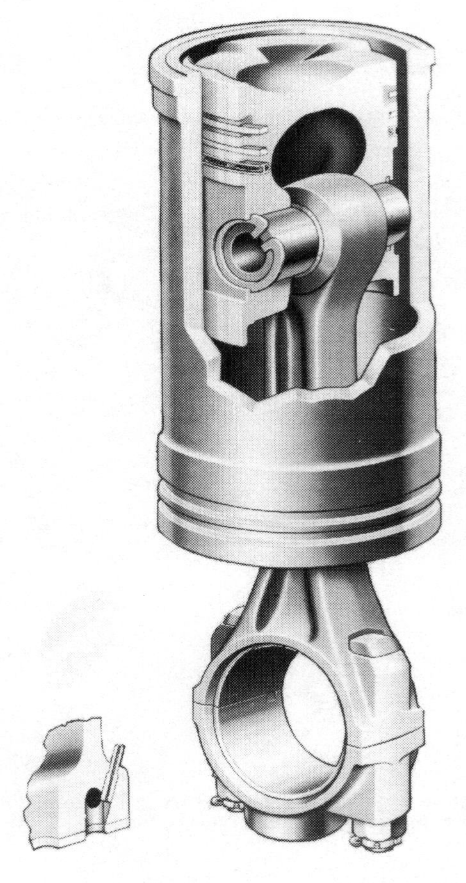

Courtesy Caterpillar Tractor Co.

Fig. 11-4. Cutaway of piston in its cylinder showing piston rings and interior of the piston.

of the heat. It sticks to the piston, the rings, and the ring grooves—which results in sluggish operation and causes overheating of the piston.

What prevents piston-ring joints from rotating and becoming aligned on the piston, forming a channel through which the gases and air might escape?

Answer: Dowel pins are set in the piston to prevent the rings from rotating in the grooves.

How can the piston rings be sealed to prevent air leakage in a badly worn cylinder?

Answer: By coating the cylinder wall with a thick lubricating oil. This can be done by heating the thick oil to make it more fluid and then injecting it into the cylinder with a squirt gun—making sure that the heavy oil reaches and coats the entire cylinder wall.

How can piston rings be removed if they stick in the ring grooves?

Answer: By boiling the piston and rings in a strong solution of lye water to remove any deposits that might cause them to stick.

How can stuck piston rings be detected in a two-stroke-cycle engine?

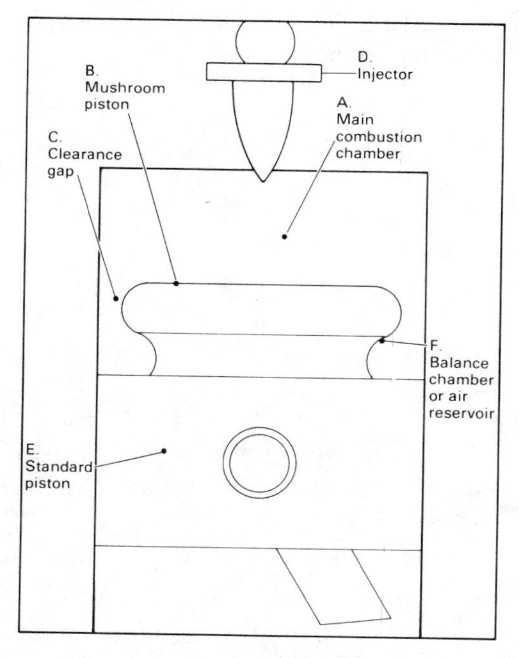

Fig. 11-5. Cross section of a mushroom piston.

Answer: By removing a plate on the exhaust manifold, opposite the exhaust port, and pushing on the rings with a stick or iron rod to determine whether there is any movement of the rings in their grooves.

Explain how rings, which are set on their pistons, may be easily compressed for insertion of the piston into the cylinders.

Answer: One method is to use a sloping pot or cylinder, which fits into the cylinder head recess and compresses the rings as the piston goes into the cylinder. Another method is to use a ring compressor, which is a round clamp that can be drawn tightly around the rings and compresses them. Still another method is to use a homemade device when the cylinder head joint recess is bored outward from the edge of the cylinder.

What is a mushroom piston?

Answer: The mushroom piston is a mechanism that can be adapted to any standard diesel piston. It enhances combustion. It also increases thermo-efficiency (Fig. 11-5).

CHAPTER 12

Governors

How are diesel engine speeds controlled?

Answer: By a speed governing device—or governor (Figs. 12-1 and 12-2.).

What is a governor?

Answer: A governor is a device that automatically governs or controls the speed of an engine. It keeps the engine speed constant regardless of changes in the engine load.

Why must engine speeds be controlled or governed?

Answer: To prevent ruining the engine when there is a sudden drop in load and to maintain required engine speed when an increased load tends to slow down the engine. Engine speeds must also be controlled because electric generators require a constant speed for maintaining a constant voltage.

How can a governor control engine speeds?

Answer: By automatically increasing or decreasing the fuel supply to the cylinders in accordance with the load requirements—the more fuel, the more power—the less fuel, the less power.

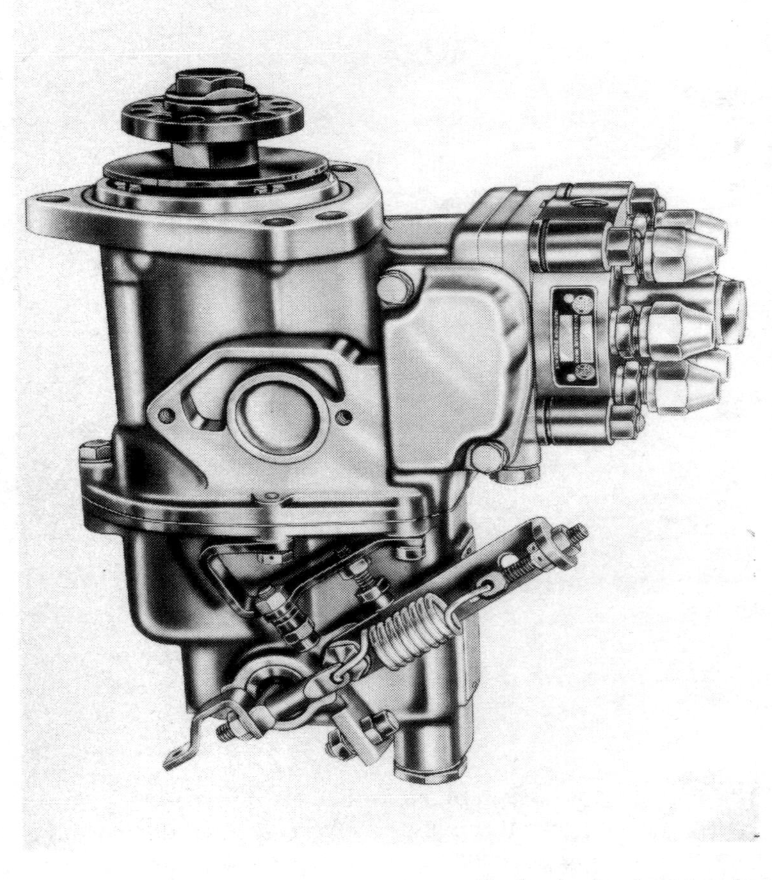

Courtesy American Bosch Arma Corp.

Fig. 12-1. American Bosch PSB fuel injection pump with external-spring governor.

How can governors increase or decrease the fuel supply to an engine?

Answer: In a number of ways; the most common way is to open the suction valve of the fuel pump at given times, permitting part of the fuel oil to be by-passed to the storage tanks. Another way is to lengthen or shorten the fuel pump plunger stroke so that either more or less fuel is injected at each stroke or movement. Still another way is to oscillate the fuel pump plunger, causing part of the fuel to be by-passed to the storage tank. The control of all these is maintained by a system of levers actuated by the governor mechanism (Fig. 12-3).

Courtesy American Bosch Arma Corp.

Fig. 12-2. American Bosch PSB fuel injection pump with internal-spring governor.

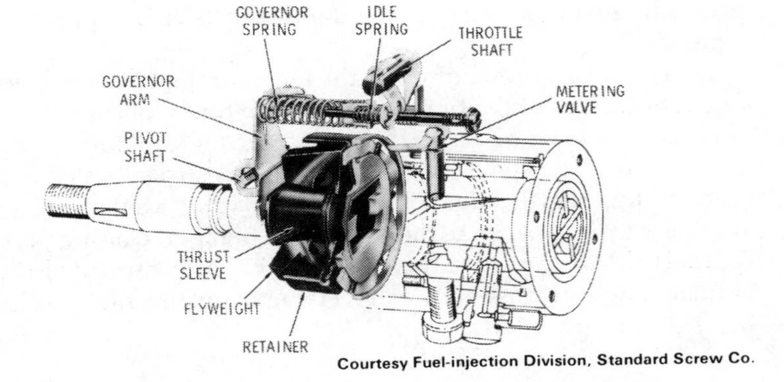

Courtesy Fuel-injection Division, Standard Screw Co.

Fig. 12-3. Roosa Master centrifugal governor. Movement of the flyweights against the governor thrust sleeve rotates the metering valve. This rotation varies the registry of the metering-valve slot with the passage to the rotor, thus controlling the flow of fuel to the engine.

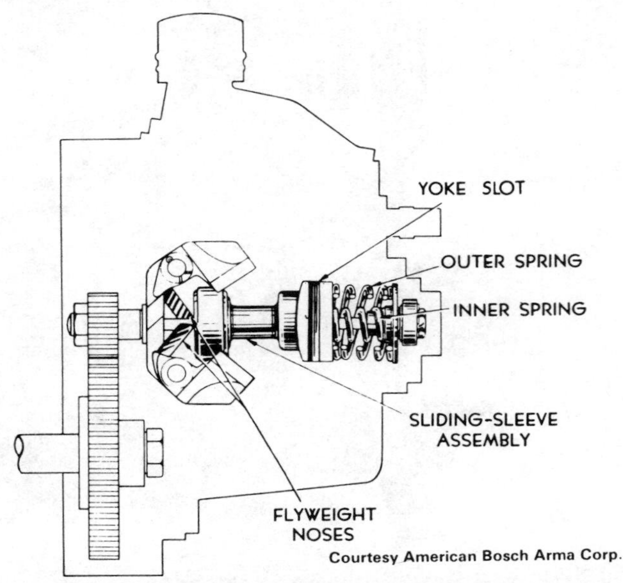

Courtesy American Bosch Arma Corp.

Fig. 12-4. Operating principle of American Bosch mechanical governor. Centrifugal force moves the weights outward and moves the weight noses longitudinally along the shaft. The governor spring opposes this movement through the sliding-sleeve assembly. Any change in speed changes the position of the sliding-sleeve assembly.

Describe the various governors.

Answer: There are several types but the most common is the centrifugal governor, centrifugal meaning action by centrifugal force (Figs. 12-4 and 12-5.).

Centrifugal force is the force that tends to make a weight move away from the center when it is revolved. The more rapidly it is revolved, the farther away it will tend to move. It is well known that when a weight on a string is swung about a central point, the string will become taut and quite a pull or force will be exerted

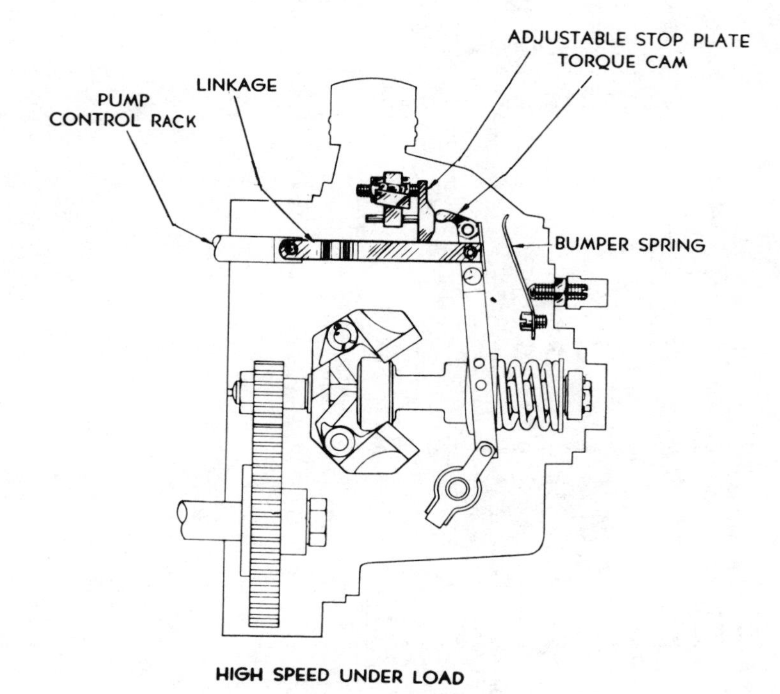

Courtesy American Bosch Arma Corp.

Fig. 12-5. Type GV American Bosch governor operating at high speed under load. The operating-lever shaft hub and the yoke hub are aligned with each other. The yoke, pivoting about the pivot screws, has forced the torque cam against the stop plate, thereby giving maximum control rack movement. Centrifugal force, caused by high speed, causes the weights to be extended.

upon it. That pull or force has the power to move an object. This power is utilized in doing work—the work being the movement of rods or levers that can open and close valves.

A centrifugal governor is such a device; but instead of having a

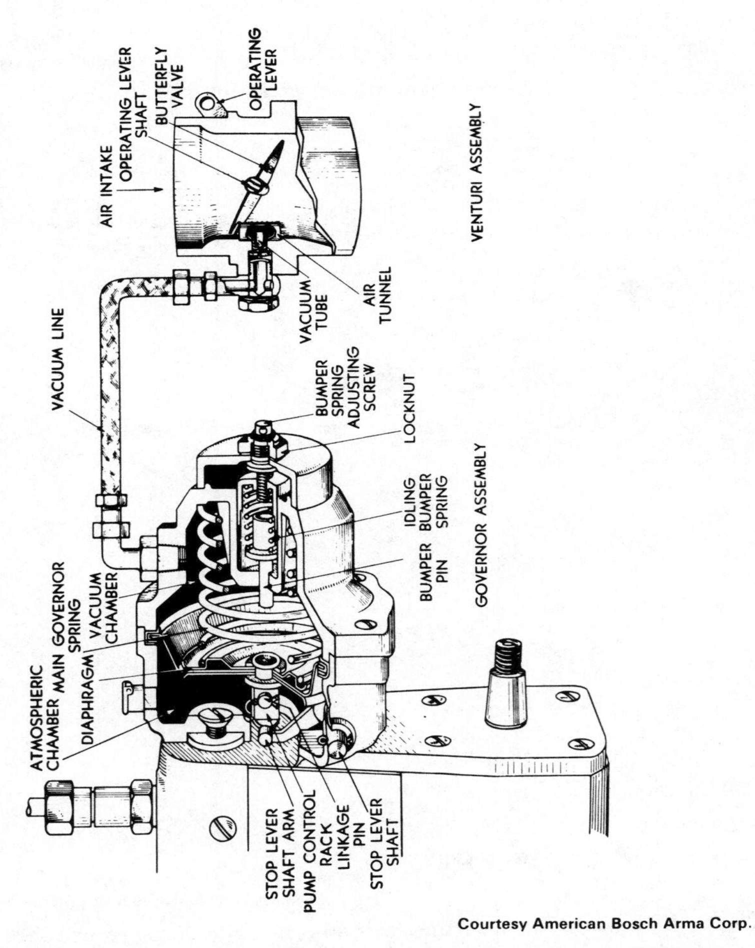

Courtesy American Bosch Arma Corp.

Fig. 12-6. American Bosch GPA pneumatic governor. The principle underlying the action of the pneumatic governor is that air passing through a tube tends to create a vacuum in another, smaller tube, entering it at right angles to the tube.

weight at the end of a string, it has two weights or fly balls fastened to the ends of movable levers so that they can swing outward when revolved and at the same time carry with them other levers that pull or push rods to actuate valves.

Other types of governors, such as shaft or inertia governors, act on the camshaft and control the cutoff time of the injected fuel. A shaft governor is connected to the end of the camshaft and an inertia governor is directly connected to, or is a part of, the flywheel. Both types may be working on the principles of centrifugal force and inertia. Inertia is the property of a weight to remain at rest when at rest and remain in motion when in motion. The variations are brought about by changes in the speed of the engine against the natural position of the governor aided by springs.

CHAPTER 13

Lubrication

Why is thorough and continuous lubrication necessary in a diesel engine?

Answer: Because of the intense heat and the high pressure present in a diesel engine.

What effect do intense heat and high pressure have on lubrication?

Answer: Heat thins the oil and destroys the lubricating quality of the oil making it useless for its function. The high pressure forces the burned oil particles into corners or crevices where they stick and eventually clog the lubricating system.

What is the real purpose of lubrication?

Answer: To provide a thin film or coating of oil between moving metal parts. The oil reduces friction and wear.

What general methods of lubrication are used in diesel engines?

Answer: The pressure system and the splash system.

What is meant by the pressure system of lubrication?

Answer: An oil pump provides the pressure required for forcing lubricating oil to the moving parts (Fig. 13-1).

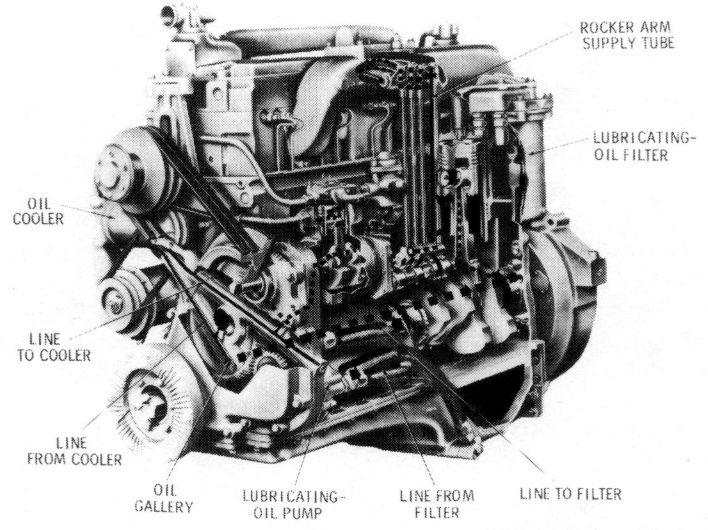

Courtesy Cummins Engine Co.

Fig. 13-1 Pressure-type lubricating oil circulation through the Cummins six-cylinder engine.

How is lubrication provided by the splash system?

Answer: A quantity of lubricating oil is maintained in the crankcase. The action of the crank journals dipping into the supply of crankcase oil splashes the oil over the moving parts.

What is meant by the gravity system of lubrication?

Answer: A system where the oil drops into the parts by the force of its own weight or gravity.

How is lubricating oil splashed into a cylinder?

Answer: By being thrown from the crank journals. The oil originally may come from a pressure pump, which forces it into the bearings, the crankshaft, the crank journals, up through the connecting rod and wrist pin, and then throws it against the cylinder wall.

What is the chief drawback of the splash system to proper lubrication?

Answer: Inability to regulate the quantity of oil splashed against the cylinder wall, or inability to keep the oil from getting past the piston head into the combustion chamber, burning with the fuel, and passing out with the exhaust gases.

What may be another effect of too much oil getting into the working cylinder?

Answer: Oil may flood the piston grooves, carbonize, and cause the rings to stick in the grooves.

What is the meaning of timed cylinder lubrication?

Answer: Lubrication, or providing oil, at a definite time conforming to the movement of the piston. In a four-stroke-cycle engine, the pump injects oil into the space between the first and second rings of the piston while the piston is at the bottom of the stroke so that, on its upward stroke, it is riding in oil. In two-stroke-cycle engines, the pump is timed to inject oil after the piston has passed the exhaust ports on its upward stroke. Timing is brought about by adjusting the forced-feed cylinder oil pumps to inject the oil at a proper time to meet the requirements.

How do forced-feed pumps lubricate the cylinders?

Answer: They force oil into spiral grooves encircling the inside of the cylinder wall. This spreads lubricating oil over the entire cylinder-wall surface and piston. They also force oil, through small holes in the cylinder wall, below the ring travel, from where the oil is spread by the piston (Fig. 13-2).

Is there any difference in the quantity of lubricating oil used in two-stroke-cycle engines and in four-stroke-cycle engines?

Answer: Yes. In two-stroke-cycle engines some of the lubricating oil is blown out through the exhaust ports or valves by the scavenging and charging air. In four-stroke-cycle engines this does not occur because air is drawn in and exhausted by the piston through the exhaust valves.

Why is more lubricating oil used in the operation of a crankcase-scavenging two-stroke-cycle engine than in other types of two-stroke-cycle engines?

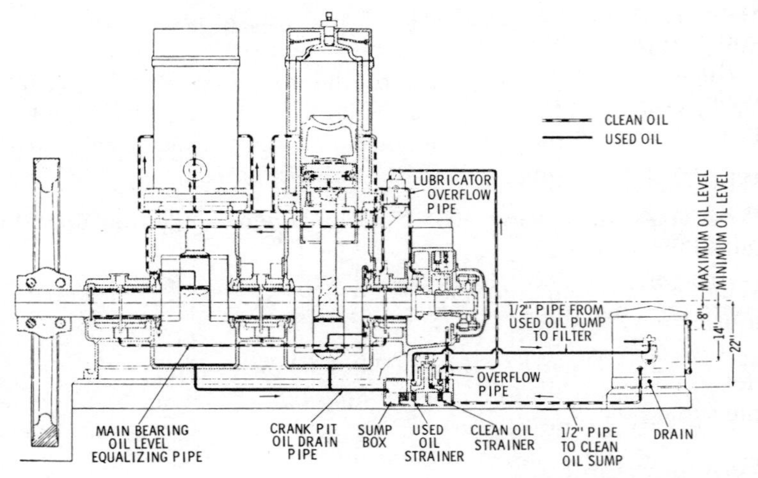

Courtesy Fairbanks, Morse & Co.

Fig. 13-2 Forced-feed lubrication system. Lubricating oil from a storage tank flows to an oil sump from where a reciprocating pump forces it through the engine. Used oil returns to another sump, where it is strained and sent back to the tank.

Answer: Because the scavenging and charging air, in its passage through the crankcase, absorbs the heated oil drippings from the bearings—some of which are burned in the cylinder and some of which are passed out through the exhaust port.

What determines the proper quantity of lubricating oil to be used in a cylinder?

Answer: The condition of the cylinder wall, the pistons, and the piston rings. The cylinder walls and pistons should appear highly polished. They should have clean polished surfaces and be free of gum and deposits of carbon.

How can one determine, without removing the cylinder head, whether a cylinder and piston are getting the required amount of lubricating oil?

Answer: In a two-stroke-cycle engine, by observation through the hole in the exhaust manifold opposite the exhaust port. In a four-stroke-cycle engine, by removal of an exhaust- or inlet-valve cage to permit inspection.

How can it be determined whether an air compressor cylinder wall is properly lubricated?

Answer: By rubbing a piece of white paper along the cylinder wall. If the cylinder is properly lubricated, the paper should show just a stain of oil.

What factors may cause gumming of piston rings?

Answer: The use of improper quality or quantity of lubricating oil; or improper combustion may cause gumming.

What is the actual amount of lubricating oil needed for cylinder lubrication?

Answer: Just a few drops per minute—the actual quantity is determined by the condition of the piston and cylinder wall, and the suitability of the oil.

How much lubricating oil should a well-designed, properly operated diesel engine consume?

Answer: About a gallon of oil for each 600 to 1,000 horsepower hours of operation. Some will operate up to 3,000 horsepower hours on a gallon of oil.

How are journal bearings usually lubricated?

Answer: By the pressure system. The oil pump forces lubricating oil into and through the journal bearings at a pressure of 10 to 60 pounds per square inch, depending upon the size and type of the journal bearings.

What is the usual course of the lubricating oil in lubricating the journals or bearings?

Answer: It enters the main journal box then passes through passages to the crankshaft, the connecting rod, and to the wrist pin or crosshead. The excess oil drops back to a sump below. From the oil sump, it goes through a filtering device, or is hand filtered, and returned to the supply tank for reuse. In lubricating systems where the oil is cooled before reusing, the oil is passed through a cooler after filtering (Fig 13-3).

How is the quantity of lubricating oil for bearings regulated?

Answer: By stopcocks in the oil lines, by holes in disks placed in the lines, and sometimes by special nozzles.

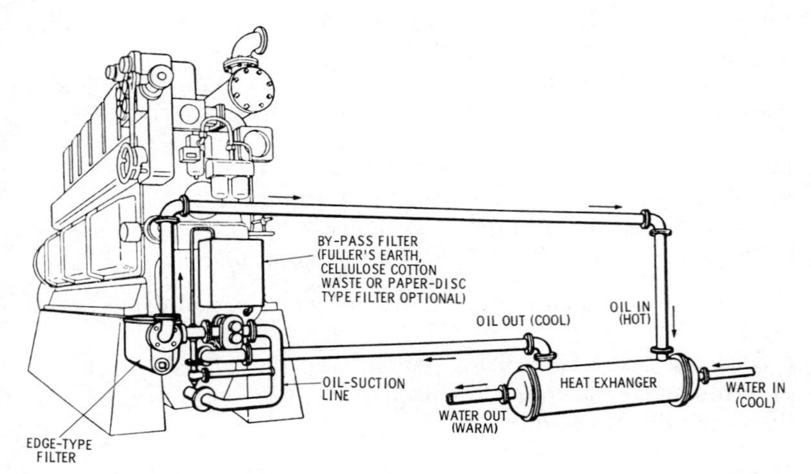

Fig. 13-3. Lubricating oil system in a four-stroke-cycle diesel engine. A positive displacement pressure oil system is used to lubricate all bearing surfaces in this totally enclosed engine. Lubricating oil is taken from the built-in oil sump and pumped, at a pressure of 20-30 lbs. per sq. in., through edge-type filters rand an oil cooler to the main header cast in the engine base.

What kinds of oil pumps are used for providing pressure for lubricating systems?

Answer: Usually rotary pumps, although plunger-type pumps have been found satisfactory.

What are the general properties of good diesel engine cylinder lubricating oil?

Answer: Good diesel engine cylinder lubricating oil does not become gummy from use under heat and pressure. The oil adheres well to metal surfaces, has an oily feel, flows freely, spreads readily to all parts, and can be separated easily from water when reduced to an emulsion.

What care is necessary for lubricating oil?

Answer: Lubricating oil must be clean, free from water, and used efficiently. The circulating system must be given proper care and also kept free from water.

How does water get into the lubricating system?

Answer: Through leaks in the cooling water system and by

carelessness in filling oil storage tanks. These conditions must be watched and either avoided or eliminated.

What is the result when water gets into the lubricating system?

Answer: It forms emulsions which clog small openings and oil passages, prevent proper circulation, and thereby cause the bearing to overheat.

How is water kept from the lubricating system?

Answer: By keeping the lubricating system, the water circulating system, and the crankcase watertight.

What is meant by an emulsion?

Answer: Combination of liquids which do not mix—or combine chemically. One stays with the other in the form of bubbles or small films, such as oil and water. The two liquids form layers.

Is the water and oil emulsion harder to circulate than either the lubricating oil or the water?

Answer: Yes, because it forms a sort of sticky or elastic substance which adheres to the surface of parts.

Is there any difference in the result when fresh water gets into the oil and when salt water gets into the oil?

Answer: Yes. A fresh water and oil emulsion does not hold together as firmly as does a salt water and oil emulsion, and fresh water can be separated from the oil by letting it stand for 24 hours at a temperature of approximately 180° Fahrenheit. The two liquids separate into layers because they do not combine chemically.

Can the oil be separated from salt-water emulsion into oil and water?

Answer: Yes, but not easily—especially where it has been in contact with rusty pipes or iron chips.

How can the presence of salt water in lubricating oil be detected?

Answer: Drop a few drops of silver nitrate solution into a glass of the emulsion. If salt water is present, the mixture will become milky in appearance.

What action is necessary when the lubricating oil indicates the presence of salt water?

Answer: The water leak must be located and stopped. The oil must be drained and new lubricating oil substituted.

How often is it necessary to test the lubricating oil in the system for water content?

Answer: At every watch.

If lubricating oil pressure fails to rise when the engine is started, what is likely to be the cause?

Answer: The strainer on the pump suction may be clogged or the pump itself may not be in proper working condition. There may be a lack of sufficient lubricating oil in the system.

Can too much stress be placed on proper lubrication for an engine?

Answer: No, because the operating results and the maintenance cost are dependent upon proper engine lubrication.

How important is the use of proper lubricating oil?

Answer: Very important, so much so that a laboratory test is not always sufficient to indicate the proper oil to be used. It must be tried in actual practice. An oil may meet specific tests in a laboratory, yet break down in service. This breakdown may be due to local conditions that had not been taken into account when laboratory determinations were made.

What is meant by the viscosity of an oil?

Answer: The fluidity, the thickness or stickiness of it, or the rate at which it will flow at a given temperature. Viscosity is measured by the Saybolt Universal Viscosity Test, which measures the number of seconds required for a definite quantity of the oil to flow into a definite measure.

What do the flash point and the fire point indicate?

Answer: The temperature at which the vapors of the oil will just catch fire or flare up is the flash point; and the temperature at which the oil vapors will continue to burn is the fire point.

What is meant by API degrees?

Answer: API, or American Petroleum Institute, degrees, is an accepted specific gravity hydrometer scale for measuring the specific gravity of petroleum products. It was adopted and accepted by the American Petroleum Institute. The hydrometer is placed in the liquid and the degrees read off at the surface level. A hydrometer is a weighted tube marked off into definite measurements. It will sink to a definite level depending upon the thickness or weight of the liquid.

What effect has temperature on the viscosity of an oil?

Answer: The higher the temperature, the less viscous or less thick is the oil. Higher temperatures have a tendency to thin an oil and make it flow more freely and quickly.

What is a Saybolt viscosimeter?

Answer: A measure through which a definite quantity of oil (60 milliliters) will flow out and the duration of the flow can be timed. A 60-second flow at 210 degrees Fahrenheit means that it would take 60 seconds for the 60 milliliters of oil to flow out when the temerature of the oil is 210° Fahrenheit. If the oil were heavier, it would take a longer for that quantity to flow out. If it were lighter, it would take a shorter time. If the temperature were lower, the oil would be thicker, more sticky, and it would take longer for it to flow out, and vice versa. Temperature is very important in the test of viscosity.

How important is the use of oil of proper viscosity?

Answer: Very important, because the high temperatures of diesel engines will break down or thin an oil and make it useless for proper lubrication. The high temperature may also change its characteristics or qualities and reduce its lubricating effect (Fig. 13-4).

What makes the lubricating oil requirement different for diesel engines than for gasoline engines?

Answer: The intense heat, the high pressure, and the manner in which the oil must be used in diesel engines.

What factor, other than the engine itself, governs the selection of lubricating oil?

Answer: The type of lubricating system in which it must be used.

Should the same kind of lubricating oil be used continuously?

Answer: Yes, until a better or more suitable kind is found and proven by test in service. A lubricating oil that on test and trial has

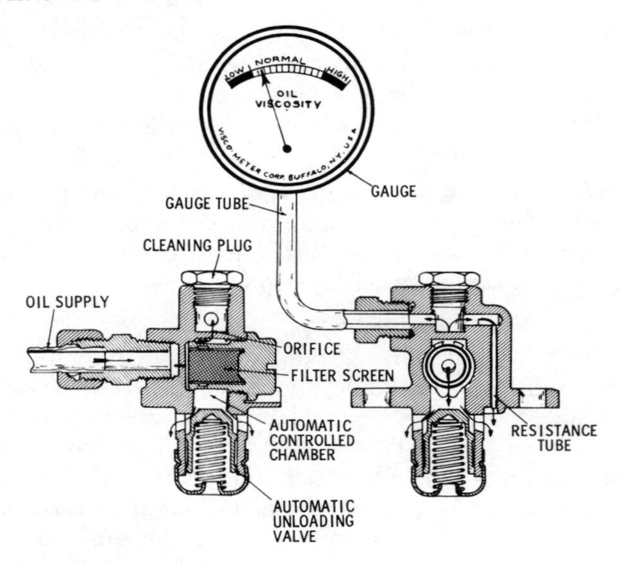

Fig. 13-4. A viscosimeter, an instrument for indicating the viscosity of the oil just before it enters the main bearings. The indicator hand registers higher with thicker oil, as the pressure in the gauge line is higher. The gauge reading is lower for thinner oil, as there is less resistance to flow through the resistance tube.

proven to be a good lubricator, free from ruinous substances, does not subject the engine to undue maintenance, and is economical, is the proper oil to contnue to use.

Do impurities in lubricating oil have a detrimental effect upon the system?

Answer: Yes. Acid causes rusting and pitting of metal parts that, in time, have to be replaced. It makes crankshaft journals rough and requires larger quantities of oil to keep the bearings running cool. Oil adulterated with soap provides ash when heated, and therefore is injurious to the system. It is well to adhere to the use of an oil which has proven satisfactory in practice.

What factors spoil a good lubricating oil?

Answer: Good oil is spoiled and its lubricating qualities are lessened when it becomes contaminated or mixed with foreign matter such as carbon, sludge, dirt, small particles from bearings, or rust; along with leakage of fuel oil and water into the lubricating system.

How will a good lubricating oil react to water?

Answer: The oil should be readily separated from water by the ordinary heat treatment.

What is the heat treatment for removing water from oil?

Answer: The water and oil mixtures are heated to a temperature of about 180° F. and retained at that temperature for several hours. Then, the oil should be separated from the water and the impurities can be readily poured off. Because oil is lighter than water, it will float on top and form almost a complete separation. If a centrifuge is at hand, the water and oil emulsion can be passed through it for more complete separation.

What is a centrifuge?

Answer: A device or machine for separating heavier liquids from lighter liquids by centrifugal force, or by revolving them—the heavier liquids being thrown outward and carried off in one container and the lighter ones in another container.

Centrifugal force is the force that acts upon a revolving body tending to force it farther away from the center.

If the water will not separate from the oil by heating the emulsion, what other method may be used?

Answer: A heavier solution may be made of the emulsion by adding fine salt in the proportion of three pounds of salt to 100 gallons of emulsion. Then heat to 180° F. for about two hours and pass through a centrifuge.

If the above methods will not separate the water from the oil, is there another method?

Answer: Yes. Add five pounds of soda ash to 100 gallons of oil-water emulsion and heat for one hour at 180°F.; then pass it through the centrifuge. If the water content is heavy, add soda ash up to 15 pounds per 100 gallons of oil-water emulsion and heat for one hour.

Give two important rules for the proper operation of continuous-pressure lubricating systems.

Answer: Keep the oil in the lubrication system and keep the water out. This is the best practice to follow because it saves considerable work and maintenance cost.

How long should a quantity of lubricating oil be used?

Answer: Lubricating oil could be used indefinitely if it were not for leakage and spoilage because it seldom wears out. If the impurities are removed by filtration, by centrifuging, or by some other method, it can be used repeatedly. Good oil is economical if it is retained in the system and properly used.

How often should lubricating oil be cleaned?

Answer: It is a good practice to clean lubricating oil frequently. In fact, it should be kept clean at all times regardless of how often it must be done. If it must be cleaned often, the cause should be determined and the fault rectified. The directions of the manufacturers of the oil and its use in the cleaning apparatus should be carefully followed.

What attention should be given sump-tank drains?

Answer: Sump-tank drains should be checked frequently for water or impurities in the system. Considerable work may be saved by keeping the old adage in mind: "An ounce of prevention is worth a pound of cure."

Can acid form in lubricating oil?

Answer: Yes. Lubricating oil, after long use, may become sufficiently acid to form a caky substance—mineral soap. This substance may cause considerable trouble if it is not removed.

How are wicks used in lubrication systems?

Answer: In about the same manner as wicks are used in a lamp. They attract and feed the lubricating oil to the parts with which they are in contact.

Under what conditions are wicks for oil feed suitable?

Answer: Under normal engine room conditions and temperatures, they will feed regularly.

What precautions must be taken in using wicks for oil feed?

Answer: The wicks must be kept free from dirt by frequent washing. Wicks filter the oil as well as feed it, and as they filter it, they retain impurities and soon become clogged. The wicks should also be lifted from time to time to prevent foreign substances, that have been filtered, from clogging the tubing and stopping the flow.

How much oil can be fed with a wick?

Answer: With the oil in the container at a distance of 3/8 to 3/4 inch from the top of the wick tube, a strand of wool yarn should feed oil at a rate of about one drop of oil per minute.

What daily routine should be followed in maintaining the lubricating system?

Answer: To maintain a lubricating system properly, it is good practice to look at the pressure gauges, indicators, and oil feeds on tanks and reservoirs every half hour.

Also, while the engine is operating, open the drains on the sump tank every six hours and take from them a small sample of the oil in the palm of the hand, in a shallow vessel, or in a glass for examination. Test for water and at the same time take a small sample of oil from the drain of the oil pump and test for water.

How often should oil tanks and sump tanks be cleaned?

Answer: The main lubricating oil tanks and sump tanks should be cleaned at least every six months and more often if the oil indicates that it is necessary.

What other care is necessary for the lubricating system?

Answer: The feed-regulating devices placed in the various lines to regulate the quantity of oil for each part should be examined periodically and adjustments made where needed. Pipe lines that supply lubricating oil to various engine parts should be blown out periodically with compressed air at about 50 pounds pressure.

Why is it important to clean the oil strainers and examine the deposits?

Answer: Because the strainers will not permit a sufficient quantity of oil to go through them if they are clogged. The deposit in the strainers gives important information. If there are indications of metal or mineral substances in the deposit, the substances should be washed in kerosene to determine their origin.

If the metal particles are Babbitt or bronze, it is an indication that a journal or bearing is cutting. If they are iron, it may be rust from a pipe, a tank, or a casting.

Sand may also appear in the deposit. This may be from the casting on which it was left after being drawn from the mold. Sand is often covered by paint or enamel during manufacture and later released by the paint chipping off.

This is often experienced in new engines. Many crankshaft journals and bearings have been ruined by the failure of the operator to examine the oil deposit and to care for the lubricating system properly.

After a lubrication system has been cleaned, what procedure should be followed before again starting the engine?

Answer: All lubricating oil lines should be disconnected at the bearings, and the oil pumped through by hand to be sure that the oil reaches all parts properly. The lines should then be reconnected, the engine started, and the system watched closely until the pressure gauge indicates a steady oil pressure.

If the gauge fails to show oil pressure after five minutes, the engine should be stopped again and the cause investigated.

What temperature should be the maximum to which lubricating oil is permitted to rise?

Answer: To not more than 100° F. If it rises higher than 100° F., the oil should be run through a cooler because it will be too thin and will flow too freely.

What should be the oil pressure reading in the lubricating system compared to the pressure reading in the oil cooling system?

Answer: The pressure in the lubricating system should be higher than that in the cooling system so that if a leak occurs in the cooling system, the oil will enter the water rather than the water entering the lubricating system. The drain cock of the oil cooler should be opened, at each watch, to determine whether there is oil in the water.

CHAPTER 14

Cooling Systems

Why is a cooling system necessary for a diesel engine?

Answer: Because a diesel engine operates at intense heat, and the cylinders require cooling to enable them to retain proper lubrication.

What means is generally employed for cooling the cylinders of diesel engines?

Answer: A water circulating system. This cools the cylinders, pistons, and occasionally the main journals.

How are the cylinders constructed to enable water to cool them?

Answer: They are provided with a water jacket, which is really a larger cylinder outside the smaller cylinder with a space between them in order to permit a wall of water to be continually in contact with the cylinder lining. The water is led into the water jacket by piping and then out again after it has absorbed the heat.

What becomes of the water after it has left the hot cylinder jacket and is heated?

Answer: It is either wasted or cooled and reused, or it is used in its warm state for heating purposes.

Are the cylinders and the pistons the only parts of the engine that require cooling?

Answer: No, besides the cylinders, the lubricating oil and the air-compressing system require cooling (Fig. 14-1).

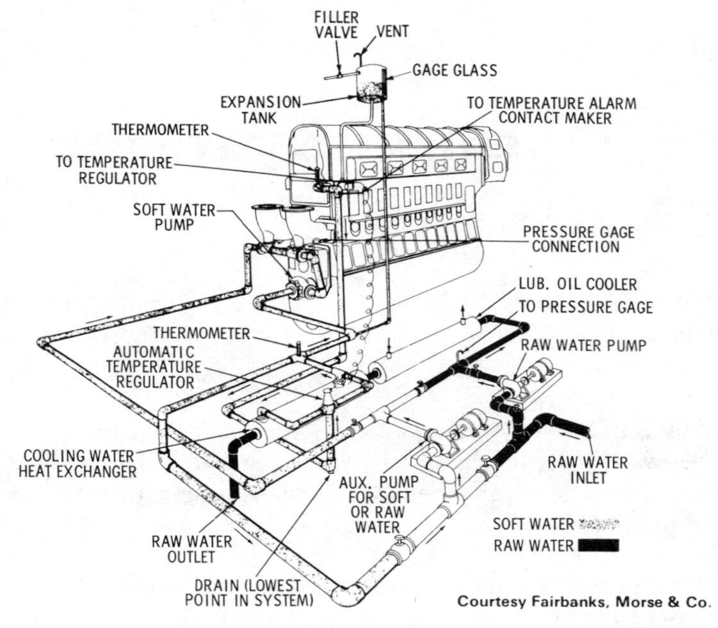

Fig. 14-1. Fresh-water piping. Water circulates in a closed system; the same water is used repeatedly. It is piped out of the engine, cooled in a heat exchanger, then returned to the engine.

Why does lubricating oil need cooling?

Answer: Because lubricating oil, when it leaves the sump, is hot and thin and cannot be used again until it is cooled and thickened to restore its lubricating quality. Most engines have a lubrication system in which the oil is forced through the various parts by pump pressure and reused after being cooled.

Why does the air-compression system need cooling?

Answer: Because air heats upon compression and air pressure cannot be increased to the required high-pressure levels without passing through intercoolers to reduce the intense heat of compression.

Why is it important to use clean water in the cooling systems of diesel engines?

Answer: Because mud or dirt deposited on the surfaces of the cylinder liners and other parts forms a coating that prevents heat absorption by the water. In addition, the narrow passages may become clogged and stop circulation of the cooling water.

Can salt water be used in the circulating system?

Answer: Yes, although fresh water is to be preferred. Salt water leaves a deposit on the heated surfaces in the form of a scale that hinders the transfer of heat to the cooling water.

What is the damaging effect of a scale and dirt coating on cylinder liners?

Answer: Unabsorbed heat may reach a temperature high enough to destroy the lubricant.

Describe the means by which water is cooled.

Answer: By means of towers over which the water is made to flow in thin streams or sheets through currents of air or in contact with the air (Fig. 14-2). Also by passing the circulating water through a series of tubes around which cold water is circulated (Fig. 14-3).

At what temperature should cooling water for power cylinders be maintained?

Answer: When it leaves the cylinders, cooling water should be at a temperature of 110° to 120° Fahrenheit.

Can the water be too cold when it enters the cylinder jackets?

Answer: Yes, because too low a temperature of the cooling water interferes with ignition and combustion in the cylinders.

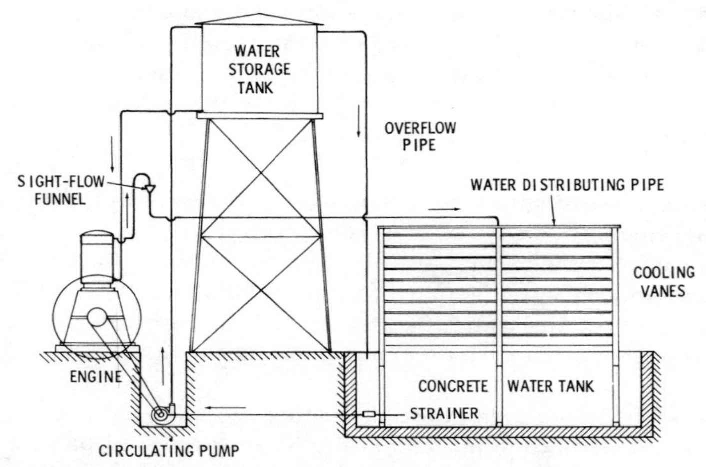

Fig. 14-2. Open-type water-cooling system. This system is preferred where the water supply is not limited. Water from the overhead tank flows through the engine water jackets and then out to be cooled. Water must be free from scale-forming substances.

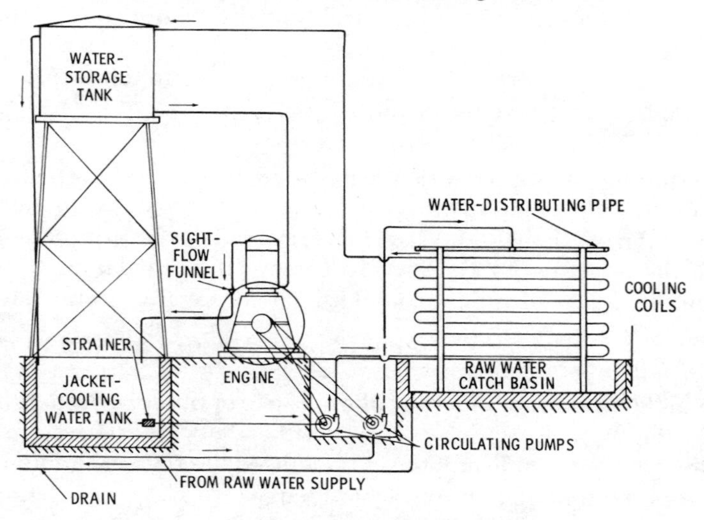

Fig. 14-3. A closed-type water-cooling system is preferred where there is a limited supply of water. A tank is filled with soft water or rain water, which is circulated by a pump through cooling coils containing raw water. The water is cooled, delivered to an overhead tank, and recirculated into the engine water jackets.

What are the air pockets in the water circulating system?

Answer: Pockets or spaces containing air under pressure which the circulating water cannot displace.

Where are the pockets in which air may be trapped likely to occur in the circulating system?

Answer: In the cylinder head, in the discharge manifold, in a pipe connection, and sometimes in the water jacket.

How do trapped air pockets affect operation?

Answer: They cause parts of the engine to remain hot in spite of the circulating water of the system. Air pockets, principally in the cylinder head, may develop regardless of the engine design and may require valves or other means to clear the trapped air from the water circulating system.

How can the existence of air pockets be detected?

Answer: By checking the cylinder heads and other suspected parts for hot spots. The spots immediately above the air pockets will be hotter than the surrounding parts.

In what part of the engine is proper circulation of the cooling water of special importance?

Answer: Around the exhaust valves—the valves or openings through which the burned fuel gases are driven out at every exhaust stroke. The temperature at these points is extremely high (Fig. 14-4).

What indicating instruments are used on a water cooling system?

Answer: Thermometers, pyrometers, and pressure gauges. They preferably should be on the discharge side of the system because the heat of the water, after it leaves the engine, tells whether the engine is cooled sufficiently. The pressure gauge indicates whether the circulating water pump is working properly.

What types of pumps are used for circulating water?

Answer: The plunger, piston, and rotary pumps. They are generally connected directly to the engine.

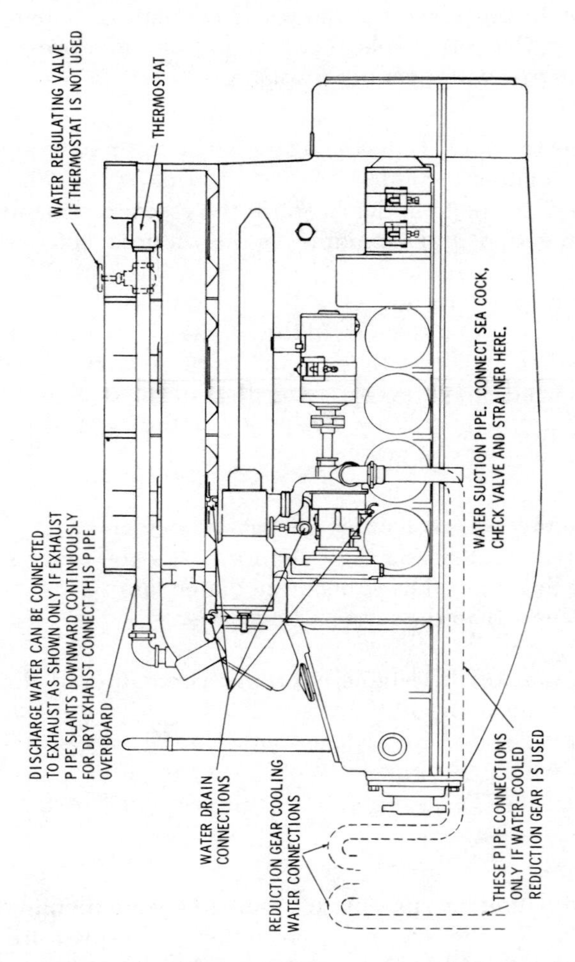

Fig. 14-4. Cooling-water piping of a marine engine. A gear-type water pump driven from one of the shafts forces water through the cylinder jackets, the cylinder head, and the exhaust manifold jacket. A thermostat on the manifold regulates water temperature.

What are auxiliary water circulating pumps?

Answer: Pumps operated independently of the engine to maintain the temperature at fair limits after the engine is stopped. Independently operated pumps are also useful for circulating warm water through the engine before starting it. They also can be used in event of breakdown of the directly connected pumps.

What provision is made to regulate the amount of water which the pump circulates?

Answer: A valve that controls the quantity of water flowing through the pump when the water is taken in under pressure is placed in the intake or suction line. A valve controlling the intake prevents hammering and probable wrecking of the pump. Also, there is a by-pass valve connecting the discharge side of the pump with the suction side to control the quantity of water delivered to the system.

Why is it important to keep the temperature of circulating water within the prescribed limits—110° to 120° Fahrenheit?

Answer: In order to have efficient operation. Low temperatures tend to cool the cylinders too much and retard combustion. They make the exhaust gases smoky, reduce the horsepower or output. High temperatures increase the internal temperatures, thereby destroying lubrication and causing pounding.

What care should the cooling water jackets and water spaces be given?

Answer: The cooling water jackets and water spaces should be washed out frequently and any existing scale removed.

How can the scale be removed?

Answer: By dissolving it. Analysis shows it can be dissolved without injuring the metal parts. Scale formed by different substances in the water requires different solutions for dissolving. After dissolving, the scale and solution may be driven out by the water.

Another way of removing scale is to dry the surfaces and apply kerosene flakes, which can then be removed through any suitable opening in the water jacket or cylinder head. Scale can also be removed by scraping.

What is the danger in forcing cold water through an engine while it is hot?

Answer: Cold water admitted to hot cylinders can cause piston seizure—besides the effect of chilling the hot metal. Piston seizure may wreck the engine.

How does experience help the operator?

Answer: Experience can develop a sense of touch which enables the operator to know exactly at what temperature each part works best and enables him to tell when there is a difference in operating conditions.

CHAPTER 15

Starting Diesel Engines

Can a diesel engine start on its own power?

Answer: No. It usually requires an outside source of power to start it.

Why can a diesel engine not start on its own power?

Answer: Because it must have air compression in the cylinders to create the heat that ignites the fuel.

What means are generally used for starting diesel engines?

Answer: Compressed air, electric motors, and hand cranking.

How is a diesel engine started by compressed air?

Answer: Compressed air is admitted to the working cylinder and exerts a pressure on the piston the same as burning fuel exerts during the power stroke (Fig. 15-1).

Courtesy Fairbanks, Morse & Co.

Fig. 15-1. Air-starting system (for 4- to 10-cylinder engines). The engine is started by the action of compressed air on the pistons in their proper firing order. Air is stored in tanks and is required at between 150 and 250 lbs. per sq. in. (preferably 250 psi) at the engine.

What air pressure is needed for air starting a diesel engine?

Answer: About 250 pounds per square inch.

Where is the air for starting the engine obtained?

Answer: From bottles or tanks containing compressed air, or from outside sources.

How are the bottles or tanks supplied with compressed air?

Answer: By auxiliary air compressors operated by an outside source of power or by compressors attached to the main engine.

Why is air more generally used for starting large engines rather than other sources?

Answer: Because the starting air can be compressed and stored in tanks while the engine is running. Also it is used because of its convenience and suitability.

What are air compressors?

Answer: They are pumps that compress air to definite pressures.

Where do air compressors get their power?

Answer: Auxiliary compressors are driven by electric motors, by gasoline engines, or by other sources of power (Fig. 15-2). Main compressors are attached to the main engine which drives them while the engine is running.

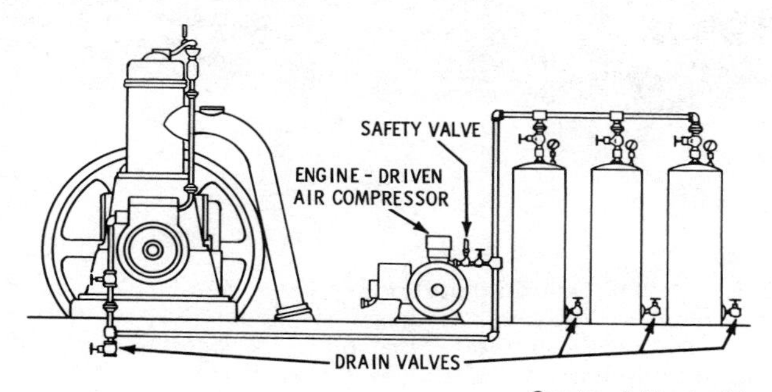

Fig. 15-2. A simple air-starting system with a separately driven compressor. Many engines drive their compressors as part of their equipment.

How does either an electric motor or hand cranking start an engine?

Answer: By rotating the engine until it attains sufficient speed to compress the air in the cylinders to provide enough heat to ignite the fuel (Fig. 15-3).

During what period of the cycle, in a four-stroke-cycle diesel engine, is the starting air admitted?

Answer: The air is admitted while all the valves are closed except the air-inlet valve and it follows the piston to about half its stroke during the starting operation.

For how long a period is the starting air supplied to the engine?

Answer: Just long enough to get up engine speed which will cause the pistons to compress the intake air to a high enough temperature to ignite the fuel oil.

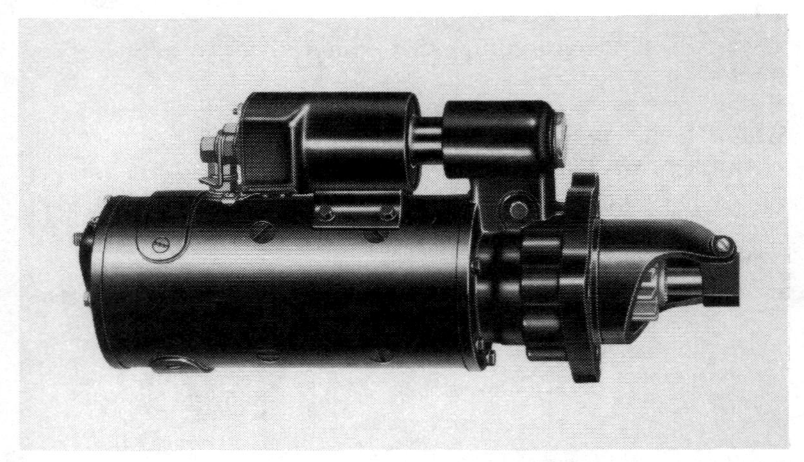

Courtesy Delco-Remy Division, General Motors

Fig. 15-3. Cranking (starting) motor for a diesel engine. An electric motor similar to those used on gasoline engines and powered by storage battery current rotates the engine. Air compression in the cylinders ignites the fuel and starts the engine.

How should an air-starting valve be handled for starting?

Answer: The valve should be opened completely in order to give the engine a quick rotation. If the engine does not start firing promptly, the valve should be shut, then opened a number of times, successively, until the engine does start. If the valve is left open, the expanding air has a cooling effect on the cylinder and retards fuel ignition (Fig. 15-4).

If the engine does not begin to rotate promptly upon injection of the starting air, what may be the cause?

Answer: Some of the starting gear or a line shaft may be out of adjustment, or the engine may be jammed.

Is fuel oil injected at the same time as the starting air?

Answer: No. The fuel is injected at a different time in the cycle.

Is starting air admitted to all cylinders of a multiple-cylinder engine?

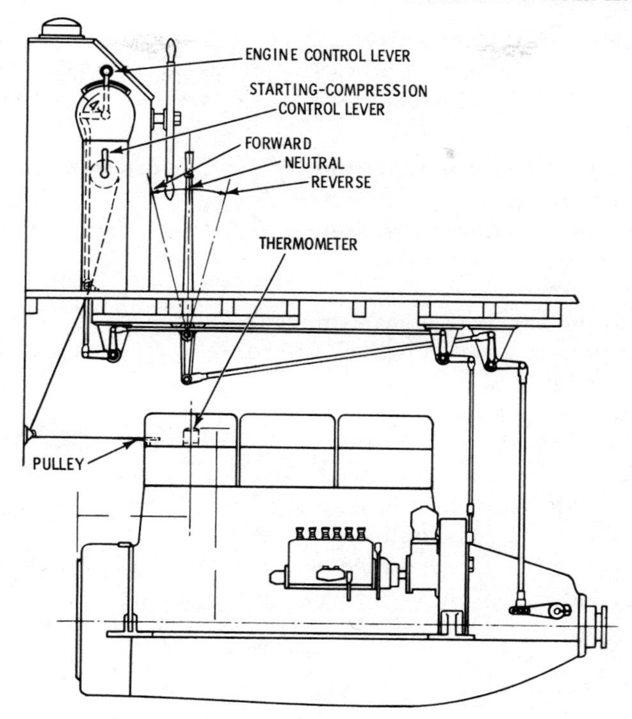

Fig. 15-4. Pilot house or bridge control of a marine diesel engine. Remote control of engines is accomplished by levers. Reversal is brought about by a reverse gear and clutch actuated by a lever.

Answer: No, generally to only three cylinders of a six-cylinder engine.

What may cause an engine to fail to ignite its fuel oil promptly while the engine is being rotated?

Answer: The engine may be too cold, compression may be too low, the fuel may not be properly atomized, or the fuel oil may not be injected at the proper time.

What action should be taken when the engine fails to start?

Answer: Check the fuel supply, the atomization, and the compression.

What is an air-starting valve?

Answer: A valve that admits compressed air to the working cylinders of a diesel engine in order to start the engine rotating.

How is the admission of starting air timed in order to bring about proper starting?

Answer: In a direct air-admission system, it is timed by starting valves actuated by cams on the camshaft.

How is the air-start mechanism controlled?

Answer: A lever or wheel controls the air-start and reverse mechanism.

When the lever is moved to "Start" position, the air cutoff valve opens and admits compressed air to a distributor; at the same time the cams under the air-starting valves move out of position. The valves are lowered to their seats and brought into contact with the starting-air cam. As the engine rotates, the starting-air cam lifts the valves in the required sequence and air is admitted to the respective cylinders; this flows during a portion of the downstroke of the piston.

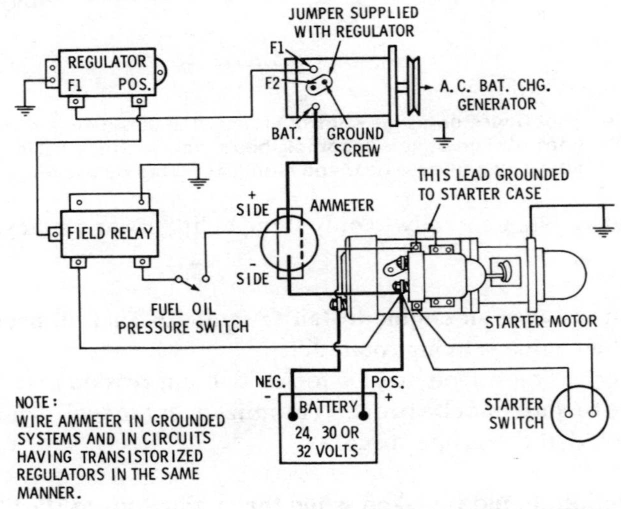

Courtesy Detroit Diesel Engine Division, General Motors

Fig. 15-5. Typical AC battery-charging electrical system.

When the engine is rotating, the lever is moved to the "run" position which closes the air cutoff valve, stopping the admission of compressed air, raising the air-starting valves off their seats, and restoring the cams to their normal operating position. Fuel oil is then admitted and the engine runs. The valves and cams are interlocked so that compressed air cannot enter the cylinders while the valves are off their seats.

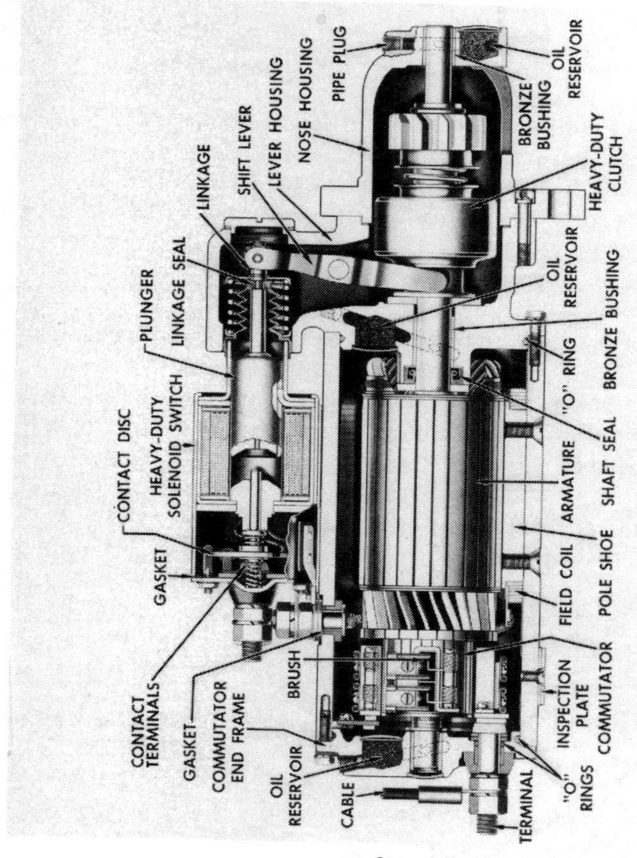

Courtesy Delco-Remy Division, General Motors

Fig. 15-6. Cutaway of cranking (starting) motor for diesel engine. Diesel engines require stronger motors and higher voltages because of the higher compression pressures to be overcome.

In another method of air starting an engine, a small rotary valve controls and directs the compressed air through tubes to the starting cylinders; this valve can be operated in either direction so that reversal of the engine can be brought about easily by the operator of the engine.

How is starting air admitted to the working cylinders in a pilot or pneumatic valve starting system?

Answer: By auxiliary valves actuated by cams on the camshaft that admit air at the proper time to small cylinders that open the main air-starting valve to the working cylinders.

If the heat of compression is not sufficient to start an engine, how can ignition be brought about?

Answer: By heating the outside of the cylinder heads with a blow torch—thus adding to the internal heat—or by circulating hot water through the water jackets around the cylinders and throughout the engine block.

What other methods may be used to start balky or hard-starting engines?

Answer: Balky, hard-starting engines may be started by the insertion of saturated cartridges or punks in the cylinders to help to ignite the fuel oil.

How are the cartridges or punks made?

Answer: By saturating short pieces of braided absorbent mate-

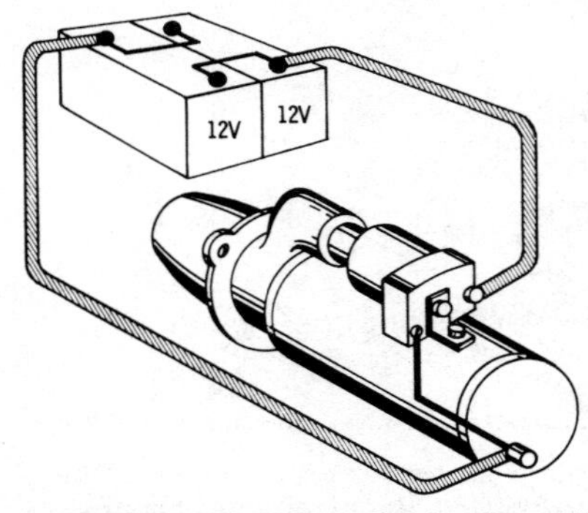

Courtesy Detroit Diesel Engine Division, General Motors

Fig. 15-7. Typical Battery-to-starting-motor circuit.

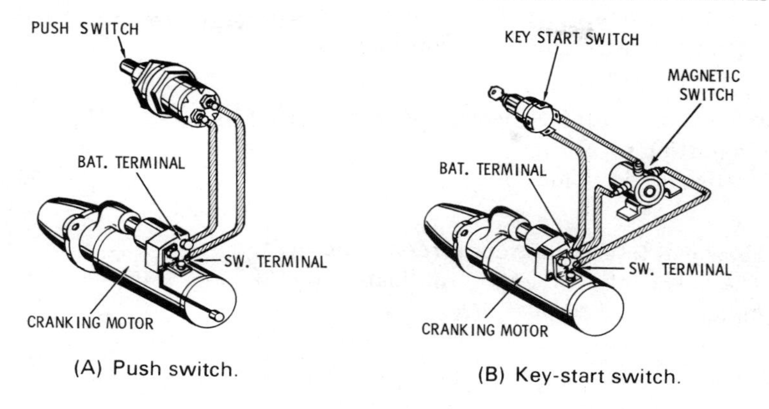

(A) Push switch.

(B) Key-start switch.

Courtesy Detroit Diesel Engine Division, General Motors

Fig. 15-8. Typical starting-switch wiring.

rial or rags with either kerosene or fuel oil. Pieces of rolled-up blotting paper may also be used.

How is saltpeter used for making the cartridges?

Answer: Soak the absorbent material in a saturated solution of saltpeter for six hours, then remove and thoroughly dry. The solution is made by dissolving as much saltpeter in fresh water as it will absorb during shaking or mixing for a half hour at a moderate temperature. After removing and drying the cartridges, they should be put into airtight containers and kept there, when not in use, to prevent loss of strength.

What is a self-igniting cartridge?

Answer: One which ignites by the heat of compression in the cylinder of the engine.

How are self-igniting cartridges made?

Answer: By soaking blotting paper in a solution of 8 ounces of saltpeter and 5 ounces of copper nitrate dissolved in 3¾ pounds of water. After sufficient saturation, the cartridges are removed and set out to dry for 48 hours at a temperature below 170° Fahrenheit. They must be kept in airtight containers to prevent their absorbing moisture and in a cool place to prevent spontaneous ignition.

At what temperature will self-igniting cartridges ignite?
Answer: At about 190° Fahrenheit.

What is the most frequent cause of an engine failing to start by the air-starting method?
Answer: Low air compression.

How can tests for air compression be made?
Answer: By the engine indicator. (See Chapter 18 on engine indicators.)

CHAPTER 16

Engine Reversing

How can a diesel engine be reversed?

Answer: By a reversing mechanism that automatically cuts off the fuel supply and brings the engine to a stop; then it admits starting air and fuel oil and starts the engine in the reverse direction.

How is the direction of engine rotation changed?

Answer: By stopping the engine and placing the reversing gear in the opposite running direction. This changes the admission time of the starting air and the fuel. In principle it is as follows: when the crank throw is, say, on the left side of the crankshaft and a pressure is applied to the piston, it will move the crank throw downward on the left-hand side thereby causing left-hand rotation. When the crank throw is on the right side, the downward pressure will cause right-hand rotation of the crankshaft.

How can a change of admission time be brought about?

Answer: By the cams which open and close the valves—there being two sets—one for one direction and one for the other. Another method is by means of a loose coupling.

How are the cams moved to bring about reversing?

Answer: By an endwise movement of the camshaft, which brings one or the other set of cams into position to operate the rocker arms which control the opening time of the valves; or by means of a set of swinging rollers that come in contact with the cams. The latter method is used mainly on two-stroke-cycle engines.

How can the engines be reversed by the loose-coupling method?

Answer: In this method, one set of cams is used for valve control. Their position is altered by the motion of a loose coupling with a slot that permits a 30-degree change of the cam angle. This coupling is located on the camshaft between the cams and its driving mechanism. The change in cam angle is brought about by moving the crankshaft to which the camshaft is connected by gears.

What is meant by endwise movement of the camshaft to bring about reversal of the engine?

Answer: An endwise shifting of the camshaft upon which are located two sets of cams, one for one direction of operation, the other for the other direction. The cams engage the valve levers.

State the sequence followed in reversing an engine.

Answer: One movement of a handle or wheel lifts the valve levers off the cams, moves the camshaft endwise, and replaces the levers on the cams for reverse (Fig. 16-1).

The second movement—to "start," admits starting air to the cylinders and starts the engine rotating.

A third movement—to the "running" position, cuts off the starting air, admits fuel to the cylinders, and the engine operates in reverse.

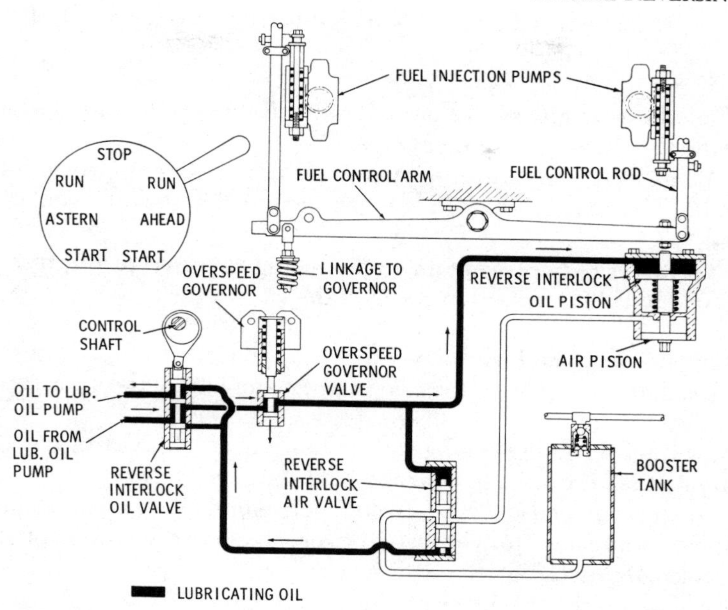

Courtesy Fairbanks, Morse & Co.

Fig. 16-1. Overspeed governor and reverse interlock in "ahead run" position marine engines. When engine rotation is reversed, the cam on the control valve.shifts the reverse interlock valve to maintain constant direction of oil flow.

Are all the reversing gears moved by hand?

Answer: No. Some have a hand control but the movements are made either by an electric motor or by compressed air.

Is there any danger of getting the valve movements wrong—that is, both air and fuel oil coming in at the same time?

Answer: Not when the original cam adjustments are kept intact.

Is the operation for reversing two-stroke-cycle engines different from that for reversing four-stroke-cycle engines?

Answer: Only in the timing and in the valve arrangement. The principle of stopping and starting the engines is the same.

NOTE: It is well to follow directions of the manufacturers for each particular type of engine. This holds true for operation of the engine as well.

When diesel engines have large flywheels, how is the momentum of the flywheels overcome?

Answer: By applying a pneumatic brake—a compressed-air brake. This may be applied at the same time that the fuel is cut off.

What factor may prevent an engine being reversed promptly?

Answer: Air trapped in the cylinders.

How are the cylinders relieved of trapped air?

Answer: By opening the relief valves and letting the air escape.

Why does it require a longer time to reverse a four-stroke-cycle engine than it does a two-stroke-cycle engine?

Answer: Because a four-stroke-cycle engine has only one compression and one power stroke every second revolution of the crankshaft. A two-stroke-cycle engine has one power stroke at every revolution of the crankshaft.

Is it difficult to bring a diesel engine to a stop?

Answer: No. When the fuel is cut off, the engine is still compressing air in the cylinders and their compression causes a back pressure—which helps the engine to come to a stop.

What precautions should be taken in reversing an engine?

Answer: Make sure that the engine has come to a full stop before placing the reversing gear in the opposite running direction. Otherwise, the engine may continue to run in the original direction.

CHAPTER 17

Valve Timing

What is meant by valve timing?

Answer: The adjustment of the valves to open and close at the proper time for smooth and efficient operation of the engine.

Why is valve timing important?

Answer: Because the opening and closing of all valves necessary to make the engine run properly must be done in proper sequence or order.

What are the various functions of valves?

Answer: They must admit air for compression, fuel oil for ignition, clean air for scavenging, air for starting; they must open and close the inlet and exhaust manifolds which lead to and from the power cylinders (Fig. 17-1).

How frequently must the valves operate in relationship to the engine cycle?

Answer: Once in each engine cycle or in each series, because a definite series of events must take place and be repeated in regular order if the engine is to run continuously (Fig. 17-2).

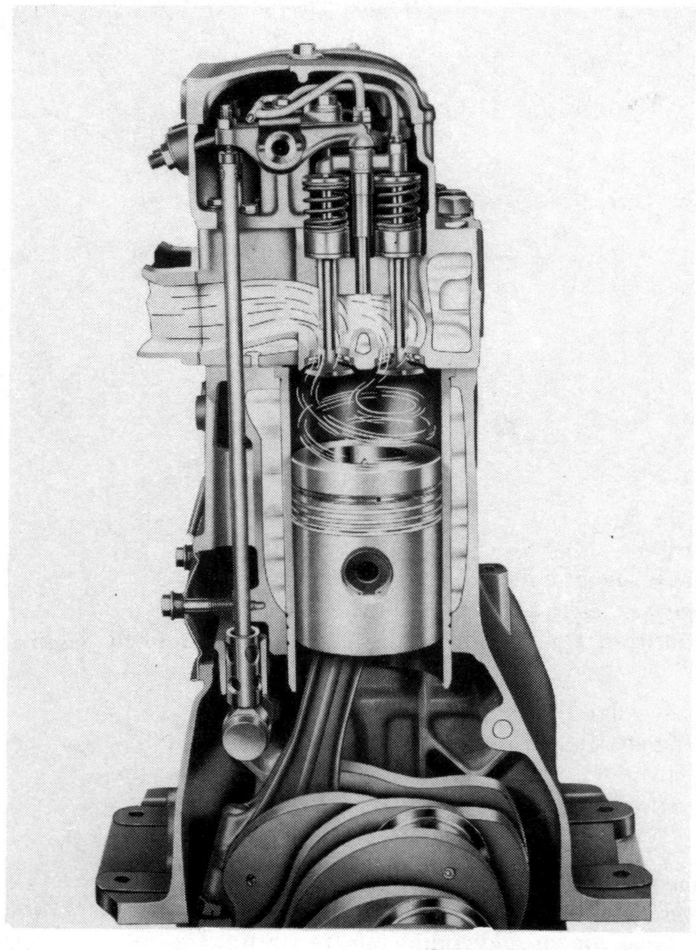

Courtesy Allis-Chalmers Mfg. Co.

Fig. 17-1 Cutaway of Allis-Chalmers Model 25000 diesel engine showing valve mechanism.

What is the length of the time interval for valve operation?

Answer: Just a fraction of a second. The interval of time is dependent upon the speed of the engine.

How are the valves timed?

Answer: By adjusting their actuating mechanisms so that the valves open and close a definite number of degrees before and after the piston has reached top dead center or bottom dead center (Fig. 17-3).

What is meant by top dead center?

Answer: The extreme top end of the piston stroke.

What is meant by degrees in circular distance?

Answer: One 360th division of a circle. A circle is usually divided into 360 parts called degrees. This is standard division. Each of these spacings is again divided into 60 parts called minutes, so that it is possible to designate points along a circle in a definite way. The readings may be given as so many degrees, minutes, and seconds in circular distance from a certain point.

Where are the timing divisions marked on a diesel engine?

Answer: On the flywheel and on the gear train. These are in degrees. Also shown are the points of opening and closing of the inlet, exhaust, spray, and air-starting valves, as well as the top and the bottom dead centers of the piston strokes (Figs. 17-4 and 17-5).

Why are the dead-center markings important?

Answer: Because both the air and fuel entrances and the exhaust must be accurate for each cylinder. They are timed from both dead-center points.

What is the result if the timing for some cylinders is inaccurate?

Answer: The power strokes, and therefore the actions in the cylinders, may retard rather than help the other cylinders. Thereby improper timing prevents full power and smooth running of the engines.

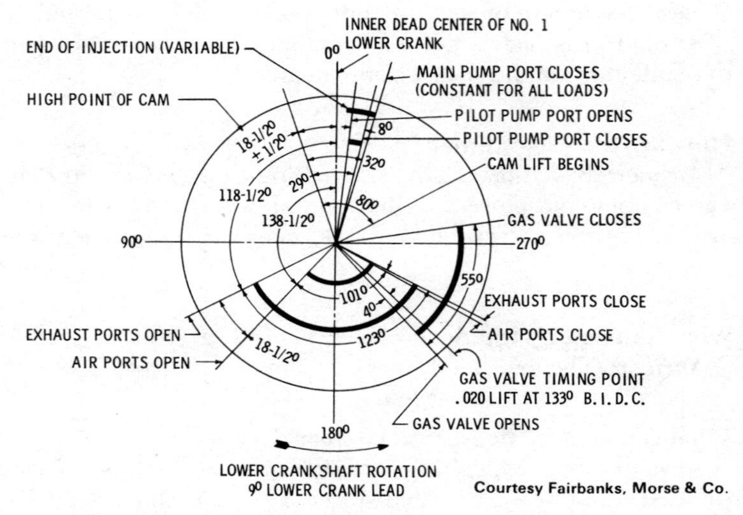

Fig. 17-2. Sequence and timing of events in Fairbanks-Morse opposed-piston diesel engine.

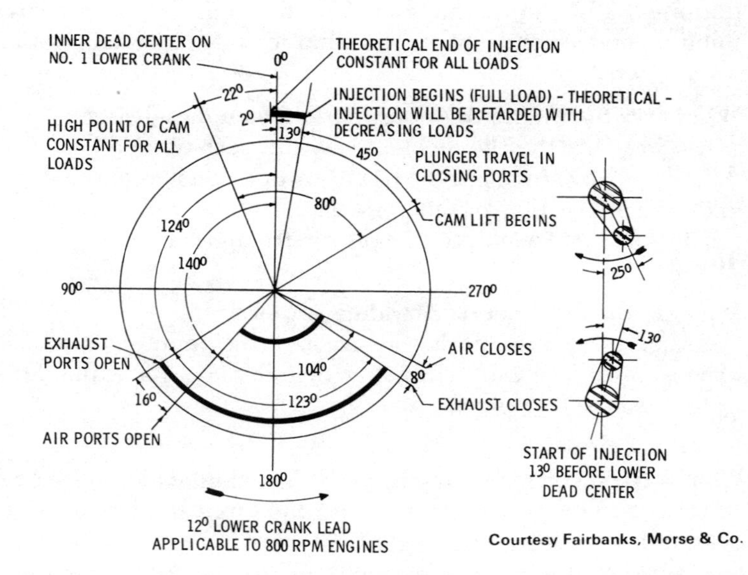

Fig. 17-3. Sequence and timing of events in Fairbanks-Morse opposed-piston, dual fuel engines.

Courtesy Cummins Engine Co.

Fig. 17-4. Gear-train timing marks on Cummins diesel engine.

Is it wise to check the existing markings on the flywheel?
Answer: Yes, to make sure that they are correct.

How can the dead centers of an engine be determined?
Answer: Remove a fuel inlet or exhaust valve from the head of the cylinder. Place the crank throw off dead center and measure from the top side of the cylinder head to the piston. Make a mark on the flywheel at the indicator point. Turn the engine past dead center until the distance from the top of the cylinder to the piston is the same. Measure from the mark previously made on the flywheel to the indicator point. Halfway between the mark on the flywheel and the indicator point is the dead center.

The positions or relative positions of the crank throw depend on what factors?

Courtesy Cummins Engine Co.

Fig. 17-5. Valve-set timing marks. Bar engine in direction of rotation until 1-6VS mark on the fan drive pulley is in line with timing mark on side of the gear-case cover. In this position both intake and exhaust valves will be closed for No. 1 cylinder.

Answer: On the cycle of the engine and on the number of cylinders of the engine.

The timing of an engine depends on what factor?

Answer: On the position of the crank throws because timing is based on the degrees before or after dead center.

Is the firing order of the cylinders in direct order—1, 2, 3, 4, 5, etc.?

Answer: No. The firing order is dependent upon the cycle and upon the number of cylinders. A two-stroke-cycle engine has a

Courtesy Caterpillar Tractor Co.

Fig. 17-6. Partial view of crankshaft of Caterpillar D342 engine.

power stroke in each crankshaft revolution while a four-stroke-cycle engine has a power stroke in each two revolutions of the crankshaft. It is obvious that the firing order must be such that certain cylinders will draw in air, compress, and exhaust while other cylinders are firing.

What important arrangement of the fuel injection, in multiple-cylinder engines, must by made to ensure proper balance?

Answer: The power strokes must balance throughout the entire engine to avoid undue strains and vibration.

How can the firing order be arranged in a two-stroke-cycle diesel engine for proper balance?

Answer: Practically in alternate or balanced form—depending upon the number of cylinders—because the firing of one cylinder compresses the air in the next.

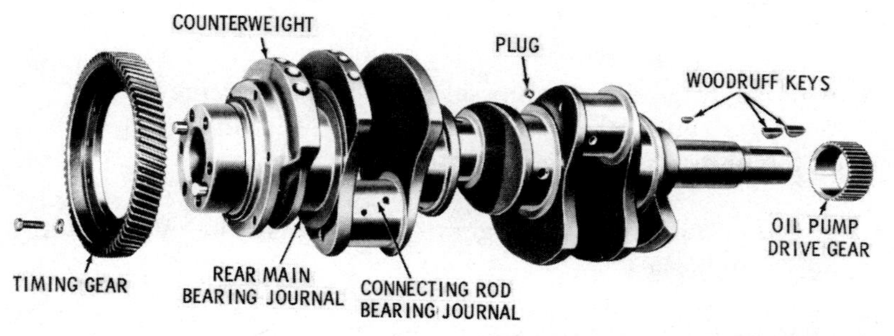

Courtesy GMC Truck & Coach Division, General Motors

Fig. 17-7. Crankshaft and gears of General Motors diesel engine.

What is the firing arrangement or firing sequence of two-stroke-cycle multicylinder engines?

1 cylinder 1;
2 cylinders 1, 2;
3 cylinders 1, 2, 3;
4 cylinders 1, 4, 2, 3;
5 cylinders 1, 3, 5, 2, 4;
5 cylinders 1, 5, 2, 3, 4;
6 cylinders 1, 4, 5, 2, 3, 6;
6 cylinders 1, 6, 2, 4, 3, 5;
8 cylinders 1, 7, 5, 4, 2, 8, 6, 3;
8 cylinders .... 1, 8, 6, 4, 2, 7, 5, 3.

What is the firing arrangement or firing sequence of four-stroke-cycle multicylinder engines?

1 cylinder 1;
2 cylinders 1, 2;
3 cylinders 1, 3, 2;
4 cylinders 1, 2, 4, 3;
5 cylinders 1, 3, 5, 4, 2;
6 cylinders 1, 5, 3, 6, 2, 4;
6 cylinders 1, 2, 3, 6, 5, 4;
8 cylinders 1, 6, 2, 4, 8, 3, 7, 5;
8 cylinders .... 1, 5, 2, 6, 4, 8, 3, 7.

At what angles are the cranks set for uniform rotation?

Answer: Uniformly around the circle—the angle depending upon the number of cylinders.

Give the crank throw arrangement of two-stroke-cycle multicylinder engines.

2 cylinders .... 180° apart;
3 cylinders .... 120° apart;
4 cylinders .... 90° apart;

5 cylinders.............................. 72° apart;
6 cylinders.............................. 60° apart;
8 cylinders.............................. 45° apart.

Give the crank throw arrangement of four-stroke-cycle multi-cylinder engines.

2 cylinders.............................. 360° apart;
2 cylinders.............................. 180° apart;
3 cylinders.............................. 120° apart;
4 cylinders.............................. 180° apart;
4 cylinders.............................. 90° apart;
5 cylinders.............................. 72° apart;
6 cylinders.............................. 120° apart;
8 cylinders.............................. 90° apart.

What may cause excessive vibration in an engine?

Answer: Unbalance in firing arrangement, unbalance in mechanical construction, and operating at the vibratory critical speed.

What is meant by the critical speed or the vibratory critical speed of an engine?

Answer: The speed at which the power stroke impulses tie in with, or match, the crankshaft vibrations, causing excessive engine vibrations. This critical speed is dependent upon the construction of the engine, the number of power strokes per minute, and the arrangement of the parts.

What action should be taken if an operating engine begins to vibrate excessively?

Answer: If the vibration is due to no developed fault in the engine, it is undoubtedly due to the critical speed. Slightly increasing or decreasing the engine speed until the vibrations stop will get it past the critical speed.

What effect will continued operation at the critical speed have on the engine?

Answer: It may shake loose the nuts and bolts, crack engine parts, and communicate its vibrations to other equipment.

How is the crankshaft affected by the engine's vibration through continued operation at the critical speed?

Answer: It receives most of the jolts and strains—these will in time get the crankshaft out of alignment, will wear and crack the bearings, and will necessitate repair or replacement. The crankshaft is an expensive part of the engine and should receive care and attention regularly.

How can a crankshaft be checked for wear and alignment?

Answer: By gauges, or micrometers made for the purpose. One such device is known as a bridge gauge and is used in conjunction with a thickness gauge. A properly designed gauge should be made to show accurately the original factory alignment of the crankshaft. The gauge should be carefully handled and preserved for future use. The crankshaft should be checked at least once a year and on suspicion of misalignment.

How is a bridge gauge used to check wear in the main bearings?

Answer: Remove the main bearing caps, set the gauge on the parts of the frame intended for it; and, by use of a thickness gauge or feelers, note the distance between the gauge and the top of the crankshaft. Compare the gauge readings with the original figures as stamped on the gauge. Any variation from the original measurements indicates wear in the lower half of the bearing.

What method can be used for testing misalignment of the crankshaft?

Answer: By stretching a fine wire parallel with the axis of the crankshaft and measuring from this wire to the crankshaft at each bearing. The measurements should be taken at four equal points of the crankshaft diameter at each bearing.

How can the interval between the firing strokes be calculated?

Answer: To determine the interval between firing strokes in two-stroke and four-stroke-cycle engines for various numbers of cylinders, divide the number of degrees of travel for the cycle by the number of cylinders.

For example:

1. A two-stroke-cycle engine makes its complete cycle in one crankshaft revolution—which is 360 degrees.
2. A four-stroke-cycle engine makes its complete cycle in two crankshaft revolutions—twice 360 degrees—or 720 degrees.

To determine, therefore, the distance or angle between firing strokes, divide 360 degrees by the number of cylinders for two-stroke-cycle engines, and divide 720 degrees by the number of cylinders for four-stroke-cycle engines.

Example: What is the interval between firing strokes on a six-cylinder two-stroke-cycle engine—and on a four-stroke-cycle engine?

$$\text{two-stroke-cycle} = \frac{360^\circ}{6} = 60^\circ$$

$$\text{four-stroke-cycle} = \frac{720^\circ}{6} = 120^\circ$$

How can inch marks on the flywheel be converted to degrees?

Answer: Inches marked on a flywheel rim are readily converted to degrees by the formula:

$$\text{degrees} = \frac{\text{inches} \times 360^\circ}{\text{circumference}}$$

Example: A 30-inch diameter flywheel is turned 24 inches from the center index point for a valve setting. How many degrees has it been turned?

Substituting, we get:

$$\text{degrees} = \frac{24 \text{ in.} \times 360^\circ}{\text{circumference}}$$

$$\begin{aligned}\text{circumference} &= \text{diameter} \times 3.1416\\ &= 30 \times 3.1416\\ &= 94.248 \text{ in.}\end{aligned}$$

$$\text{degrees turned} = \frac{24 \times 360}{94.248} = 91.67$$

How can degrees marked on the flywheel rim be converted to inches?

Answer: Knowing the number of degrees which a flywheel is to be turned for a valve setting, the number of inches on an inch-marked flywheel may be obtained by:

Multiplying the circumference of the flywheel by the degrees turned and dividing by 360.

Example: A 30-inch diameter flywheel marked off in inches is to be turned 90 degrees; how many inches is it to be turned?

$$\begin{aligned} \text{circumference} &= \text{diameter} \times 3.1416 \\ &= 30 \times 3.1416 \\ &= 94.248 \text{ in.} \end{aligned}$$

$$\text{inches to be turned} = \frac{94.248 \times 90}{360} = 23.56$$

CHAPTER 18

Engine Indicators

What is an engine indicator?

Answer: An engine indicator is an instrument that registers compression pressure, combustion pressure, mean effective pressure, exhaust pressure, time and duration of fuel injection, and time of opening and closing the inlet and the exhaust valves; all as they occur in the cylinder at the time indicated on the indicator card or diagram.

Why is the use of an engine indicator important?

Answer: The importance of using the engine indicator cannot be too strongly stressed. Only by its continued use can efficient operation be maintained. It enables the operator to know the performance of each cylinder of the engine and thereby make adjustments so that an equal amount of power can be developed by each cylinder. Distributing the power equally to each cylinder means smoother running, more efficient operation, and less vibration. It also shows defects in adjustments or leaks that may be present. The indicator is the tell-tale device of the engine and should be used frequently (Fig. 18-1).

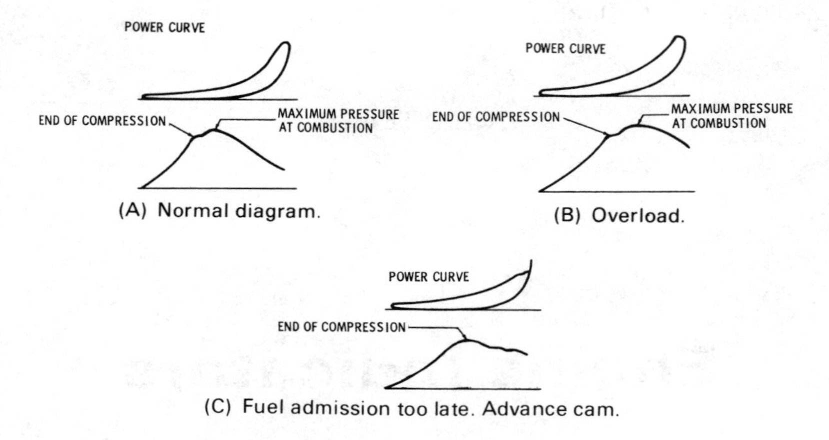

Fig. 18-1. Specimen indicator cards.

What is the most important use of the engine indicator?

Answer: For determining air compressions and firing conditions. Air compressions in the various cylinders should not vary by more than ten pounds for smooth engine operation (Fig. 18-2).

On what principle does an engine indicator work?

Answer: On the principle of a piston, in a cylinder, forced by pressure against a coiled spring that corresponds in tension to the working pressure to be measured.

How does an indicator show the pressure it measures?

Answer: By a pencil actuated by the piston through a system of levers. To record the cylinder pressure, the pencil registers, on a paper or card on the indicator drum, the movement of the indicator piston, which corresponds to the movement of the piston in the working cylinder.

How are the movements of the indicator and the piston in the working cylinder made to correspond?

Answer: By means of a cord attached to the crosshead or piston of the working cylinder through a reducing motion which moves the indicator drum.

How is an indicator attached to the working cylinder?

Answer: By means of an indicator cock connection on the cylinder.

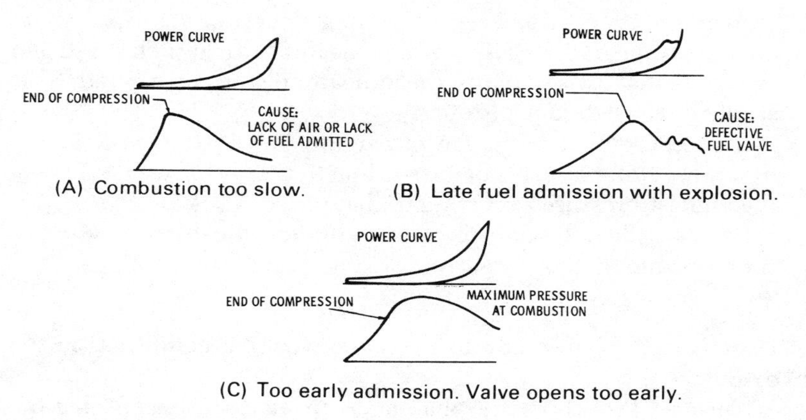

Fig. 18-2 Specimen indicator cards.

What is the method of determining air compression pressure?

Answer: Run the engine for half an hour to warm it up and assume normal operating condition. See that the air-starting, the air-inlet, the spray, and the exhaust valves are pressure tight.

Cut off the fuel oil from the cylinder to be tested. Blow out the indicator cock. Close the indicator cock and attach the indicator. Place a paper on the indicator drum. Adjust the pencil point so that it will not bear too hard on the paper.

Open the indicator cock and let the pencil arm move up and down a few times before marking the paper. Then press the pencil point lightly against the paper, permitting it to make a few strokes. This will register a few compressions.

Close the indicator cock, press the pencil to the paper again, and pull the cord to make an atmospheric or horizontal line from which the pressure is to be measured.

Remove the paper and the indicator from the cylinder and

arrange to test the next cylinder. Proceed until all the cylinders have been tested.

The pressure is determined by measuring the height or length of the vertical lines from the horizontal or atmospheric line by a scale that corresponds to the indicator spring number.

How is a card made to indicate the firing condition in a cylinder?

Answer: Proceed as given for testing air compression except do not cut off the fuel. Pull the cord on the indicator at such speed and time as to make the stroke of the indicator piston correspond to the stroke of the cylinder piston.

A measurement from the horizontal line to the peak of the diagram with the scale corresponding to the spring used will give the highest pressure occurring in that power stroke.

By carefully noting the movement of the indicator arm, you will soon be able to make a perfect "firing card."

How is a diagram made to show the working conditions in the cylinder?

Answer: Proceed as for making a "firing card" except that the cord must be attached to the crosshead or working piston in such a manner that it will cause the indicator drum to revolve in a movement corresponding to the movement of the working piston.

This diagram is used for determining the horsepower of the engine.

NOTE: It is necessary to make sure that the cord attaching the indicator to the working piston is of such length or so adjusted that the movement of the indicator drum exactly corresponds to the movement of the working piston.

How is horsepower determined by the indicator diagram?

Answer: By calculating or by measuring the mean effective pressure in the working cylinder from it.

What is meant by effective pressure in a diesel cylinder?

Answer: The average pressure exerted upon the piston by the expanding gases of combustion which produce the power.

How is mean effective pressure applied to the horsepower calculations?

By providing the figure required in the formula for engine horsepower which is $\frac{P \times L \times A \times N}{33{,}000}$ where P is the mean effective pressure. (The calculations for horsepower are given in Chapter 19 on horsepower.)

How is mean effective pressure determined from the diagram card?

Answer: Either by calculating the area in square inches or by obtaining the area by means of a planimeter. Then divide by the length of the diagram and multiply by the scale or number of the spring.

What is a planimeter?

Answer: It is a measuring device that traces the outline of the diagram and records on a scale the accurate area of the diagram.

What scale spring should be used on a first test?

Answer: Always use a 500-pound spring. Then if it is found that the compression is low enough to permit the use of a smaller spring to get a larger curve or diagram, the change may be made. *Caution:* The piston of the indicator should never be permitted to reach the top of the cylinder.

What is meant by a 500-pound spring?

Answer: A spring calibrated to make a diagram one inch in height at 500 pounds cylinder pressure; hence the need for using a high-pressure spring when making a first card so that the indicator piston will not reach its extreme height. A 250-pound spring on a 500-pound cylinder pressure would make a diagram two inches high, which is the usual size made to determine conditions.

How should an indicator be handled for different readings?

Answer: An indicator should always be removed from the cylinder as soon as the readings of that cylinder have been completed. The pencil arm should not be permitted to work while changing cards or while doing other things. The heat of the engine may destroy the lubricant in the indicator cylinder and damage either the piston or the barrel.

What care should be given an engine indicator?

Answer: The same as any other accurately made and delicately adjusted instrument. It must be used and handled carefully. It should be kept clean, and when not in use, should be placed where it will not be subject to vibration and injury. The accuracy of determining the operating conditions of the engine are dependent upon the accuracy of the indicator and upon the care with which the readings are made.

What care should be given the pencil arm?

Answer: The best possible care. If the pencil arm becomes bent, the accuracy of its movement will be destroyed and the card diagram will be incorrect. When through with the readings, wipe off the instrument with a clean cloth, oil it thoroughly, and return it to its box or container.

How often should an indicator be oiled?

Answer: An indicator should be oiled for each new cylinder on which it is to be used.

CHAPTER 19

Horsepower

What is meant by the horsepower of an engine?

Answer: Horsepower is defined as the amount of energy or work required to raise a weight of 33,000 pounds a height of one foot in one minute of time; or to overcome or create a force that is equivalent to doing that amount of work. Therefore, in simplified terms, one horsepower is 33,000 foot-pounds of work done in one minute. However, horsepower may be required at various power points of the engine, and the calculation must be made accordingly. For instance, the horsepower generated by an engine, as indicated by the power in the cylinders, may not be the actual power that can be utilized for work because there is the friction of the engine to be considered. Then there is brake horsepower, which can be used for work. There is also the effective horsepower and indicated horsepower.

What is meant by indicated horsepower?

Answer: Indicated horsepower represents the power deve-

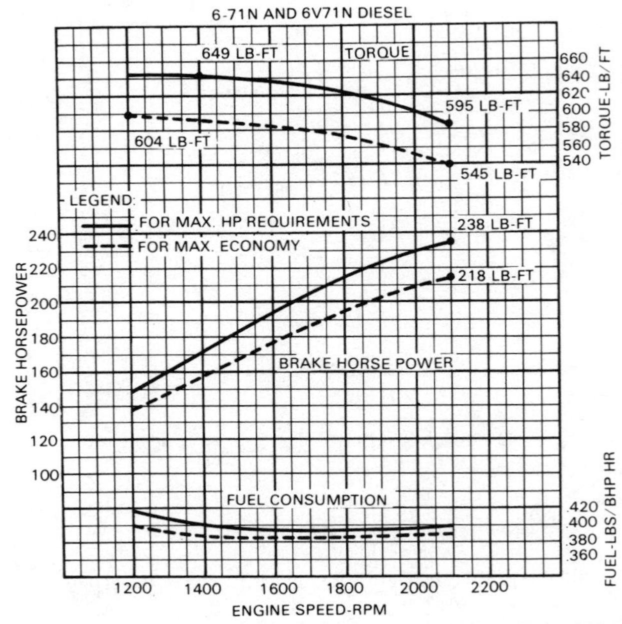

Courtesy Detroit Diesel Engine Division, General Motors Corp.

Fig. 19-1. Performance curves for General Motors 6-71N and 6V-71N diesel engines used in truck applications.

loped in the engine cylinder as obtained from the pressure in the cylinder. This is pressure obtained from an indicator card which shows cylinder pressures. Indicated horsepower does not represent the actual useful power delivered by the engine.

What is meant by brake horsepower?

Answer: Brake horsepower is the actual horsepower delivered by the engine to the drive shaft; it is equal to the indicated horsepower less the friction of the engine. It is the useful horsepower (Fig. 19-1).

What is meant by effective horsepower?

Answer: Effective horsepower is the final horsepower delivered to the equipment. An engine may be operating compressors, pumps, and auxiliary equipment as part of its own power production needs; the remaining power is therefore the effective power for drive. The difference between indicated horsepower and effective horsepower may be as much as 25 percent. The formula for indicated horsepower is:

$$\text{ihp} = \frac{\text{PLAN n}}{33{,}000} = \frac{\text{P} \times \text{L} \times \text{A} \times \text{N} \times \text{n}}{33{,}000}$$

in which:

P = mean effective pressure in cylinder (lbs. per sq. in.).
L = length of piston stroke (feet).
A = area of piston (sq. inches).
N = number of power strokes (per minute).
n = number of cylinders.

How many power strokes are delivered per crankshaft revolution in a two-stroke-cycle engine?

Answer: In a two-stroke-cycle engine, there are two piston strokes per revolution—one power stroke and one return; therefore, the number of power crankshaft strokes equals the number of revolutions of the crankshaft.

How many power strokes per crankshaft revolution are delivered in a four-stroke-cycle engine?

Answer: In a four-stroke-cycle engine, the number of power strokes equals one-half the number of revolutions of the crankshaft. If the engines are double acting, each figure for number of power strokes, is multiplied by two; i.e., a two-stroke-cycle, double-acting engine cycle would have two power strokes per revolution; and a four-stroke-cycle double-acting engine would have one power stroke per revolution.

Opposed-piston engines would be calculated the same as double-acting engines. A double power stroke is obtained by the two pistons traveling in opposite directions—power forces them apart.

How is mean effective pressure obtained?

Answer: Mean effective pressure is obtained via the engine indicator card and equals the average pressure exerted on the piston throughout one power stroke. This may be measured either by a planimeter (described in the engine indicator chapter) or by dividing the indicator card diagram into ten ordinates and obtaining the average height of the pressures by adding their lengths together and dividing by ten.

How is the length of a piston stroke obtained?

Answer: Length of piston stroke may be obtained by measuring the distance between the center of the crankshaft and the center of the crank throw and multiplying by two. (If in inches, divide by 12 for feet.)

How is the area of the piston calculated?

Answer: Area of the piston equals the diameter of the piston squared, times .7854 ($D^2 \times .7854$).

How is the number of power strokes obtained?

Answer: Number of power strokes is obtained from the speed of the engine, by a speed indicator (the number of revolutions per minute). The ratio is given above for various types of engines.

Calculate the indicated horsepower (iph) for the following engine.

Answer: A four-cylinder, two-stroke-cycle diesel engine, with cylinder diameter, 14 inches; length of piston stroke, 18 inches; mean effective pressure, 80 pounds per square inch; and speed, 300 rpm.

Use the formula for indicated horsepower (ihp).

$$\text{ihp} = \frac{P\,L\,A\,N\,n}{33{,}000}$$

$$= \frac{\text{pressure} \times \text{length stroke} \times \text{area piston} \times \text{speed} \times \text{no.cylinders}}{33{,}000 \text{ ft.-lbs. per minute (one horsepower)}}$$

$$= \frac{80 \times 1.5 \times (14^2 \times .7854) \times 300 \times 4}{33{,}000}$$

Calculate as follows:

$$\text{length of piston stroke (in feet)} = \frac{18 \text{ in.}}{12 \text{ in.}} = 1.5 \text{ ft.}$$

$$\text{area of piston} = (\text{diam. piston or cylinder diam.})^2 \times .7854$$
$$= (14)^2 \times .7854$$
$$= 196 \times .7854$$
$$= 153.9384 \text{ sq. in.}$$

Substituting in the formula:

$$\text{ihp} = \frac{80 \times 1.5 \times (153.9384) \times 300 \times 4}{33{,}000}$$

$$= \frac{22{,}167{,}129.60 \text{ ft.-lbs.}}{33{,}000 \text{ ft.-lbs.}}$$

$$= 671.73 \text{ horsepower}$$

If the engine were a four-stroke-cycle engine, the ihp would be 671.73/2 or 335.865 horsepower.

Shorter and briefer methods may be devised, using constants for obtaining the indicated horsepower of engines, but the principles are given here.

What equipment is used to measure brake horsepower?

Answer: Brake horsepower (bhp) is measured usually by the manufacturers, or on test, and is determined either by a dynamometer, or a Prony brake or friction brake applied to the flywheel as shown in the illustration.

How is the mechanical efficiency of an engine determined?

Answer: The mechanical efficiency of an engine is the ratio for the delivered power of the engine (brake horsepower) to the power developed by the engine (indicated horsepower). This is obtained by dividing the brake horsepower by the indicated horsepower.

Example: If the indicated horsepower (ihp) is 300, and the brake horsepower (bhp) is 225, the mechanical efficiency is:

$$\frac{225 \text{ (bhp)}}{300 \text{ (ihp)}} = 75 \text{ percent}$$

Therefore, knowing the mechanical efficiency of an engine, the brake horsepower, or delivered horsepower, may be readily obtained from indicated horsepower measurements and the percentage of efficiency.

How is mean piston speed calculated?

Answer: Mean piston speed, used in calculations, is the product of piston stroke and speed; it equals length of stroke in feet, times revolutions per minute, times two. (L × R × 2).

How is the Prony brake used to determine the horsepower of an engine?

Answer: The Prony brake is a device for determining engine horsepower by the mechanical method of measuring directly at the shaft. (Hence, often called shaft horsepower, this is the actual horsepower delivered by the engine.)

The Prony brake usually consists of a friction band ring that may be placed around a pulley or flywheel and attached to a lever bearing upon a weighing scale. The friction between the surfaces in contact will tend to rotate the arm in the direction in which the shaft is revolving. This motion is resisted and measured in pounds. In setting up the Prony brake, the distance between the center of the shaft and the point of contact with the scales, must be accurately measured—the point being placed at the same elevation as the center of the shaft. By this arrangement the amount of friction between the brake band and the revolving pulley or wheel is weighed upon the scales.

Since the brake fits tightly and could be carried around by the pulley, but for the arm bearing upon the scale, the amount of frictional power exerted by the revolving pulley in turning free within the brake band, may be measured, just as would a load or a machine attached to the wheel or shaft.

The distance between the center of the shaft and the point of contact with the scales, is called the lever arm.

The information that must be had in determining brake horsepower is:

1. The length of the lever arm in feet.
2. The pressure exerted upon the scales.
3. The speed—in revolutions per minute.

The formula for determining brake horsepower is:

$$bhp = \frac{2 \pi L N P}{33,000}$$

in which:

π = 3.1416 (for circumference).
L = length of lever arm from center of shaft to point on scale (in feet).
N = number of revolutions (per minute).
P = weight or pressure on scale (in pounds).

Example: If on the test, length of lever arm is six feet; reading onscale, 90 pounds; speed of pulley, 1,00 revolutions per minute; what is the brake horsepower of the engine?

Substituting:

$$bhp = \frac{2 \times 3.1416 \times 6 \times 1000 \times 90}{33,000} = 102.8 \text{ horsepower}$$

How is the Prony brake designed to test the horsepower of an engine?

Answer: The Prony brake, as shown in Fig. 19-2, consists of a band of rope, leather, or strap iron to which are fastened a number of wooden blocks—some of which carry shoulders to prevent the band from slipping from the wheel rim. The band is drawn tightly so that the blocks press against th surface of the rim all around. The brake band is restrained from revolving with the flywheel or pulley by two arms attached to it near the top and bottom centers of the wheel and joined at the outer ends to form a lever or contact point, which bears upon a platform scale. The contact point is maintained at a height level with the center of the shaft. This is done by a block or leg set upon the scale platform. Circulating water passes through a channel on the interior of the band to prevent overheating while the wheel is revolving against the friction.

How can brake horsepower be measured with the rope and the spring-scale form of Prony brake?

Answer: A Prony brake of rope and spring-scale form can be

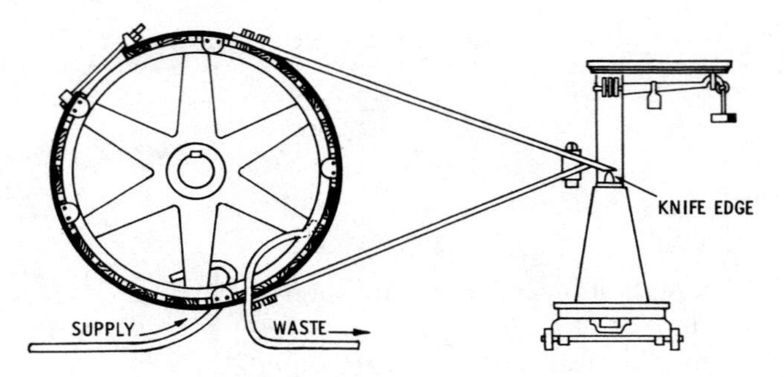

Fig. 19-2. Prony brake for testing brake horsepower.

used for measuring brake horsepower. This form of brake is readily constructed and set up; and being self-adjusting, needs no accurate fitting. For heavy power tests, the number of ropes or turns of rope around the wheel may be increased as needed (Fig. 19-3).

To calculate the brake horsepower, subtract the pull registered by the spring balance B, from the weight W. The radius of the wheel, plus one-half the diameter of the rope, is the lever arm length for the calculation.

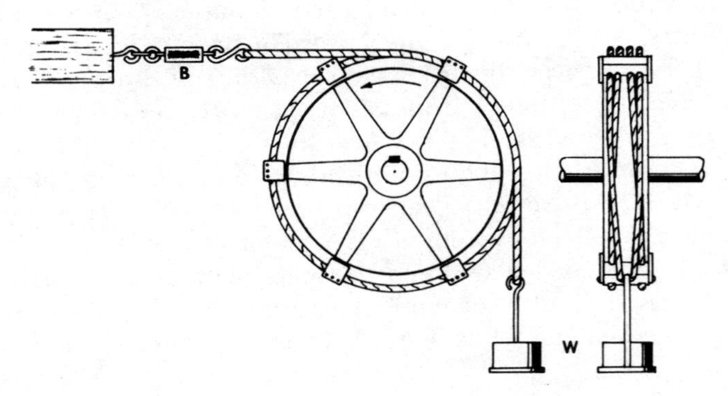

Fig. 19-3. Rope and spring-scale form of Prony brake. A spring balance, B, is shown horizontally, although it may be attached vertically. The ropes are held to the wheel by blocks, O, and a weight, W, or a spring balance attached to the rope.

The formula is:

$$bhp = \frac{2\pi R N (W-P)}{33{,}000}$$

in which:

π = 3.1416 (for circumference).
R = Radius (distance from center of shaft to center of rope) in feet.
N = number of revolutions per minute.
W = weight in pounds.
P = pull on spring balance.
NOTE: If P is greater than W, the wheel is turning in the reverse direction and the formula should read (P−W).

Give a brief description of the various kinds of horsepower.

Answer: Nominal horsepower is of the original calculation by Watt; the efficiency of engines has greatly increased.

Indicated horsepower (ihp) is the horsepower developed by an engine as a result of the generated power in the cylinders, as shown by an engine indicator card; the mean effective pressure being utilized in the formula:

$$ihp = \frac{PLAN\ n}{33{,}000}$$

in which

P = the mean effective pressure (lbs. per sq. in.).
L = the length of piston stroke (feet).
A = area of piston (sq. ins.).
N = number of power strokes (per minute).
n = number of cylinders.
33,000 = the foot-pounds per horsepower of work.

Brake horsepower is the horsepower actually delivered to the drive shaft and is equal to the indicated horsepower less the friction of the engine. It is measured by a Prony brake applied to the flywheel.

A Prony brake consists of a friction band or belt placed around the flywheel and attached to a lever bearing upon the platform of a scale, or drawing on a spring scale, and balancing the pull of the engine.

The formula for determining brake horsepower is:

$$\text{bhp} = \frac{2\pi LNP}{33{,}000}$$

in which:

π = 3.1416.
L = length of lever arm (feet).
N = number of revolutions (per minute).
P = balancing weight or pull shown on scales.
(This is more fully explained in the text.)

Hydraulic, or pump, horsepower is the horsepower required to elevate or force water against pressure. It is usually calculated as the indicated horsepower at the water end of the pump.

The formula for pump horsepower is:

$$\text{php} = \frac{V \times W \times (L + H)}{33{,}000}$$

in which:
V = volume in cubic feet per minute.
W = weight of one cubic foot of water.
L = lift in feet.
H = head in feet.

Boiler horsepower—One boiler horsepower is the power required or consumed in evaporating 34½ pounds of water in one hour when the water is at 212° Fahrenheit. The heat consumed is 33,465 Btu.

Electrical horsepower is the electrical equivalent of mechanical energy and is 746 watts. One watt = one volt × one ampere. The equivalent is determined from the work done, or heat generated, by passing a current of one ampere at a pressure of one volt through a resistance of one ohm during one second of time; and is known as the joule. One watt equals one joule per second; therefore 60 joules per minute. One joule equals .7374 foot-pounds, hence 60 joules equals:

$$60 \times .7374 = 44.244 \text{ ft.-lbs.}$$

$$33{,}000 \div 44.244 = 746 \text{ watts}$$

Effective horsepower is the actual effective horsepower for the driven equipment. As, in a train, the effective horsepower for the train of cars would be:

$$\frac{\text{pull on tender drawbar} \times \text{speed}}{33{,}000}$$

CHAPTER 20

Pulleys, Belts, and Gears

What is meant by a driver pulley and by a driven pulley?

Answer: When two pulleys are working together, connected by a belt, the one that transmits the motion is called the *driver* and the one that receives the motion is called the *driven*.

How can the size of the driver pulley be determined?

Answer: To find the size of the driver pulley, multiply the diameter of the driven pulley d, by its number of revolutions r, and divide by the revolutions of the driver pulley R.

$$\text{diameter of driver pulley } D = \frac{d \times r}{R}$$

in which:

D = diameter of driver pulley.
d = diameter of driven pulley.

r = revolutions per minute of driven pulley.
R = revolutions per minute of driver pulley.

Example: What is the diameter of the engine pulley required to drive a generator at a speed of 2,000 revolutions per minute, when the generator pulley is 4 inches in diameter and the engine speed is 500 rpm?

Substituting values:

$$D = \frac{4 \times 2000}{500} = \frac{8000}{500} = 16 \text{ inches}$$

How can the size of the driven pulley be determined?

Answer: To find the size of the driven pulley multiply the diameter of the driver pulley D, by its number of revolutions R, and divide by the revolutions of the driven pulley r.

$$\text{diameter of driven pulley } d = \frac{D \times R}{r}$$

Example: What size pulley must be used on a generator to be driven at a speed of 2,000 rpm, if the engine speed is 500 rpm and the engine pulley is 16 inches in diameter?

Substituting values:

$$d = \frac{16 \times 500}{2000} = \frac{8000}{2000} = 4 \text{ inches}$$

How can the speed of the driver pulley be determined?

Answer: To find the speed of the driver pulley multiply the diameter of the driven pulley d, by its number of revolutions r, and divide by the diameter of the diver pulley D.

$$\text{speed of driven pulley } r = \frac{D \times R}{d}$$

Example: At what speed will a generator run if it has a 4-inch pulley and is driven by an engine having a 16-inch pulley and running at 500 rpm?

$$r = \frac{16 \times 500}{4} = \frac{8000}{4} = 2000 \text{ rpm}$$

How can the speed of the driver pulley be determined?

Answer: To find the speed of the driver pulley, multiply the diameter of the driven pulley d, by its number of revolutions r, and divide by the diameter of the driver pulley D.

$$\text{speed of driver pulley } R = \frac{d \times r}{D}$$

Example: At what speed must an engine run to drive a generator at 2,000 rpm when the generator has a 4-inch pulley and the engine a 16-inch pulley?

Substituting values:

$$R = \frac{4 \times 2000}{16} = \frac{8000}{16} = 500 \text{ rpm}$$

What is the relationship of size to speed in pulley calculations?

Answer: The proportion is as follows:

1. Diameter of driver pulley D, is to diameter of driven pulley d, as the revolutions of driven pulley r, are to the revolutions of the driver R.

$$D : d : : r : R \text{ or } DR = dr$$

$$\text{from which } D = \frac{dr}{R},\ d = \frac{DR}{r},\ R = \frac{dr}{D},\ r = \frac{DR}{d}$$

2. The ratio of diameters between two pulleys should not ordinarily exceed 1 to 5 without some means of securing belt contact to the smaller pulley (Fig. 20-1).

How can the required size of belt be determined?

Answer: A single belt traveling 1,000 feet per minute will

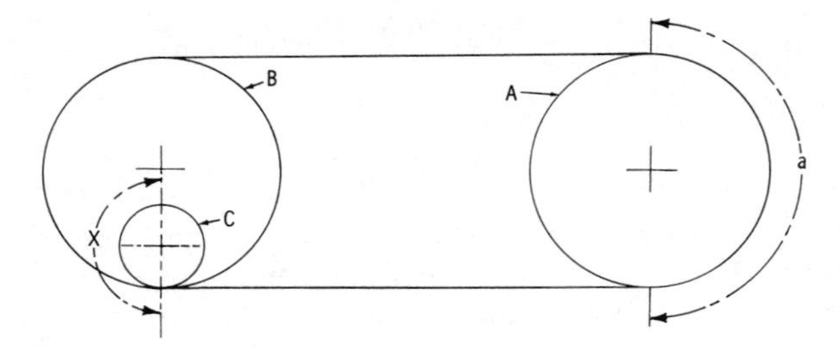

Fig. 20-1. **Variation in arc of belt contact caused by difference in sizes of driver and driven pulleys. If the two pulleys, A and B, are of equal size, the arc of contact, a, will be 180° (slightly more because of the sag of the belt). When the driver A is larger than the driven pulley C, the arc of contact x is reduced, depending on the difference in pulley diameters and their center-to-center distances.**

transmit one horsepower per inch of width; a double belt will transmit twice this amount (Fig. 20-2).

Example: What size of double belt is required to transmit 50 horsepower at a belt speed of 4,000 feet per minute, and what diameter pulley must be used for 954 rpm at a belt speed of 4,000 feet per minute?

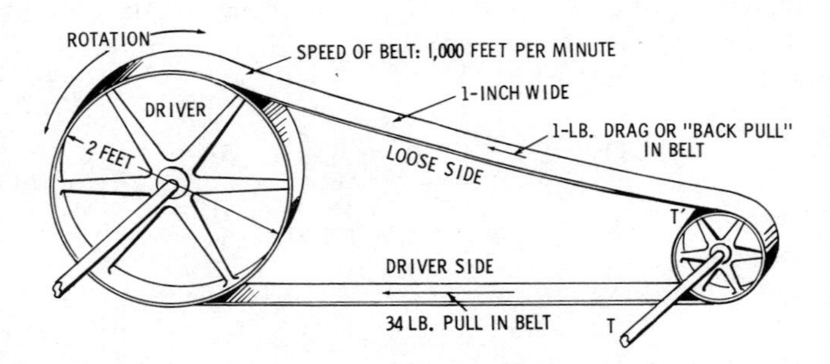

Fig. 20-2. **Belt transmission of one horsepower. A pulley is driven by means of the friction between the belt surfaces in contact. The driving force of the belt is equal to the tension on the driving side T, minus the tension on the loose side T'. If the belt travels at 1,000 feet per minute, T = 34 lbs., and T' = 1 lb., then 33 lbs. × 1,000 feet = 33,000 ft.-lbs. per minute = 1 horsepower = the power transmitted by the belt.**

The horsepower transmitted per inch is:

$$\frac{4000}{1000} \times 2 = 8$$

Accordingly, the width of belt required to transmit 50 horsepower is:

$$50 \div 8 = 6.25, \text{ or } 6 \text{ inches.}$$

For a belt speed of 4,000 feet per minute, the distance in inches traveled by the belt per revolution of the pulley is:

$$\text{circumference} = \frac{4000 \times 12}{954} = 50.31 \text{ inches}$$

This is equal to the circumference of the pulley, and the corresponding diameter is:

$$\text{circumference} \div \pi, \text{ or}$$
$$50.31 \div 3.1416 = 16.1 \text{ inches}$$

Does it matter if the driver or tight side of the belt is on the top or on the bottom?

Answer: Whenever possible, belts should be run with the tight side on the bottom; the upper or loose side will then form a concave arc, which increases the contact on both pulleys. A belt used in this manner may be kept more slack than a belt put on the reverse way.

How can the number of teeth required on a gear be calculated?

Answer: Gear-wheel calculation is similar to that for pulleys, except that number of teeth instead of diameter is used in the calculation as follows:

1. $\text{rpm of driver} = \dfrac{\text{rpm of follower} \times \text{teeth on follower}}{\text{teeth on driver}}$

$$2.\ \text{rpm of follower} = \frac{\text{rpm of driver} \times \text{teeth on driver}}{\text{teeth on follower}}$$

$$3.\ \text{teeth on driver} = \frac{\text{rpm of follower} \times \text{teeth on follower}}{\text{rpm of driver}}$$

$$4.\ \text{teeth on follower} = \frac{\text{rpm of driver} \times \text{teeth on driver}}{\text{rpm of follower}}$$

How can the speeds of trains of gears be calculated?

Answer: As with pulleys, marked speed differences are made with trains of gears. Each pair constitutes a driver and a follower. If the shafts are parallel and two gears of different diameters are keyed to each shaft, except the first and last, considerable speed difference can be obtained (Fig. 20-3).

Rule: The product of the number of teeth of all the drivers divided by the product of the number of teeth of all the followers, is the velocity ratio.

Example: Assume that a train of gears has three drivers A, B, C, and three followers L, M, and N, as shown. A has 14 teeth and drives L having 70 teeth. Pinion B, on same shaft with L, has 13 teeth and drives M, having 104 teeth. Pinion C has 15 teeth, and is on the same shaft with M; C drives N, having 75 teeth. What is the velocity ratio of A to N?

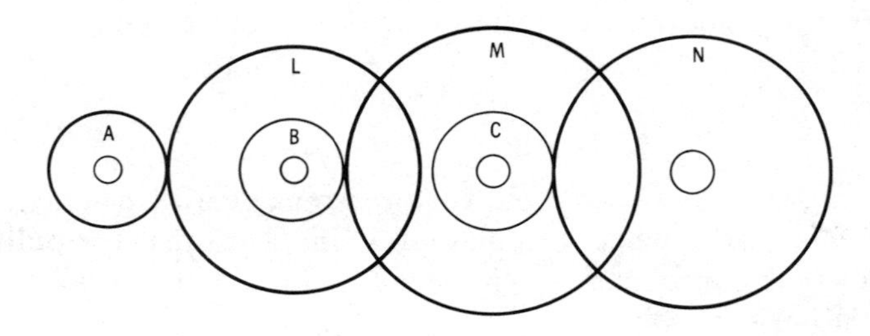

Fig. 20-3. Velocity ratio of gears in a gear train. The product of the number of teeth on all the drivers divided by the product of the number of teeth on all follower gears is the velocity ratio.

$$\text{velocity ratio} = \frac{\text{teeth on A} \times \text{teeth on B} \times \text{teeth on C}}{\text{teeth on L} \times \text{teeth on M} \times \text{teeth on N}}$$

$$= \frac{14 \times 13 \times 15}{70 \times 104 \times 75} = \frac{1}{200}$$

Knowing the velocity ratio of the train, the speed of N may be readily determined from the speed of A as follows:

If A runs at 1,800 rpm; then N will make

$$\frac{1800}{200} = 9 \text{ rpm.}$$

CHAPTER 21

Tank Calculations

How can the capacity of fuel and oil tanks be calculated?

Answer: To find the capacity of fuel and oil tanks, determine their cubical contents (cu. ft. or cu. in.) and convert the measurement to gallons or barrels as follows:

1. One barrel (bbl.) = 31.5 gallons (gal.).
2. One gallon occupies 231 cubic inches of space (cu. ins.).
3. One cubic foot (cu. ft.) of space is occupied by 7.5 gallons.
4. One cubic foot = 1728 cubic inches.

Example: How many barrels of oil are required to fill a cylindrical tank, 3 ft. in diameter and 9 ft. in length?

Find the cu. ft. in the tank by the formula:

$$\begin{aligned}\text{cu. ft.} &= \text{area of cross section} \times \text{length}\\ &= (D^2 \times .7854) \times 9\\ &= (3 \times 3 \times .7854) \times 9\end{aligned}$$

$$= 7.0686 \times 9$$
$$= 63.6174 \text{ cu. ft. in the tank}$$

One cu. ft. holds 7.5 gallons:

Therefore 63.6174 cu. ft × 7.5 gallons per cu. ft. = 477.13 gallons or the capacity of the tank. Since there are 31.5 gallons in 1 barrel, 477.13 ÷ 31.5 = 15.15 barrels or the capacity of the tank in barrels.

If the above tank has a square base, 3 ft. × 3 ft., and is 9 ft. in length, the cubical contents (cu. ft.) is the product of 3 ft. × 3 ft. × 9 ft., or 81 cu. ft. is the capacity of the tank.

How can the capacity of a tank be calculated if the measurements are in inches, or in feet and inches?

Answer: The measurements are reduced to inches and divided by 1728 (1728 cu. in. = 1 cu. ft.) to give the capacity in cubic feet.

Example: Tank base is 3 ft. 2 in. × 3 ft. 4 in. and the height is 9 ft. 3 in.

Change to inches as follows:

$$3 \text{ ft. } 2 \text{ in.} = 3 \times 12 + 2 = 38 \text{ inches.}$$
$$3 \text{ ft. } 4 \text{ in.} = 3 \times 12 + 4 = 40 \text{ inches.}$$
$$9 \text{ ft. } 3 \text{ in.} = 9 \times 12 + 3 = 111 \text{ inches.}$$

Then the area in cu. ft. is calculated:

$$\text{cu. ft.} = \frac{38 \times 40 \times 111}{1728}$$

$$= \frac{168720}{1728}$$

$$= 97.63 \text{ cu. ft. in the tank}$$

If the tank size is small, the cubical contents may be obtained in inches and divided directly by 231 (one gallon occupies 231 cu. in.) for the number of gallons in the tank.

Example: How many gallons will a tank 10 inches in diameter and $2^1/_2$ ft. (30 ins.) hold?

Find the cubic inches in the tank:

$$\text{cu. ins.} = \text{cross-sectional area} \times \text{length}$$
$$= (D^2 \times .7854) \times 30 \text{ ins.}$$
$$= (10 \times 10 \times .7854) \times 30 \text{ ins.}$$
$$= 78.54 \times 30 \text{ ins.}$$
$$= 2356.2 \text{ cu. in. in the tank}$$
$$\text{gallons} = \frac{2356.2}{231} = 10.2 \text{ gallons capacity of tank}$$

How can the capacity of a tank be calculated, if the tank is irregular in shape?

Answer: Determine the average measurements and proceed as above.

Example: What is the cubical content of a tank 4 ft. wide at the bottom, 6 ft. wide at the top, 3 ft. in height, and 10 ft. in length?

Find average width:

$$\frac{4 \text{ ft. (bottom)} + 6 \text{ ft. (top)}}{2} = 5 \text{ ft.} = \text{average width}$$

Then find cu. ft. in tank as:

$$\text{cu. ft.} = \text{width} \times \text{height} \times \text{length}$$
$$= 5 \text{ (average width)} \times 3 \times 10$$
$$= 150 \text{ cu. ft. capacity of tank}$$

The gallons or barrels capacity may be determined from the cubical contents as above. Note that one cubic foot actually holds 7.481 gallons of liquid, but 7.5 is sufficient for average calculations.

How can the size of tank required to hold a definite quantity of oil be determined?

Answer: Convert the oil quantity in barrels or gallons to cubic space and reduce to proper lineal dimensions.

Example: What size tank is required to hold 10 gallons of oil?

Since one gallon of oil occupies 231 cubic inches of space, ten gallons will occupy (10×231) or 2,310 cubic inches of space.

The tank may be any shape in combination of width, height, length, or cross section, provided it contains 2,310 cu. in. of space. Therefore, a figure such as space available for manufacture may govern the physical dimensions.

Assuming that 2.5 feet of space is available for the length, and the tank is to be round in cross section, the calculation for the diameter of the tank is as follows:

$$\text{cu. in.} = \text{cross-sectional area (diameter}^2 \times .7854) \times \text{length}$$

$$2310 = \text{area (diameter}^2 \times .7854) \times \text{length (30 in.)}$$

Then:

$$\text{Area (diameter}^2 \times .7854) = \frac{2310}{30}$$

$$= 77 \text{ sq. in. of cross-sectional area in the tank}$$

Therefore,

$$\text{diameter} = \text{the square root of } 77 \div .7854$$

$$= \sqrt{\frac{77}{.7854}}$$

$$= \sqrt{98.04}$$

$$= 9.9 \text{ inches as diameter of tank or}$$

10 inches

Thus, the tank required to hold 10 gallons of oil will have to be 10 inches in diameter and 30 inches in length.

If a tank is to be entirely filled, a small allowance must be made for the expansion of the oil with rise in temperature.

CHAPTER 22

Engine-Room Procedure

What action should be taken upon assuming charge of a diesel power plant or a diesel-operated ship?

Answer: Make a general inspection of the plant and a careful examination of all parts and equipment and report to the owners or superiors any damage or irregularities found.

Why is it necessary at all times to leave the fuel oil supply line from the service or day tanks open to the fuel pumps?

Answer: In order to prevent the accumulation of air in the supply line between the spray nozzle valves and the pump. This ensures a solid and continuous supply of fuel oil. Air in the fuel oil line hinders operation and must be eliminated. It is better therefore to keep air out of the system.

What tests and examinations should be made before starting a new engine—or an engine that has been out of service?

Answer: Tests for water in the oil, for leaks, for fuel supply, for pump pressures, for lubrication, and for air. The procedure should be about as follows:

1. Check the lighting circuits to see that there is light in all places.
2. Open the drains on the service and storage tanks and examine the oil for presence of water. If none appears and there are no impurities visible, open the valves to the pipe lines and test for leaks.
3. Check the cooling water supply and check for leaks in the pipe lines.
4. Start the water circulating pump and check for leaks.
5. Test the auxiliary equipment.
6. Check the air-starting and air-injection equipment.
7. Note the level of the lubricating oil in the tanks and in the sump.
8. Pump lubricating oil to normal pressure, by hand or otherwise, and note operation.
9. Open fuel by-passes, either at spray nozzle valves or at other convenient points, and operate the fuel pump to clear the lines of air, water, etc.; then close the lines.
10. Rotate the engine a few times either by hand or by the jacking gear.
11. Open the drains of the air-starting tanks or bottles and clear them of oil and water.
12. Prime the fuel oil lines with the hand pump.
13. Rotate the engine with starting air to clear the cylinders of impure air before admitting fuel.
14. Check the fuel oil supply, the lubricating oil supply, the compressed air supply, the cooling water supply—and make certain that all are at proper level and that the valves are operating properly—start engine per starting instructions.

What should be done before turning over the watch to another operator?

Answer:

1. Check the temperatures by sight and by feel.
2. Check the fuel oil and lubricating oil tanks.
3. Check the sump for proper quantity of oil.

4. Check the working of the mechanical feed pump.
5. Check the starting air.
6. Feel the accessible parts of the oiling system.
7. Oil all moving parts not served by either pump or mechanical means.

How can an engine be kept running when the exhaust rocker arm is broken beyond repair, and there is no spare?

Answer: By shifting the inlet arm to the exhaust and loosening the inlet valve spring enabling it to open by suction of the piston.

What is an auxiliary exhaust valve?

Answer: It is a mechanically operated poppet valve at the base of the cylinder. Its purpose is to divert some of the exhaust gases and their high temperatures from the exhaust valve at the head.

If all the bolts in a shaft coupling are broken and there are no spare bolts, what can be done to continue operating?

Answer: Half the number of bolts of another coupling can be removed, placed in the coupling without bolts, and operation continued at reduced speed.

How can a tight pulley, a wheel, or coupling be removed from a shaft when the ordinary methods fail?

Answer: By building a box around it and expanding it by heat. The pulley will then become larger due to expansion and slide off the shaft.

Describe the method of removing a coupling by expanding it with hot lead.

Answer: Prepare a pulling rig and a pounder for immediate use. Build a wooden box around the coupling and stop it with clay, flour dough, or other convenient material to prevent the heat reaching the shaft or the center of the coupling before it affects the coupling. Pour red-hot lead into the box around the coupling. When the coupling has expanded, pull or pound it off the shaft.

How can a smoking cylinder be detected?

Answer: By noting the color of the exhaust at the try cock or by holding a piece of moistened white paper in front of an opened indicator cock or the relief valve. If the cylinder is smoking, the paper will become blackened.

What is a try cock?

Answer: It is a round-way stopcock placed in the connection between the cylinder and the exhaust manifold which, on being opened, permits the exhaust gases to escape and become visible.

What does a slapping sound in a cylinder indicate?

Answer: That either the piston or the cylinder is worn and is needing attention.

What does pounding in a cylinder indicate?

Answer: That there may be either a loose piston pin or a loose crank or main journal. Also, the crank journal may be worn out-of-round or fuel oil is being injected too early.

What is a fuel knock?

Answer: It is a knock caused either by too early fuel injection or by too cold fuel ignition temperature.

What does the term "pinking" mean?

Answer: It is a name given to the sound that indicates that fuel injection and combustion are occurring in an efficient manner and that the spray-valve needle is seating properly during its operation.

What is a precaution when cutting oil grooves in bearings?

Answer: That the grooves are not cut too close to the edge of the bearing and permit too much lubricating oil to escape.

Is heating of the bearings an indication that they do not have a sufficient number of oil grooves?

Answer: Not necessarily. Heating is more likely to be due to insufficient lubricating oil reaching the bearings either because of a stoppage in the lubricating oil lines or because of a lack of oil pressure in the engine.

What is a good daily routine for maintaining the lubricating system?

Answer: To maintain a lubricating system properly, it is a good practice to look at the pressure gauges, indicators, and feeds on the tanks and reservoirs every half hour.

Also, while the engine is operating, open the drains on the sump tank every six hours and remove a little oil either in the palm of the hand or in a shallow vessel or glass for examination. Test for water and at the same time take a little oil from the drain of the oil pump and test for water.

Why are most diesel engine bearings lined with Babbitt metal?

Answer: To reduce friction and to prevent damage to the crankshaft in the event of heating or bearing seizure.

What is the peculiar property of Babbitt metal?

Answer: It expands upon cooling. This is opposite to the action of other metals—they expand upon heating.

What is the best method of pouring Babbitt metal into a shell?

Answer: Pour the metal and revolve the shell at high speed. This drives out the air and condenses the metal before it solidifies.

To what temperature should Babbitt metal be heated for it to be suitable for pouring?

Answer: To a temperature no higher than is necessary for it to flow readily—it should be no hotter. The shell into which it is poured should also be heated to keep the metal from chilling upon contact and to enable it to reach all parts.

What is an easy way to tell when Babbitt metal is hot enough for pouring?

Answer: Insert a soft pine stick into the metal. When the stick turns black, the metal is ready for pouring.

What may cause a cylinder head to crack?

Answer: Overheating due either to insufficient cooling water or to accumulation of mud or scale on the cooling surfaces which prevents the heat escaping and being absorbed by the cooling water as it circulates.

How can a cracked cylinder head be detected?

Answer: By the cylinder misfiring and by the presence of water in the cylinder.

How can cooling water be kept from entering a cracked cylinder head?

Answer: Insert a piece of tin between the cylinder flange and the gasket of the inlet and outlet of the cooling water at the head.

Is it desirable for an engine operator to develop a sense of touch for temperatures?

Answer: Yes. The best of gauges and thermometers get out of order; and a sense of touch, which can be readily cultivated, can aid one considerably in determining temperature conditions. Some operators can judge temperatures as closely as five degrees.

How can air compression be increased or decreased in a diesel cylinder?

Answer: By raising or lowering the piston in the cylinder—this reduces or increases the space in which the air is compressed. The smaller the space into which a definite amount of air is compressed, the greater will be the pressure of the air, and therefore, the higher the temperature or heat of compression for ignition; the greater the space, the lower the heat of compression for ignition.

How can a piston be raised or lowered in a cylinder to change the compression and heat?

Answer: By adding to or removing shims from between the crankshaft and the connecting rod to lengthen or shorten the connecting rod. This increases or decreases the distance a piston enters a cylinder.

What are shims?

Answer: Shims are pieces of metal or other material used for either increasing or decreasing a clearance or a diameter to provide proper clearance.

Describe a method for increasing or decreasing air compression by a definite amount.

Answer: To increase or decrease the air compression in a diesel cylinder, as may be shown necessary by readings on an indicator card, the following method will be found useful.

In Table 22-1 the changes of presssure given are produced by raising or lowering the piston .010 (10/1000) of an inch, for various pressures. The table also shows the total amount which a piston must be raised or lowered for each 50 lbs. change of pressure.

Table 22-1. Increasing or Decreasing Cylinder Compression

Cylinder compression in lbs. per sq. in.	Pressure increase or decrease by raising or lowering piston .010 of an inch	Rise or drop required for 50 lb. change in pressure
700-650	14.0 lbs.	.0375 inch
650-600	11.7 lbs.	.0425 inch
600-550	10.0 lbs.	.050 inch
550-500	8.3 lbs.	.060 inch
500-450	7.1 lbs.	.070 inch
450-400	6.05 lbs.	.082 inch
400-350	5.05 lbs.	.092 inch

Example: What is the ratio required to lower the compression in a diesel cylinder for 600 lbs. to 500 lbs. per sq. in.?

The table shows the drop in pressures in 50-lb. steps. Therefore two steps are necessary—600 to 550 lbs. and 550 to 500 lbs. The drop between 600 and 550 lbs., produced by lowering the piston .010 of an inch, is 10 lbs. Between 550 and 500 lbs., it is 8.3 lbs.; the average, therefore, is 10 lbs. plus 8.3 lbs. divided by 2, or 9.15 lbs.—average pressure for each .010 of an inch change.

To lower the pressure 100 lbs. would require the removal of as many .010 inch (10/1000) shims as 9.15 will go into 100, or approximately 11 shims—a total of .110 (110/1000) of an inch.

To raise the pressure, shims are added.

Column 3 shows directly the amount that a piston must be either lowered or raised for each 50 lbs. change; 600 to 550 lbs. shows .050 (50/1000) of an inch; and 550 to 500 lbs. shows .060 (60/1000) of an inch.

Shims of any thickness may be used and the number required may be determined by dividing the total distance needed by the thickness of the shims and the nearest number used. For example, if the shims are .030 (30/1000) of an inch in thickness, the nearest number would be: .110 inch divied by .030 inch or 4 shims.

What are the best conditions for an engine room?

Answer: The following conditions are best for an engine room:

1. Light and dry—free from dust, dirt, water leaks, obstructions, dangerous chemicals, and fumes.
2. The room temperature, when the engine is shut down, should not fall below that required to keep the fuel oil sufficiently fluid to flow freely when ready to start up again.
3. Oil leaks or spatter should be avoided.
4. Sufficient supplies of fuel oil, lubricating oil, and needed accessories should always be on hand.
5. A rack, tool closet, and workbench are useful accessories.
6. A record book or log sheets should be on hand for notations of occurrences, conditions, and needs.
7. Air supply for the engine should preferably be drawn from

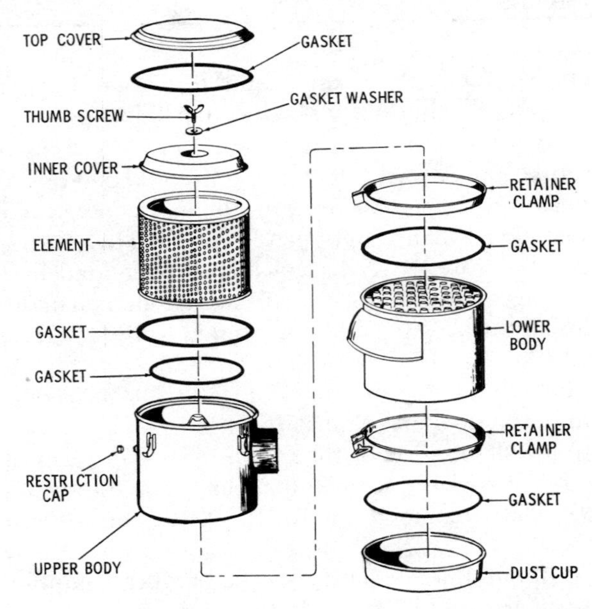

Courtesy Cummins Engine Co.

Fig. 22-1. Cummins engine composite air cleaner.

outside the engine room from a point that is not contaminated.

How does the air cleaner remove dust from the air?

Answer: Air enters through a side inlet and passes into a series of vaned tubes which give the air a cyclonic twist. Dust particles are separated from the air and thrown to the outside of the tube and fall into a removable dust cup. The clean air passes into the secondary paper filter, which removes all remaining dust.

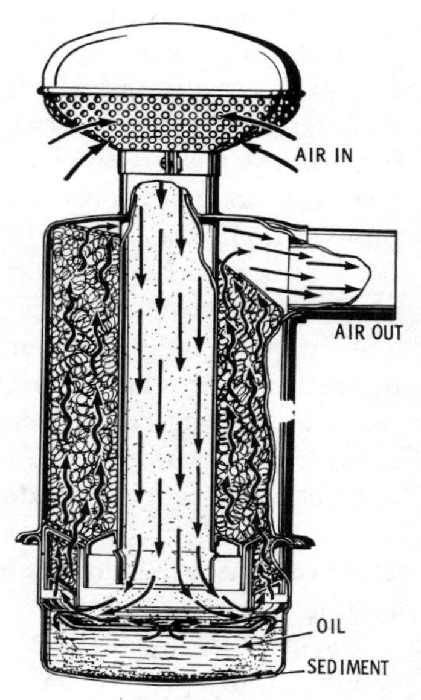

Fig. 22-2. Oil-bath air cleaner for large Waukesha diesel engine.

Courtesy Waukesha Motor Co.

What is meant by keeping a log?

Answer: Keeping a record of happenings, repairs, replacements, adjustments, lubricating oil and fuel oil used, and other matter that may be of importance for reference in the future. It is a diary—a daily record—and serves also as a reminder to do certain things at certain times, such as to make repairs, to check the fuel and oil supply, to clean a tank, to adjust a bearing, to grind a valve, to replace a part, to make an inspection, or to order supplies.

In what form is a log kept?

Answer: In any form suitable to the needs of the plant. A notebook may do, or it may require a daily diary with a page set aside for each date. It may require a printed form, on sheets, giving the engine numbers, the cylinder numbers, the pumps, the compressors, the generators, the batteries, and all other equipment that may exist in the plant.

Large plants often have separate maintenance log sheets that give the intervals during which inspections should be made and when cleaning and grinding should be done. A typical log may show that fuel valves should be removed, cleaned, and ground every three months; the governors should be examined and adjusted every twelve months; the crankshaft should be checked for alignment every twelve months; the fuel pump driving mechanism should be examined every two months; the lubricating oil piping should be cleaned out every month; and the fuel valve settings should be checked every month. The requirements are made to suit the operating conditions.

Another typical maintenance form may show the number of operating hours elapsed between changes or repairs; thus giving a record of strength and wear. This may serve as a guide later for making up a maintenance record form to show at what intervals inspections should be made.

How can smooth running, efficiency of operation, and minimum engine wear be attained?

Answer: by careful attention to details of operating requirements and by proper maintenance of equipment.

CHAPTER 23

Engine Operation

What preparation is necessary before attempting to start a diesel engine?

Answer: The following preparations should be made:

1. Read the directions furnished with each engine for preparing the engine for starting, and follow the instructions. Make sure that there are no tools, materials, or waste lying on or around the engine. They may be either caught in the moving parts or thrown into them by vibration.
2. Make sure that all tanks have sufficient fuel, lubricating oil, and air and that all equipment and parts are intact and working properly.
3. Operate the valves as given in the instructions.

If the engine fails to start, what may be the cause?

Answer: The cause may be among the following:

1. The starting-air pressure may be too low.
2. The engine may not be set on "starting" position.
3. The valves in the starting-air lines may not be open or sufficiently open.
4. Either the air-inlet or the exhaust valves may be open or leaking; permitting the starting air to escape.
5. The engine bearings may be tight. Try turning over the engine by hand.
6. The engine load may be too great. Ease it up.
7. The air-check valves may not be seating properly.
8. If electric starting—the batteries may not be fully charged, or the motor or clutch may not be in working order. Check battery, switches, motor, and clutch.

If the engine starts rotating, but the fuel does not catch fire, what may be the cause?

Answer:

1. There may not be sufficient fuel oil in tanks and pumps.
2. Fuel pumps may be air or water bound.
3. Fuel lever or handwheel may not be set in "running" position.
4. Fuel valves or spray nozzles may not be open or working properly.
5. The holes in the spray nozzles may be clogged.
6. There may be air or water in the fuel oil.
7. By-pass valves may not be in good order.
8. Fuel oil pipes may be clogged.
9. The governor linkage to the injection pumps or controls may be binding.
10. The fuel transfer pump may not be delivering fuel properly or in sufficient quantity.
11. Fuel delivery timing may be out of adjustment.
12. Circulating water may be cooling the cylinders too rapidly.
13. Check the oiling system.
14. Check feeds and temperature.
15. Fuel may be too heavy to flow.
16. Fuel screens or filters may be clogged.
17. Pistons may be binding.

18. Air compression may be low, valves may stick, or exhaust may leak.

Examine the engine and supply equipment and determine the cause or causes.

If an engine fails to start in cold weather, what may be the cause?

Answer: In addition to the causes given for ordinary failure to start, cold weather may add the following:

1. Fuel may be congealed or too heavy for cold weather and fail to flow freely. It must be either changed or preheated. The engine manufacturer prescribes the fuel oil or method to be used. Lubricating oil may be too heavy and may have to be changed or preheated.
2. Inlet air may be too cold and have to be preheated.
3. Circulating water may be too cold and have to be heated. Heating the water will warm the cylinders and aid in starting. The fuel system may become air bound and special attention must be given to venting.
4. Some engines have preheating devices and fuel ignition plugs for cold-weather starting. These should be used as per directions. Water may have gotten into the fuel tanks and frozen—thus shutting off the fuel supply.
5. Batteries may be frozen and the voltage too low for operating the starting motor at proper speed. Batteries should be kept fully charged. Additional batteries for cold-weather starting may be required, which are disconnected later. After the engine has been started, it should idle for a short time to heat the parts and attain a free flow of fuel oil and lubricating oil. The oil pressure gauge should be watched and when proper oil pressure shows, the load may be thrown on the engine.

How is an engine stopped?

Answer: The procedure for stopping an engine should be followed as:

1. By shutting off the fuel supply to the engine. However, it is a good practice to reduce or take off the load before stopping the engine.

2. The lubricating oil equipment should be stopped.
3. The condensate should be blown out of the cooling pipes of the air compressors.
4. The valve to the fuel supply tank should be closed.
5. It is well to let the cooling water continue circulating for a few minutes, after stopping the engine, to prevent undue heat and permit slow cooling.
6. If the engine is exposed to cold weather, or is to be idle for a while, all water should be drained. This prevents freezing and rusting.
7. If the engine is to remain idle for some time, it is well to turn it over once a day to keep it in proper running condition.

If an engine races, what may be the cause?

Answer: Engine racing can be caused by:

1. The governor may not be working properly.
2. There may be lost motion in the linkage between the governor and the fuel supply.
3. The valves may not be working properly.
4. The injection pump may not be working properly.

If an engine runs irregularly or slows down, what may be the cause?

Answer: One of the following may be the cause:

1. Air in the fuel pump, fuel line, or spray nozzles.
2. Water or impurities in the fuel.
3. Delivery valve in the fuel pump leaking.
4. Nozzle valve either leaking or not moving freely.
5. Screens or filters clogged.
6. Improper lubrication.
7. Governor not regulating properly.
8. Too much load.
9. Intake air insufficient.
10. Valves and ducts clogged.
11. Air filter clogged.
12. Air compression insufficient.
13. Piston rings leaking.
14. Pistons not moving freely.

15. Bearings overheating and binding.
16. All cylinders not firing. One cylinder at a time can be cut out for a check on the cylinders. If a good working cylinder is cut out, the engine speed will drop materially. If a poor one is cut out, it will make little difference.
17. Back pressure in exhaust lines due to clogged exhausts.

If an engine knocks or pounds, what may be the cause?

Answer: Pounds or knocks in an engine may be caused by:

1. Faults in operation or mechanical trouble.
2. Too early fuel injection—or too late. The timing should be adjusted.
3. Fuel injection valves may leak or stick.
4. Air may be entering the fuel lines.
5. The air compression may be insufficient.
6. Use of a poor-grade or unclean fuel oil.
7. Pistons may be worn or working loose.
8. Bearings may be loose.
9. Valves may be rattling.
10. Flywheel working loose.
11. Too much clearance in working parts.

What may cause an engine to overheat?

Answer: Causes of an engine overheating are:

1. Cooling water stoppage or not circulating properly and freely. The cooling system should be checked.
2. Cylinder water jackets may be clogged because of sediment deposits.
3. Piping may be clogged. All pipes may need flushing out.
4. Water pump or supply system may be supplying insufficient water.
5. Engine parts may be too tight and causing friction.
6. Lubrication may be improper or insufficient.
7. Fuel injection, into cylinders, may be late.
8. The temperature of the cooling water may be too high.

NOTE: If the cooling water system fails, cold water from another source should not be sent into the engine immediately. Shut down the engine and let it cool slowly, because sending cold water into a hot engine may crack the cylinder heads and liners.

What may cause white smoky exhaust?

Answer: One or more cylinders may not be getting sufficient fuel. This is often noticed on light loads. This may be due to:

1. Plugged fuel valves.
2. Too high fuel pressure.
3. Improper setting of fuel valves.
4. Too low compression at high altitudes.
5. Leaks past piston rings.
6. Too much lubricating oil.
7. Water in cylinders and exhausts.

What may cause black smoky exhaust at slow speed?

Answer: Causes of black smoky exhaust at slow speeds are:

1. Too light fuel.
2. Overload.
3. Plugged spray-nozzle valves.
4. Too high fuel pressure.

What may cause black smoky exhaust at normal speed?

Answer: Black smoky exhaust at normal speeds is caused by:

1. Too high fuel pressure.
2. Too low compression pressure.
3. Too light fuel.
4. Spray valve open too long.
5. Spray-nozzle valve opening worn and too large.
6. Carbon in exhaust pipes.
7. Piston rings sticking.
8. Cylinders scored.
9. Dirt in air filters.
10. Overload on engine.

What are some good practices in operating an engine?

Answer: Some good practices to observe in engine operation are:

1. An engine should be run at the speeds for which it is designed and carry the loads intended.
2. They should be slowed down when not using the full power in order to avoid racing.
3. Just enough fuel pressure should be maintained to develop the power required.
4. The engine should have sufficient lubricating oil at all times for proper lubrication.
5. The lubricating oil gauge should be watched and the proper pressure maintained.
6. The strainers and filters should be cleaned from time to time, depending on operating conditions.
7. The lubricating oil should be changed when needed.
8. The engine should be studied and its characteristics known.
9. Parts should be examined while the engine is running, to make sure that they are functioning properly.
10. Batteries should be kept fully charged and electrical connections kept tight and clean.
11. Motor and generator commutators and brushes should be kept in good order.
12. Water sediment should be removed from the system.
13. Carbon residue should be removed from oil.
14. Engine should be kept clean at all times.
15. Fuel and lubricating oil should be kept clean.
16. Tanks and strainers should be kept clean.
17. Water should be kept circulating at all times when the engine is running.

What should be the "daily" care of engines?

Answer: Steps in "daily" care to be observed are:

1. Fuel and lubricating oil should be checked.
2. Sediment from water and oil should be drained.
3. Filters should be kept clean.
4. All parts should be kept lubricated.
5. Engine should be kept clean.

After 50 hours of operation, what procedure should be followed?

Answer: At 50 hours of engine operation:

1. Valves and valve clearances should be examined, checked, and reset if necessary.
2. Valves should be lubricated with a mixture of fuel and lubricating oil in equal parts or with a light machine oil.
3. Bearings should be checked for lubrication.

After 200 hours of operation, what procedure should be followed?

Answer: At 200 hours of engine operation:

1. Lubricating oil should be drained and the pans, sump, and piping flushed to prevent oil contamination.
2. Filters and screens should be cleaned.
3. Water should be drained and sediment removed.
4. Operation of fuel oil pump and feeds should be checked.
5. Batteries should be filled with distilled water and the connections and charge checked.
6. Nuts and bolts should be tried for tightness.

After 1,000 to 1,500 hours of operation, what procedure should be followed?

Answer: After the engine has operated 1,000 to 1,500 hours:

1. The fuel injection pumps should be examined and cleaned. Carbon and dirt should be removed from manifolds, mufflers, cylinder water jackets, and other places where sediment or dirt may lodge. If necessary, cylinder heads should be removed, cylinders should be cleaned, and valves should be ground.
2. Water circulating pump and system should be checked.
3. The engine and all accessories should be given a general inspection and overhauling if necessary.
4. Defects may be remedied before they cause the engine to be shut down.

If an engine is to be shut down for a lengthy period, what precautions should be taken to keep it in good operating condition?

Answer: Preparation for a lengthy shutdown of an engine should be as follows:

1. The engine should be run, to warm it, and then the lubricating oil should be drained from the system.
2. All fuel should be removed from the fuel pumps, spray nozzles, filters, transfer pump, and piping to avoid fuel system difficulties when the engine is used again.
3. Rust-preventive oil, to prevent rusting of parts, may be sent through the system by turning over the engine by hand or with the starter.
4. The lubricating oil pump and system should be treated in the same manner.
5. Surfaces of valves, stems, springs, rocker arms, shafts, and other parts that may tend to rust should be covered with rust-preventive oil.
6. The cooling system should be drained of all water.
7. Batteries should be removed and stored in a service station for proper care. A battery deteriorates when not in use and given attention.
8. The accessories and auxiliary equipment should be treated with the same care as the main engine.

What are some hints for inspection and repair of engine parts?
Answer: Hints for inspection and repair of engine parts are:

1. When engine parts are removed for inspection or repair, they should be plainly marked so that they can be readily put back in their proper places.
2. When any work has been done on an engine, the engine should be turned over by hand several times, to make sure that all parts are back and are in proper working order before turning on the power.
3. When an engine does not work properly, do not change the adjustments. Check for other causes first—as given in preceding paragraphs. The trouble may be trifling. Adjustments are made at the factory and are usually correct. They should not be changed.
4. When bearings have been tightened or adjusted, they should be examined when the engine is running, to make sure that they are not heating up.
5. Oil should be wiped off the outside of the engine and not

permitted to run down onto the foundation or other parts—it may damage them.

6. Engines should be stopped when trouble occurs, and the cause investigated—rather than waiting and having a lengthy shutdown.
7. Cleanliness and attention will save many shutdowns and costly repair bills.

What is the effect of altitude on engine power?

Answer: Altitude affects diesel engines as follows:

1. High altitude reduces engine power.
2. Engines installed or operating in locations high above sea level will have less power unless designed for increase of power.
3. High altitude means less air density because of lower air pressure.
4. Less air density means less oxygen for burning fuel—therefore less fuel is burned—and the result is less power from the engine.
5. Engines may be designed for more air supply, through supercharging methods, but it is better to have engines of higher capacity for higher altitudes.
6. An engine developing 100 brake horsepower at sea level gives only 92 horsepower at 4,000 feet altitude, about 75 horsepower at 8,000 feet, and about 65 horsepower at 11,000 feet. Therefore, it is important to consider altitude in power expectations.

How does climate affect engine power?

Answer: Hot climates make a difference in power supply because the warm air expands and a lesser quantity enters the cylinder combustion space. Intake air may have to be cooled.

CHAPTER 24

Diesel Engine Installation

What are some of the advantages of diesel installations over other types of installations powered by combustion engines?

Answer: With diesel engines, auxiliary apparatus such as steam boilers, gas generators, and the like is not required. Therefore, the whole plant is extremely simple. The foundation of a diesel engine plant is simple. There is no necessity for a specially designed basement, as in a steam turbine plant, for example, where provisions must be made for locating the condensing plants with their pumps, which generally entails a considerable enlargement of the existing basement. Nor is any other construction required as with plants powered by steam engines or by gasoline engines.

In a steam power plant suitable accommodation must be provided for the boiler, the auxiliary equipment, and the coal transport and storage facilities. There must be room for disposal of ashes or clinkers and for erecting a chimney.

In a diesel plant less room is needed for storing the fuel oil than in a coal-fired steam plant. The space ratio is approximately one to four. The fuel oil can be stored in iron tanks and piped to the engine room.

Under normal working conditions, there is no nuisance caused by exhaust gases, nor is there any danger of fire or explosions. The diesel engine plant can be erected in the immediate vicinity of dwellings, such as in the basements of hotels, hospitals, or stores.

Prior to erecting a diesel power plant, however, licensing authorities should be contacted to ascertain the necessity for proper permission. In many towns and districts, a special license is not required to install a diesel engine.

What other characteristics especially adapt diesel engines to power-plant installations?

Answer: Since a diesel engine needs no fuel to prepare it for service but is ready to start at any moment, it is particularly suitable as the cheapest source of mechanical energy in works and in power stations where extra energy has to be provided at short notice.

However, a diesel engine is not only an excellent standby plant but also gives best results in continuous service, such as in mills and cement works where long running periods make it essential to generate the power needed on the most economical basis possible.

The diesel engine is also highly suitable for plants in which great importance is attached to uniformity in the generation of energy and to sensitiveness of governing, for example, in electrical power generating stations and in paper mills. Direct coupling to an electrical generator gives a higher overall efficiency than do belt drives because transmission losses are eliminated (Fig. 24-1).

The fact that the diesel engine has been widely adopted in all kinds of ships is to be attributed to the special advantages of the marine diesel engine over the marine steam engine, among which may be mentioned the following:

1. The smaller dimensions required for a ship of the same cargo-carrying capacity in consequence of the saving in weight and the reduced space requirements for the engines and fuel bunkers.
2. Constant readiness for service.

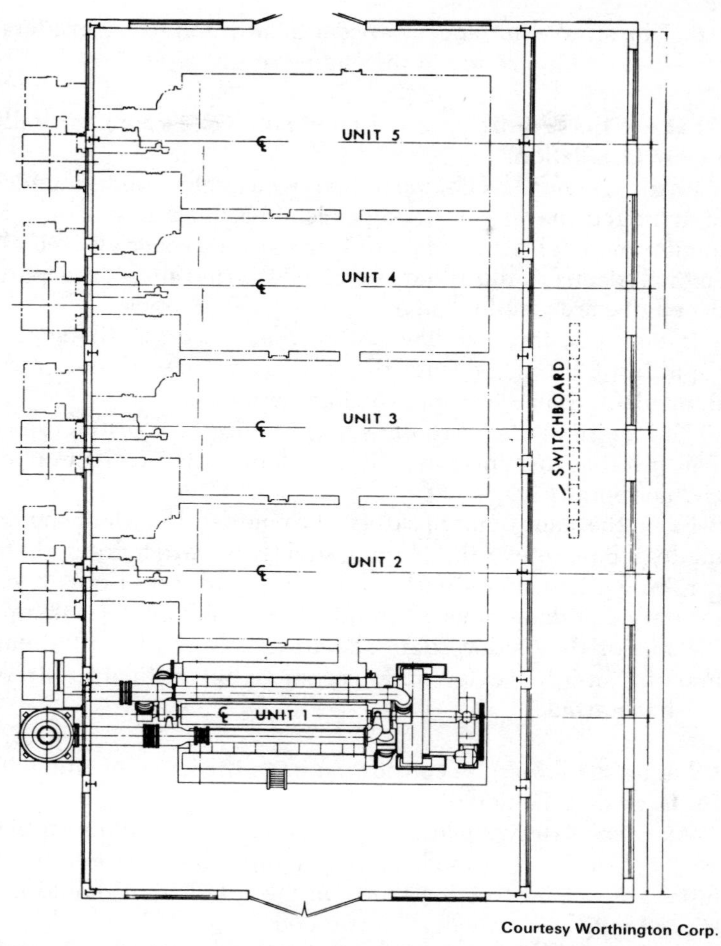

Courtesy Worthington Corp.

Fig. 24-1. Station arrangement of five Worthington SW 14-16 generating units.

3. No fuel consumption when the engines are not working, thus giving more economy.
4. Elimination of boilers and one-third to one-fourth reduction in weight of fuel.
5. Clean, easy, rapid bunkering, and simple conveyance of the fuel from the bunkers to the engine.

6. The smoke nuisance is absent in addition to a considerably lower temperature in the engine room.

What are the requirements of an ideal engine room for a diesel engine installation?

Answer: The ideal engine room is well lighted, uncramped, and so arranged that it can be kept clean easily. There should be a minimum of at least five feet of free working space all around the engine. Electric lights should be placed so that all external parts of the engine are visible at night.

It is also desirable to have an overhead crane available that is capable of lifting at least five tons. The size of the crane is dependent upon the size of the installation.

The minimum head room is that required for piston removal. This distance must be carefully calculated prior to erection of the engine room (Fig. 24-2).

From the standpoint of safety, the engine-room floor should be made of concrete or tile, sloping slightly toward a drain, so that it can be washed down with water. Iron and steel floors become slippery and dangerous when oil or water is spilled on them.

Plans of the engine room should be submitted to the engine manufacturer for review and suggestions before final construction work is started.

What points should be considered in construction of foundations for large diesel engines?

Answer: A drawing locating all foundation bolts, pipe trenches, etc., is usually furnished by the engine manufacturer for each installation. Construction work on the foundation should not be started until this drawing is examined.

If location of the engine is such that its vibration may be carried into surrounding equipment or buildings, and it is necessary to minimize this condition as much as possible, the use of a vibration-isolating device is recommended in the engine mounting. The engine foundation should be kept entirely separated from the building foundation.

What points should be considered in installation of piping for large diesel engines?

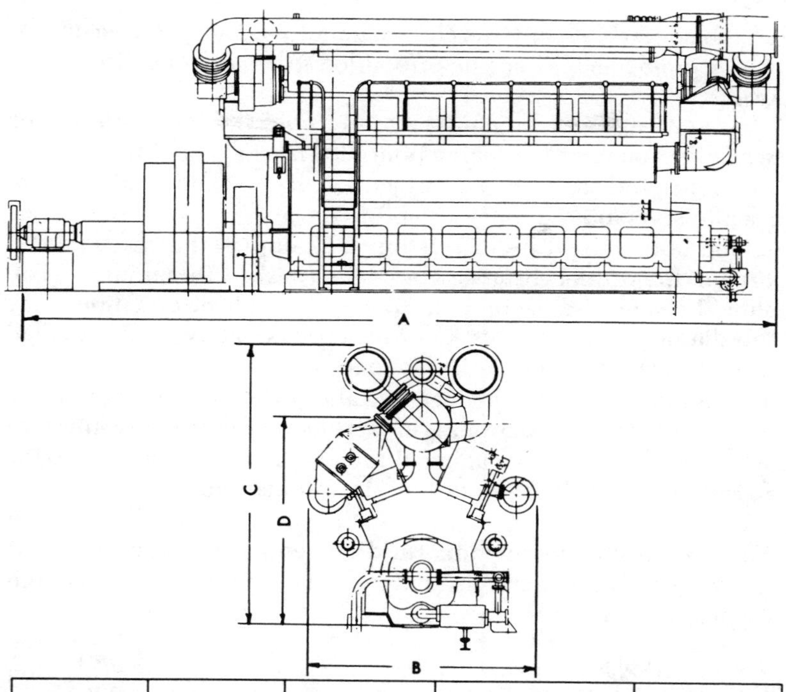

Type	Length (A)	Width (B)	Height (C)	Height (D)*
SW 14-16	38' - 6"	11' - 9"	14' - 11-3/4"	11' - 2"
SW 14-12	36' - 0"	11' - 9"	14' - 4"	11' - 2"
SW 14-8	35' - 2"	7' - 6"	11' - 10"	10' - 5"
SW 14-6	31' - 6"	7' - 6"	11' - 10"	10' - 5"

* As required to withdraw pistons, exclusive of lifting tackle.

Courtesy Worthington Corp.

Fig. 24-2. Overall dimensions of Worthington SW 14 generating units.

Answer: All piping connected to the engine should be placed in trenches. The side walls of the trenches as well as the floors should be made of concrete not less than four inches in thickness. All trenches should drain to one point. Drain pipes of about three inches in size, protected by strainers, are advisable. The engine piping in the trenches should be fastened securely to pipe racks and located accessibly for cleaning and overhauling. Exhaust or water piping should not be imbedded in either concrete or in a wall.

A template must be constructed to hold the foundation bolts while the foundation is being poured. The spacing of the foundation bolts and their correct projection above the foundation are usually given on the foundation drawing.

The bolts should be hung from the template so that they will project the proper distance above the finished foundation. They should be encased in conduit, stove pipe, or sheet-iron pipe twice the diameter of the bolts, in order to make allowance for small discrepancies in the template layout.

This bolt casing should not come above the rough surface of the concrete of the foundation, as any projection above may interfere with the aligning of the engine. Waste should be stuffed around the foundation bolts before pouring the foundation.

What is a good concrete mix for diesel engine foundations

Answer: The concrete foundation should be made of the following proportions:

1. Cement .... 1 part
2. Sand (clean and sharp) .... 2 parts
3. Broken stone .... 4 parts

Leave the top surface of the foundation rough and clean so that the grouting will adhere to it readily. Finish the surface of the flywheel and generator pit with a finely finished cement surface immediately after the forms have been removed.

What precautions and allowances must be considered in leveling the large diesel engines and the outboard bearings?

Answer: After the concrete has set for seven days, roughen and clean the foundation. Remove the waste from around the bolts and place the engine on the foundation for leveling and grouting.

Threaded holes are usually provided in the flange of the engine base for jack screws to level the engine. Place 1/2 in. × 4 in. × 4 in., or heavier, steel plates on the foundation directly under each jack screw. Use a sensitive carpenter's level lengthwise and at right angles. Place the level on the machined surfaces of the engine housing. Allow one inch between the base and the foundation for grouting.

Place the extension shaft in position with one end resting on the outboard bearing and insert the recess of the other end over the centering disk at the flywheel. Measure with feelers between the extension-shaft flange and the flywheel at top, bottom, and sides. Shift the outboard bearing pedestal to right or left and raise or lower it with wedges until all measurements taken around the extension-shaft flange are the same. Then bolt it to the flywheel.

In an installation where the outboard bearing is carried in the generator frame, the same general method is used, the extension shaft being aligned by shifting the generator frame. Jack screws are sometimes located in the feet of the generator for this purpose, but if they are not available, wedges or shims may be used. The adjustments must be made in a way to keep the clearance between the generator armature and field (air gap) equal at all points both front and back.

After the flanged coupling of the extension shaft is bolted to the flywheel, check the alignment by measuring between the crank webs of the last cylinder as described under "checking alignment."

It may be necessary to raise the outboard bearing slightly to compensate for the weight of the flywheel and the generator rotor in order to make the micrometer readings between the crank webs equal for all crank throw positions.

Do not draw up the foundation bolts at this time. The next step is to grout and place the wedges for the engine generator or the outboard bearing, etc.

What is a good mix for grouting for the foundation?

Answer: Using boards, build a dam at least three inches high around the foundation. Mark the location of the wedges so that they can be removed after the grouting has set. The grout should be mixed in the following proportions:

1. Portland cement .... 1 part
2. Clean sharp sand .... 2 parts

Make the mixture wet enough to flow freely, but not so wet that the cement will separate from the sand and float to the top. Work the grout well under the engine base and be sure that it fills in all spaces evenly. Fill the pipes encircling the foundation bolts with grout. Do not start the engine until the grout has set for at least five days.

How can the engine alignment be checked?

Answer: After the grouting under the engine, generator, or outboard bearing has set for three days, remove the aligning jack screws and wedges. Then tighten the foundation bolts. Check the alignment of the extension shaft by measuring between the crank webs of the last cylinder with an inside micrometer when the crank throw is at top and bottom dead centers and 90° to the right and to the left. When the alignment is correct, all four measurements will be the same and should not be allowed to vary more than 0.002 inch. Adjust the outboard bearing if necessary to obtain this condition.

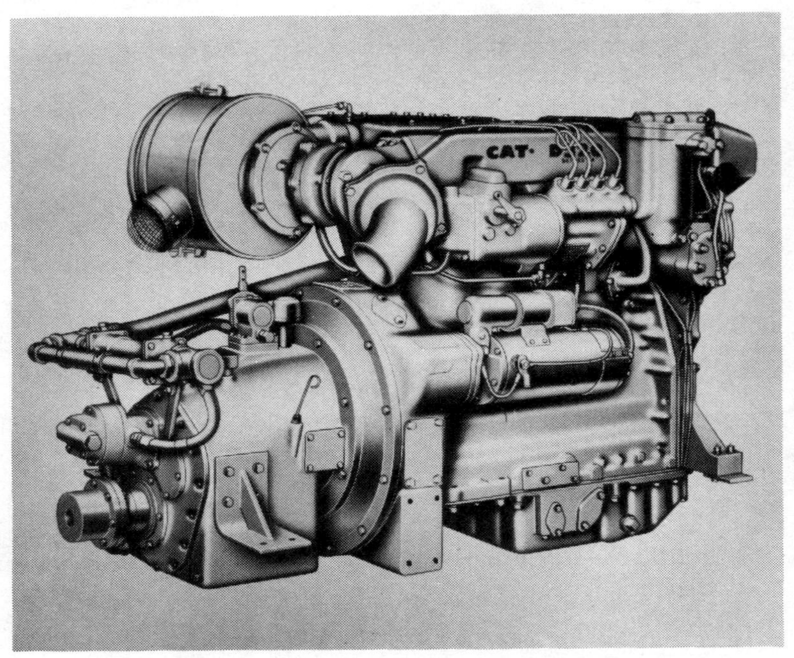

Courtesy Caterpillar Tractor Co.

Fig. 24-3. Caterpiller D320(A) marine diesel engine.

What points must be considered in providing foundations for marine engines?

Answer: In marine applications, the engine is usually mounted on a structural steel subbase which forms a rigid foundation for the engine in spite of any creep or setting in the structure of the ship. The engine and the driven unit must be lined up on the subbase in the same manner as that outlined for stationary engines (Fig. 24-3).

The foundation bolts must then be pulled up tight and the alignment checked as indicated. The flywheel end of the engine base and the generator or driven unit must be doweled to the subbase when the alignment has been checked and found to be correct.

This alignment should be checked frequently, especially when the ship is new, and whenever an accident occurs that gives a severe jar to any part of the ship structure.

CHAPTER 25

Working Principles of a Cummins Diesel

What are the four piston strokes of a Cummins diesel engine?

Answer: A good understanding of each engine part is vital. Knowing what happens in the combustion chamber during each of the four strokes is also important. The four strokes and the order in which they occur are intake stroke, compression stroke, power stroke, and exhaust stroke.

What is the intake stroke? How does it fit into an engine's operation?

Answer: When the intake stroke occurs, the piston travels downward; the intake valve is open, the exhaust valve is closed. The downward stroke of the piston permits air from the outside to enter the cylinder through the open intake-valve port. The supercharger or turbocharger increases air pressure in the intake manifold and forces it into the cylinder. Intake air is never restricted by carburation or mixing valves such as those used on spark-ignited automotive gasoline engines. Intake charge consists of air only with no fuel mixture.

What is the compression stroke? How does it fit into an engine's operation?

Answer: At the completion of the intake stroke, the intake valve closes, and the piston starts upward on the compression stroke. The exhaust valve remains closed.

At the end of the compression stroke, the air in the combustion chamber has already been forced by the piston to occupy a space about one-fifteenth as great in volume as it occupied at the beginning of the stroke. The compression ratio is then said to be 15 to 1 (15 : 1).

Air compressing into a small space causes the temperature of the air to rise. Close to the end of the compression stroke, the pressure of the air above the piston is approximately 500 to 600 pounds per square inch, and the temperature of that air is approximately 1000°F.

During the last part of the compression stroke and the early part of the power stroke, a small metered charge of fuel is injected into the combustion chamber. Almost immediately after the fuel charge is injected into the chamber, the fuel is ignited by the hot air and starts to burn.

What is the power stroke? How does it fit into an engine's operation?

Answer: The power stroke occurs when the piston travels downward. Both intake and exhaust valves are closed. By the time the piston reaches the end of the compression stroke, the burning fuel causes a further increase in the pressure above the piston as more fuel is added and ignites; the gases get hotter and hotter and expand to push the piston downward and add impetus to the crankshaft revolution.

What is the exhaust stroke? How does it fit into an engine's operation?

Answer: During the exhaust stroke, the intake valve is closed, the exhaust valve is open, and the piston is on its upstroke. Burned gases are forced out of the combustion chamber through the open exhaust-valve port by the upward travel of the piston. The exhaust stroke completes the engine cycle needed to produce power. The engine starts again with the intake stroke once the exhausting of the burned gases is complete.

How are diesel engines lubricated?

Answer: The working parts of the Cummins J Series engines are lubricated by forced feed. The force is supplied by a gear-type lubricating pump located below the crankshaft and driven by an idler gear off the pinion gear. Oil is held in the sump in the oil pan, and is drawn from this sump by the lubricating oil pump. It is delivered to all working parts of the engine through lubricating oil lines and the oil header, the latter being drilled the full length of the block (Fig. 25-A). Various drillings through the block, cylinder head, crankshaft, and rocker levers complete the oil circulation passages.

Lubricating oil is forced through the crankshaft to lubricate the main and connecting rod bearings. In some engines, rifle drillings carry oil from the crankshaft through the connecting rods to lubricate the piston pin and bushing. Lubricating oil pressure is controlled by a regulator in the oil strainer head.

Filters, strainers, and screens are provided throughout the lubricating system for proper cleaning of the lubricating oil. The air compressor, supercharger or turbocharger, and water-pump drive all receive pressure lubrication from the engine oil supply. The turbocharger is also cooled by engine oil.

The injector plunger in the injector and the working parts in the fuel pump are lubricated by fuel oil. The fuel used for lubrication of the injector plunger is returned to the fuel tank through drain lines.

Schematic diagrams in Figs. 25-1 and 25-2 show the direction of oil flow and various units provided to clean and cool hot oil and maintain a constant pressure of 30/50 psi in the header at governed engine rpm. A by-pass valve is provided in the oil strainer as insurance against interruption of oil flow by a dirty or clogged strainer element.

How does the fuel system operate in a Cummins diesel?

Answer: The Cummins PT fuel system is a completely new

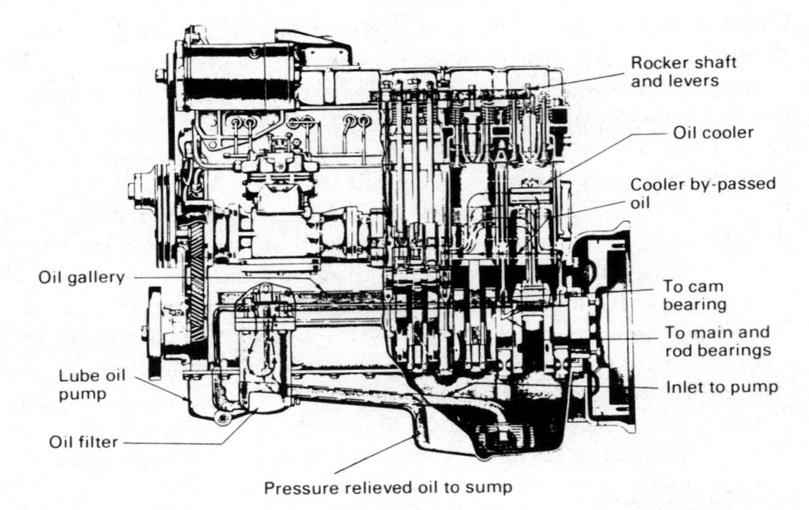

Courtesy Cummins Engine Co.

Fig. 25-1. Cross section of Cummins JS engine.

application of basic hydraulic principles to the diesel engine fuel system (Fig. 25-3). It is a Cummins design for Cummins diesels only. The identifying letters PT are an abbreviation for *pressure-time.*

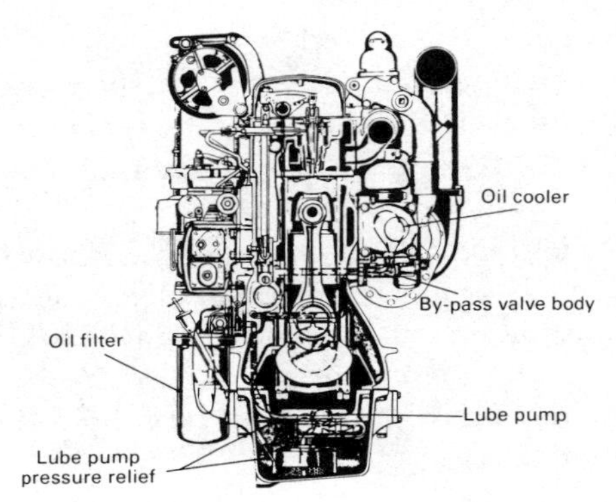

Courtesy Cummins Engine Co.

Fig. 25-2. Cross section of Cummins JS engine, end view.

On what principle does the fuel system operate in a Cummins diesel?

Answer: The PT fuel system is based on the principle that by changing the pressure of a liquid flowing through a pipe, you change the amount of liquid coming out of the open end. Increasing the pressure increases the flow or the amount of liquid delivered, and vice versa. In applying this simple principle to the diesel fuel it is necessary to provide:

1. A fuel pump to draw the fuel from the fuel tank and deliver it to individual injectors for each cylinder.
2. A means of controlling the pressure of the fuel being delivered by the fuel pump to the injectors so that the individual cylinders will receive the right amount of fuel for the power required of the engine.
3. Fuel passages of the proper size and type so that fuel will be distributed to all injectors and cylinders with equal pressure

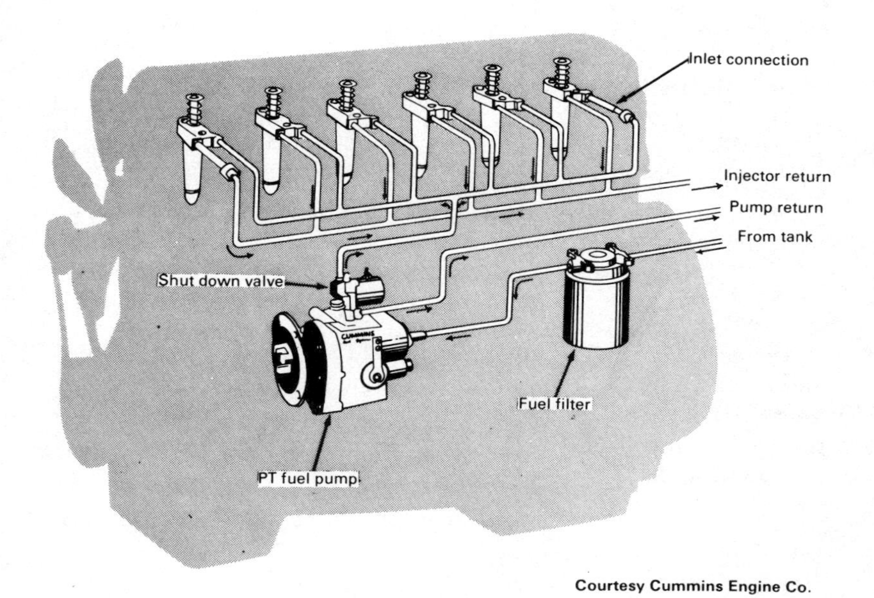

Courtesy Cummins Engine Co.

Fig. 25-3. PT fuel system. Fuel-flow diagram, pressure-regulated fuel pump.

under all speed and load conditions.

4. Injectors to receive low-pressure fuel from the fuel pump and deliver it to the individual combustion chambers at the right time in equal quantity and in proper condition to burn.

The PT fuel system consists of the fuel pump, the supply and drain lines, and the injector. Each of these is described in detail in the following paragraphs.

How many parts does the fuel pump have?

Answer: The fuel pump has three main units:

1. The gear pump, which draws fuel from the supply tank and delivers it under pressure through the pump and supply lines to the individual injectors.
2. The pressure regulator, which limits the pressure of the fuel to the injectors.
3. The governor and throttle, which act independently of the pressure regulator to control fuel pressure to the injector.

How is the fuel pump driven?

Answer: The fuel pump is coupled to the compressor or fuel-pump drive, which is driven from the engine gear train. The fuel-pump main shaft turns at engine crankshaft speed and drives the gear pump, governor, and tachometer shaft. The location of these units in the fuel-pump housing is indicated in Figs. 25-4 and 25-5.

Where is the gear pump located?

Answer: The gear pump is located at the rear of the fuel pump and is driven by the main shaft. This unit consists of a single set of gears to pick up and deliver fuel throughout the fuel system. From the gear pump, fuel flows through the filter screen and to the pressure regulator.

What is a pressure regulator?

Answer: The pressure regulator is a by-pass valve to regulate the fuel, under pressure, supplied to the injectors. By-passed fuel flows back to the suction side of the gear pump (Fig. 25-6).

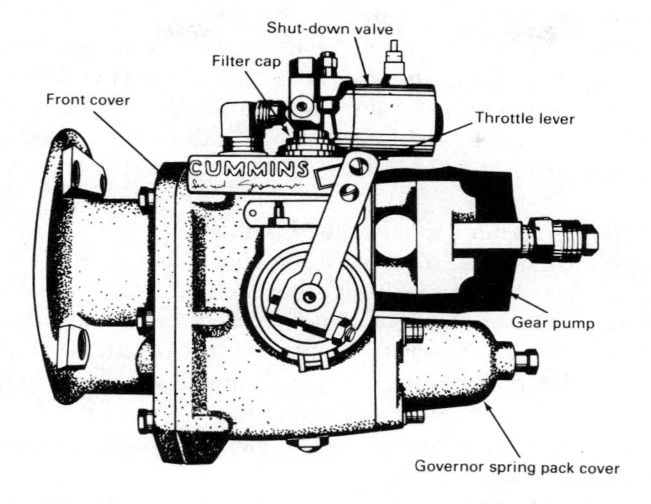

Courtesy Cummins Engine Co.

Fig. 25-4. Fuel pump units.

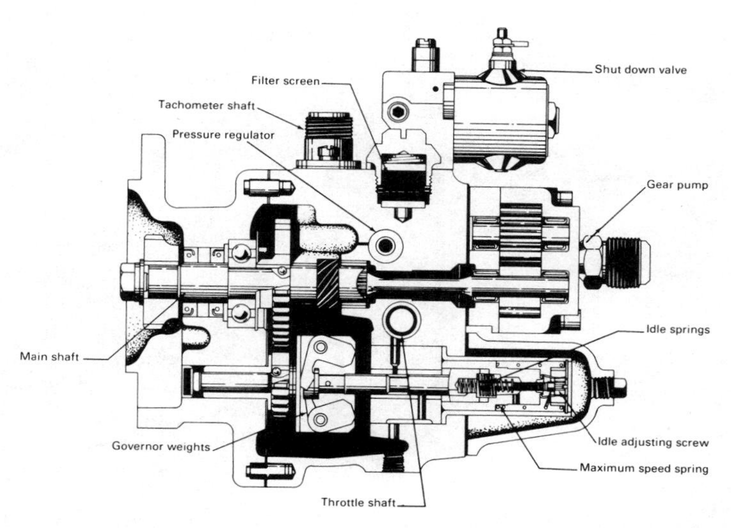

Courtesy Cummins Engine Co.

Fig. 25-5. Cross section of pressure-regulated pump with idling and high-speed mechanical governor.

How does the throttle operate?

Answer: The fuel for the engine flows past the pressure regulator to the throttle shaft. The idle fuel passes around the shaft to the idle jet in the governor. For operation above idle, fuel passes through the throttling hole in the shaft and enters the governor through the primary jets.

What is the governor's role in speed regulation?

Answer: Idling and high-speed mechanical governors have similar action. Mechanical-governor action is provided by a system of springs and weights. The governor maintains sufficient fuel for idling with the throttle control in idle position. It cuts off fuel above the maximum rated rpm. The idle springs, which are contained in the governor, pull back the governor pluger so that the idle fuel jet is open enough to permit passage of fuel to maintain engine idle speed.

During the operation between idle and maximum speeds, fuel flows through the governor to the injectors in accord with the

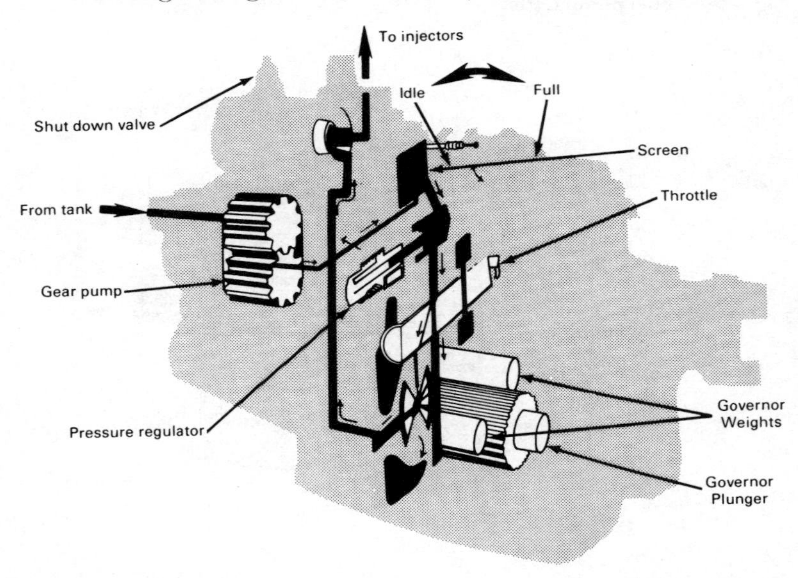

Courtesy Cummins Engine Co.

Fig. 25-6. Fuel flow through the pressure-regulated fuel pump.

engine requirements controlled by the throttle and limited by the pressure regulator. When the engine reaches governed speed, the governor weights move the governor plunger, and fuel passages to the fuel supply manifold are shut off. At the same time, another passage opens and dumps the fuel to the supply manifold back into the main pump body. In this manner, engine speed is controlled and limited by the governor regardless of throttle position. Fuel leaving the governor travels through the shutdown valve and inlet supply lines and on to the injectors.

How does the variable-speed mechanical governor work?

Answer: Variable-speed mechanical governors are designed to meet the requirements of machinery on which the engine must operate at a constant speed, but where extremely close regulation is not necessary. Adjustment for different rpm can be made by means of a lever control or adjusting screw. At full-rated speed, the variable-speed mechanical governor has a speed drop between full-load and no-load of approximately 8 percent. A cross-section of this governor is shown in Fig. 25-7.

As a variable-speed governor, this unit is suited to the varying speed requirements of cranes, shovels, and other such machines. These units use the engine both to propel the machine and to drive a pump or other fixed-speed machine.

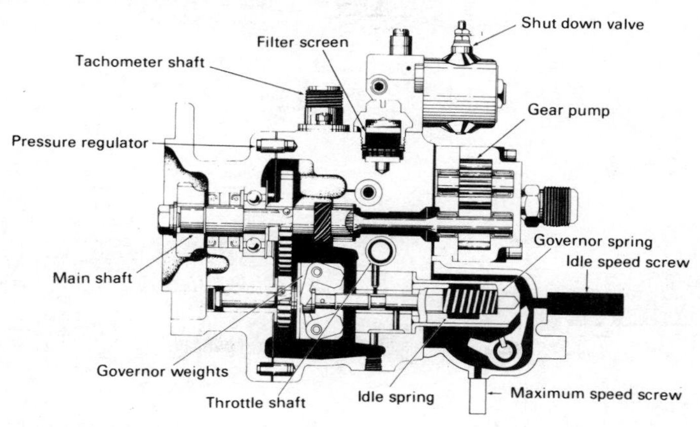

Courtesy Cummins Engine Co.

Fig. 25-7. Flange-mounted fuel pump with variable-speed governor.

As a constant-speed governor, this unit provides control for pumps, nonparalleled generators, and other applications where close regulation (variation between no-load and full-load speeds) is not required.

What is a torque-converter governor?

Answer: When a torque converter is used to connect the engine with its driven unit, an auxiliary governor may be driven off the torque-converter output shaft to exercise control over the engine governor and limit converter output shaft speed. The engine governor and the converter governor must be adjusted to work together. The PT torque converter is fundamentally two mechanical variable-speed governors in series. One is driven by the engine and the other by the converter (Figs. 25-7 and 25-8).

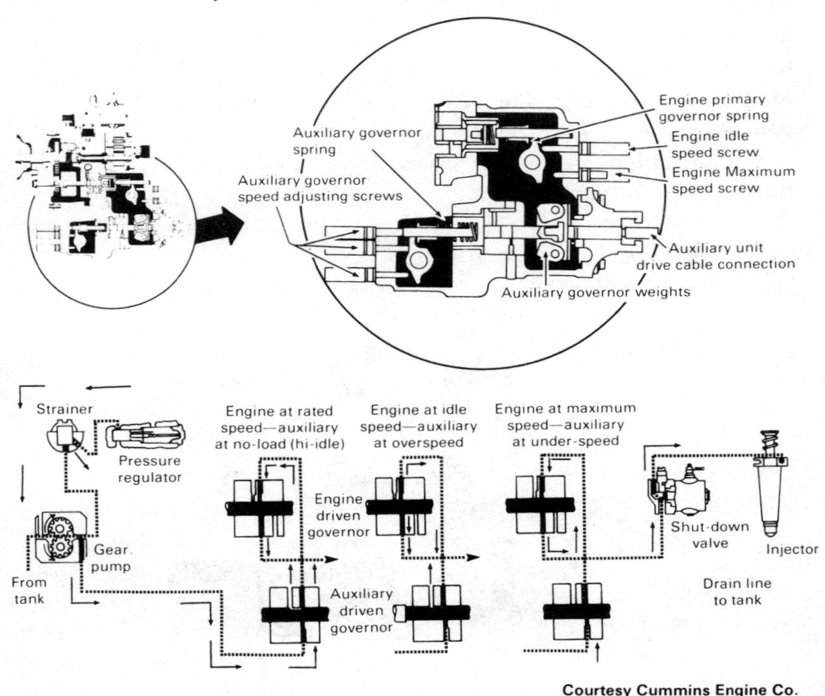

Courtesy Cummins Engine Co.

Fig. 25-8. Cross section and fuel flow through the torque-converter governor and fuel pump.

What else does the engine governor do?

Answer: The engine governor, in addition to giving a variable engine, acts as an overspeed and idle-speed governor while the converter-driven governor is controlling the engine. Each governor has its own control lever and speed-adjusting screws. The converter-driven governor works on the same principle as the standard engine governor *except* it cannot cut fuel to the idle jet in the engine-driven governor. This ensures that if the converter tailshaft overspeeds, it will not stop the engine.

Fig. 25-8 shows the position of the governor plungers under different engine and converter speed conditions.

What two types of fuel valves are used on an engine?

Answer: The manual shutdown valve is located on the top of the fuel pump and is used to shut down the engine (Fig. 25-3). This is a push-pull valve. In other words, *push* in to permit fuel to flow to the injectors, and *pull* out to shut off the fuel. It is important to keep this valve in the "Off" position at times when the engine is not running.

The electric solenoid valve permits the fuel to be shut off, or turned on, through the use of a switch key similar to that used in automotive ignition systems (Fig. 25-3).

What size should supply lines be?

Answer: Fuel supply lines must be held to a specified size to ensure an even pressure and supply of fuel to each injector. From the supply lines, the fuel enters the inlet connection to the injector (Fig. 25-3).

Where is the inlet connection made?

Answer: The inlet connection connects the supply-fuel manifold to the injector and contains a fine mesh screen at the large, or cager, end. This screen is the last protection against dirt entering the injector. There are no check valves in the inlet connection used in the PT fuel system (Fig. 25-3).

What are drain lines?

Answer: Not all the fuel entering the injector is burned in the cylinder. Part of the fuel circulates through the injector and is returned to the supply tank through the drain fittings, drain manifold, and drain line. The drain lines must be of ample size to provide freed drainage to the fuel tank (Fig. 25-3).

How does the fuel injector operate?

Answer: Fuel injectors have fuel circulating in them at all times except for a short period following injection into the cylinder. From the inlet connection, fuel flows down the inlet passage of the injector, around the injector plunger, between the body end and cup, up the drain passage to the drain connections and the manifold, and back to the supply tank. See Fig. 25-9 for a cross section of an injector.

As the plunger rises, the metering orifice is uncovered, and part of the fuel is metered into the cup. At the same time, the rest of the

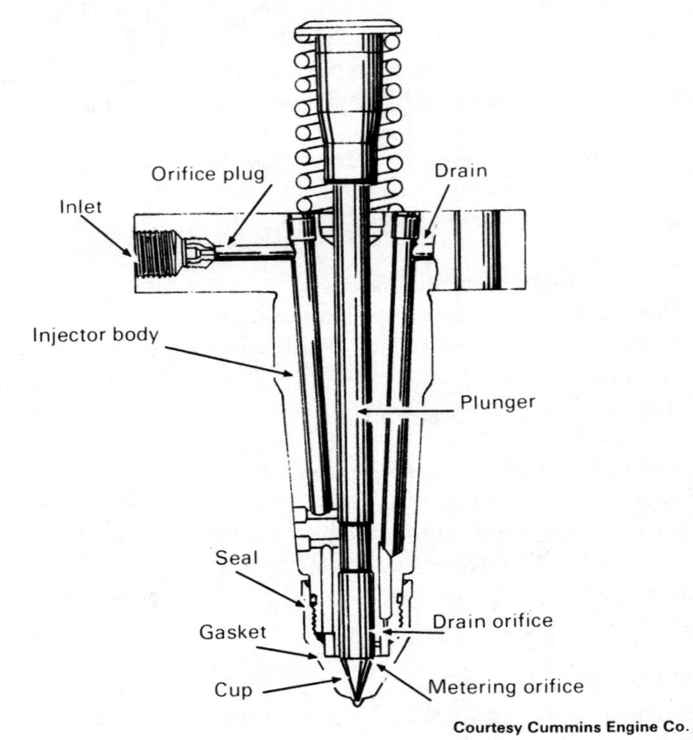

Fig. 25-9. Injector cross section.

fuel flows out of the drain orifice. The amount of fuel passing through the metering orifice and into the cup is controlled by fuel pressure, which is regulated by the fuel pump.

Some injectors contain adjustable feed orifices located in the fuel inlet passages (Fig. 25-9). The size of the orifice regulates the fuel flow to the injector.

During injection, the plunger is forced downward until the metering orifice is closed and the fuel in the cup is injected into the cylinder. While the plunger is seated the flow of fuel through the injector stops (Fig. 25-10).

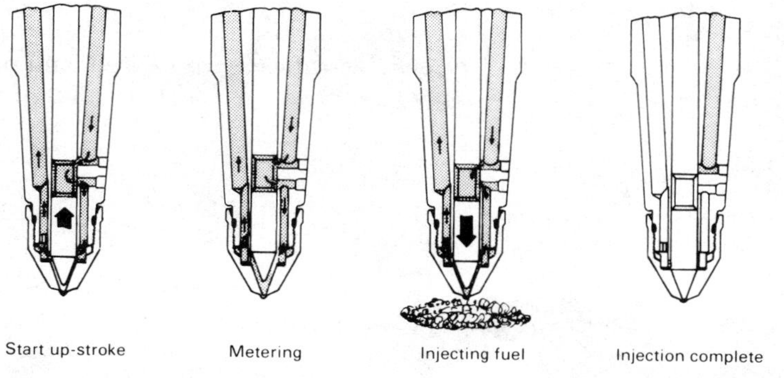

Courtesy Cummins Engine Co.

Fig. 25-10. Fuel-injection cycle.

How does the cooling system on a diesel work?

Answer: Water is circulated by a centrifugal water pump mounted on the exhaust side of the engine and driven by a supercharger or belts (Fig. 25-11). The water circulates around each of the wet-type cylinder liners, through the cylinder head, and around the injector sleeves in the cylinder head. The injector sleeves, into which the injectors are mounted, are made of copper for fast dissipation of heat.

A single large thermostat is used to control the operating temperature of the engine.

The engine cooling solution is cooled by the radiator or, on marine engines, by the heat exchanger. Where heat exchangers are used, a sea-water pump is mounted on the engine and circulates sea water inside the tubes of the heat exchanger. The engine's fresh-water supply circulates around these tubes.

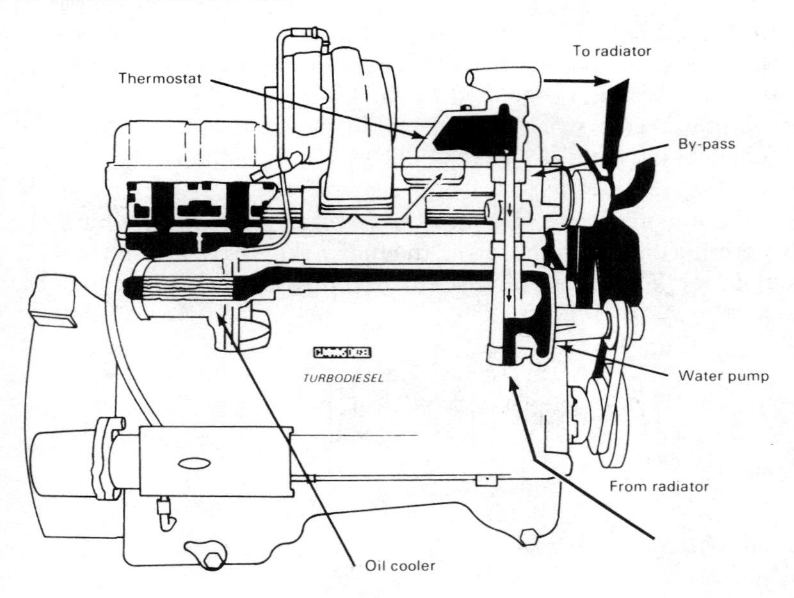

Courtesy Cummins Engine Co.

Fig. 25-11. Cummins JT-6 turbo-diesel cooling system.

What type of air system does the Cummins diesel engine have?

Answer: The supercharger and turbocharger force additional air into the combustion chambers so that the engine can burn more fuel and develop more horsepower than if it were aspirated naturally. The JS and JNS engines have superchargers; the JT engine is equipped with a turbocharger.

What is a supercharger?

Answer: A supercharger is a gear-driven pump that uses rotors instead of gears (like a gear-type fluid pump) to force air into the engine cylinders. The supercharger is driven from the engine crankshaft through a gear train turning at about 1.8 engine speed (Fig. 25-12).

What is a turbocharger?

Answer: The turbocharger consists of a turbine wheel and a centrifual blower impeller, or compressor wheel, separately

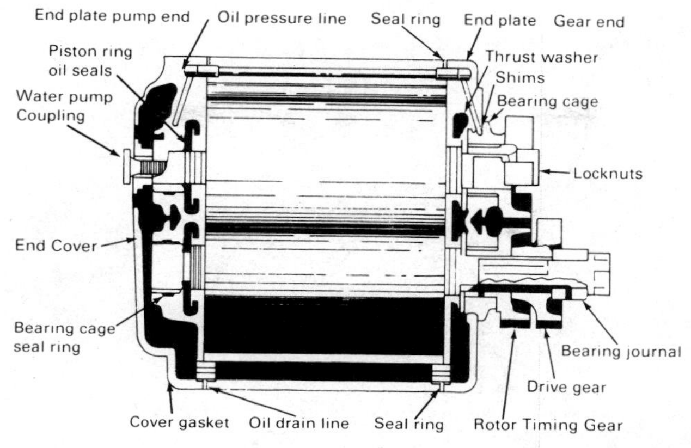

Courtesy Cummins Engine Co.

Fig. 25-12. Cutaway of supercharger.

encased but mounted on and rotating with a common shaft. The power to drive the turbine wheel (which in turn drives the compressor) is obtained from the energy of the exhaust gases. The rotating speed of the turbine changes as the energy level of the exhaust gas changes so that the engine is supplied with enough air to burn the fuel for its load requirement (Fig. 25-13).

What is a hydraulic governor?

Answer: Another engine unit is the hydraulic governor. Hydraulic governors are used on stationary power applications where it is desirable to maintain a constant speed with varying loads.

How does the hydraulic governor work?

Answer: The Woodward SG hydraulic governor uses lubricating oil, under pressure, as an energy medium. It is supplied from a sump on the governor drive housing. The governor acts through oil pressure to increase fuel delivery. An opposing spring in the governor control linkage acts to decrease fuel delivery.

For its operation to be stable, speed drop is called for. Speed drop is that characteristic of decreasing speed with increasing load. The desired magnitude of the speed drop varies with engine

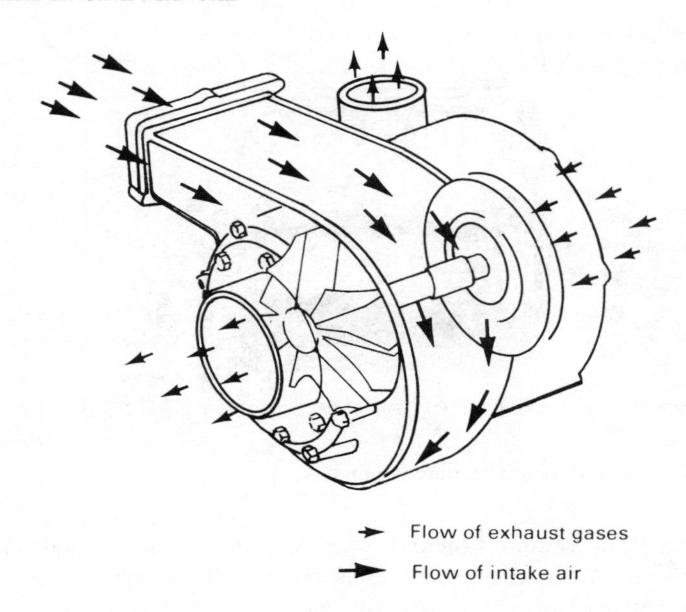

Courtesy Cummins Engine Co.

Fig. 25-13. Cutaway of turbocharger.

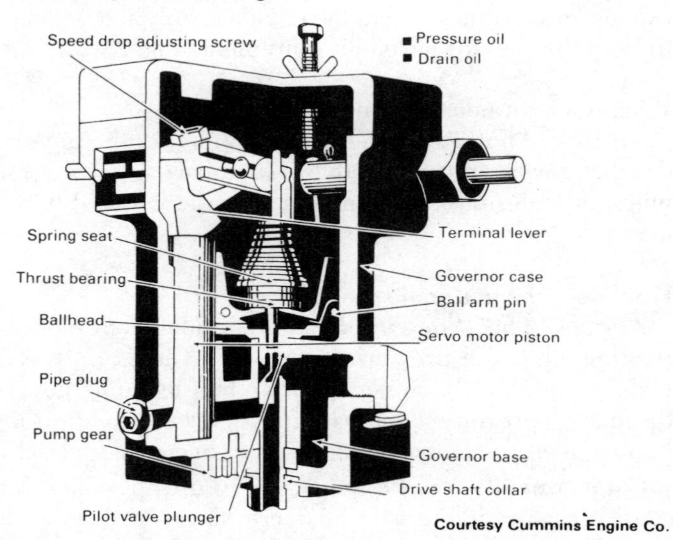

Courtesy Cummins Engine Co.

Fig. 25-14. Load-off, speed-increased position, hydraulic governor.

applications and may easily be adjusted to cover a range of approximately 0.5 percent to 7 percent.

Assume that a certain amount of load is applied to the engine. The speed will drop, and the flyballs will be forced inward and will lower the pilot valve plunger. This will admit oil pressure underneath the power piston, which will rise. The movement of the power piston is transmitted to the terminal shaft by the terminal lever. Rotation of the terminal shaft causes the fuel setting on the engine to be increased (Fig. 25-14).

If the governor is to be used for constant-speed service, speed adjustment may be made by setting the low-limit adjustment screw. Rotation of the speed-adjustment shaft increases or decreases the compression on the speeder spring (Fig. 25-5).

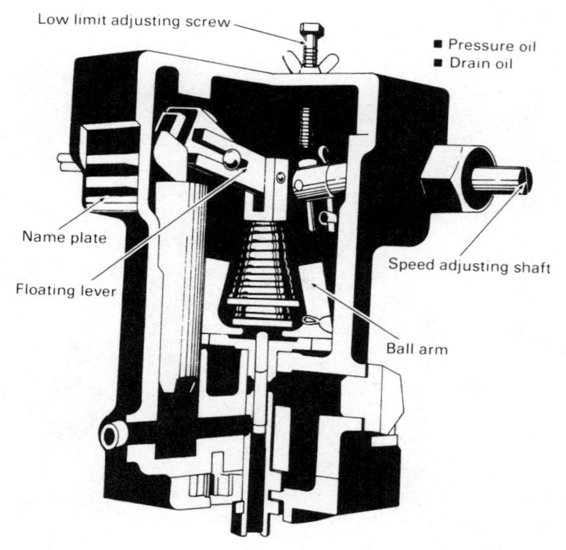

Courtesy Cummins Engine Co.

Fig. 25-15. Load-on, speed decreased position, hydraulic governor.

Rotating the speed-adjusting shaft sufficiently in the decrease-speed direction opens the area under the power piston to drain, and enables the fuel-return spring to shut off the fuel completely, thus shutting down the engine.

CHAPTER 26

Two-Stroke-Cycle Diesel Engine

What is meant by a two-stroke-cycle engine?

Answer: By definition, a two-stroke-cycle engine is one in which the four events of the cycle—scavenging or intake, compression, power, and exhaust—are performed during two strokes of the piston, as distinguished from a four-stroke-cycle engine, which requires four strokes to complete the cycle.

Since a two-stroke-cycle engine delivers one power stroke per revolution of the crankshaft, it can be made considerably lighter than a four-stroke-cycle engine, which requires two revolutions of the crankshaft per power stroke. This is one reason for the development of the two-stroke-cycle engines. Moreover, a better power flow is obtained—that is to say, one power stroke in every revolution of the crankshaft instead of one power stroke in every two revolutions of the crankshaft.

How is air provided for two-stroke-cycle diesel engines?

Answer: General Motors two-stroke-cycle diesel engines employ a blower as a part of the engine.

A specially designed blower forces air into the cylinders in order to expel the exhaust gases and fill the cylinders with fresh air for combustion as shown in Fig. 26-1. A feature of this design is the uniflow, that is, the unidirectional flow of the air from its entrance to exit from the cylinder.

Describe the cycle of events in the two-stroke-cycle General Motors diesel engine.

Answer: The cycle is as follows:

1. A series of ports cut into the circumference of the cylinder wall above the piston in its lowest position admit the air from the blower into the cylinder as soon as the top face of the piston uncovers the ports, as shown in Fig. 26-1A. The unidirectional flow of air toward the exhaust produces a scavenging effect, leaving the cylinders full of clean air when the piston again covers the admission ports.
2. As the piston continues on the upward stroke, the exhaust valves close and the charge of fresh air is subjected to the final compression, as shown in Fig. 26-1B. Since the air inlet in a diesel engine is never throttled, the high compression necessary for ignition is always maintained regardless of loads or speeds.
3. Shortly before the piston reaches its top position, the required amount of fuel is sprayed into the combustion space by the unit fuel injector, as shown in Fig. 26-1C. The intense heat generated during the high compression of the air ignites the fine fuel spray immediately, and the combustion continues as long as the fuel spray lasts. The resulting pressure forces the piston downward until the exhaust valves reopen.
4. The burned gases escape into the exhaust manifold and the cylinder volume is swept with clean scavenging air as the downward moving piston uncovers the inlet ports, as shown in Fig. 26-1D. Combustion takes place in each cylinder during each revolution of the crankshaft. The quantity of fuel burned in each cycle is controlled by the injector and is varied by the operator or by the governor.

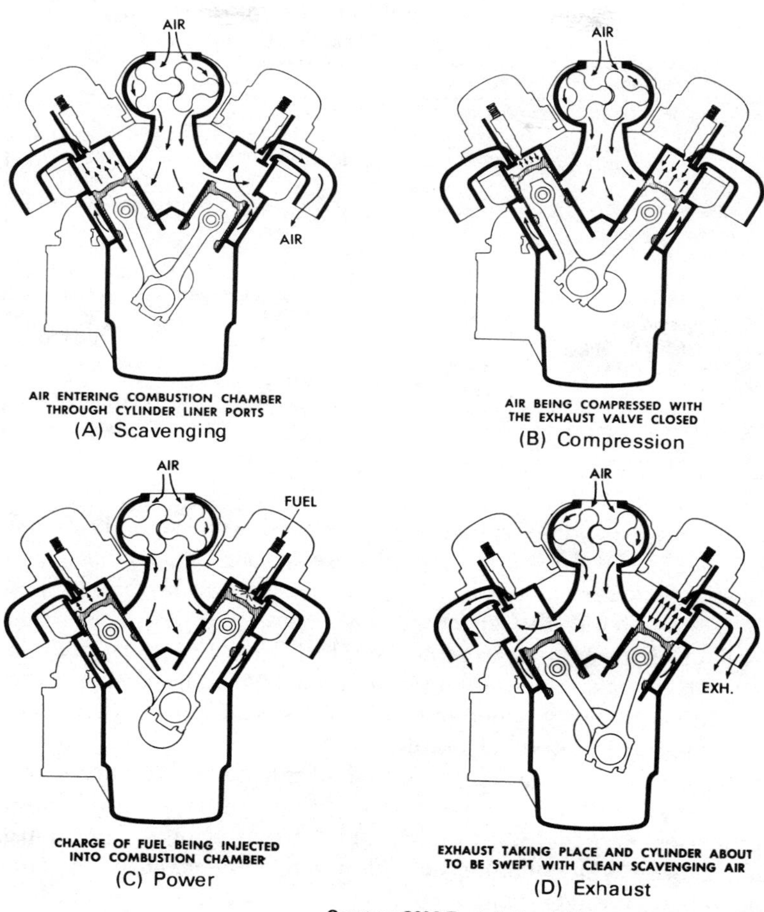

Courtesy GMC Truck & Coach Division, General Motors

Fig. 26-1. In a two-stroke-cycle engine the entire combustion cycle is completed in each cylinder in two strokes or in each revolution of the crankshaft. Intake and exhaust occur during portions of the compression and power strokes.

In what ways are several models of General Motors diesel engines similar?

Answer: General Motors two-stroke-cycle diesel engines are offered in several models having the same bore and stroke and using interchangeable parts wherever possible. For example,

Table 26-1. General Specifications for General Motors Diesel Engines

SPECIFICATION	6V-53	6-71N 6V-71N	8V-71N
Rated BHP	195 @ 2800	238 @ 2100 or 218 @ 2100	318 @ 2100 or 290 @ 2100
Maximum torque (ft.-lb.)	423 @ 1500	649 @ 1400 or 604 @ 1200	864 @ 1400 or 805 @ 1200
Engine type	two cycle	two cycle	two cycle
Number cylinders	6	6	8
Cylinder arrangement	vee	inline vee	vee
Bore and stroke	$3\frac{7}{8} \times 4\frac{1}{2}$	$4\frac{1}{4} \times 5$	$4\frac{1}{4} \times 5$
Total displacement	318.4 cu.in.	425.6 cu.in.	567.4 cu.in.
Compression ratio	17:1	18.7:1	18.7:1
Piston speed (ft./min.)	2100 @ 2800	1750 @ 2100	1750 @ 2100
No. exhaust valves	4/cylinder	4/cylinder	4/cylinder
Length	$35\frac{21}{32}$	$54\frac{3}{4}$ $48\frac{1}{4}$	$54\frac{1}{16}$
Width (in.)	$35\frac{1}{4}$	$33\frac{5}{16}$ $42\frac{19}{32}$	$38\frac{5}{16}$
Height	$38\frac{5}{8}$	$48\frac{1}{32}$ $47\frac{1}{4}$	$47\frac{3}{4}$
Approx. net wt (lbs.)	1540	2150 1960	2345

three-, four-, and six-cylinder engines having the same bore and stroke ($3\frac{7}{8} \times 4\frac{1}{2}$) are available. Also, four-, six-, eight-, and twelve-cylinder engines having the same bore and stroke ($4\frac{1}{4} \times 5$) are available (Fig. 26-2).

Thus, different power capacities (Fig. 26-3) are available in the same type of engine, in which the major working parts, such as injectors, pistons, connecting rods, all bearings, and other numerous parts are interchangeable. Engines with either direction of crankshaft rotation can be supplied to suit specific requirements.

Furthermore, the blower, water pump, oil cooker, oil filter, governor, and fuel pump (Fig. 26-4) form a group of standard accessories which can be located on either the right- or the left-hand side of the engine, regardless of the direction of crankshaft rotation (Fig. 26-5). Still further flexibility in meeting installation requirements can be obtained by placing the exhaust manifold and the water-outlet manifold on either side of the engine. This flexibility in the arrangement of parts is obtained by having both the cylinder block and the cylinder head symmetrical at both ends and also with respect to each other.

Describe the crankshaft and main bearing assembly of General Motors diesel engines.

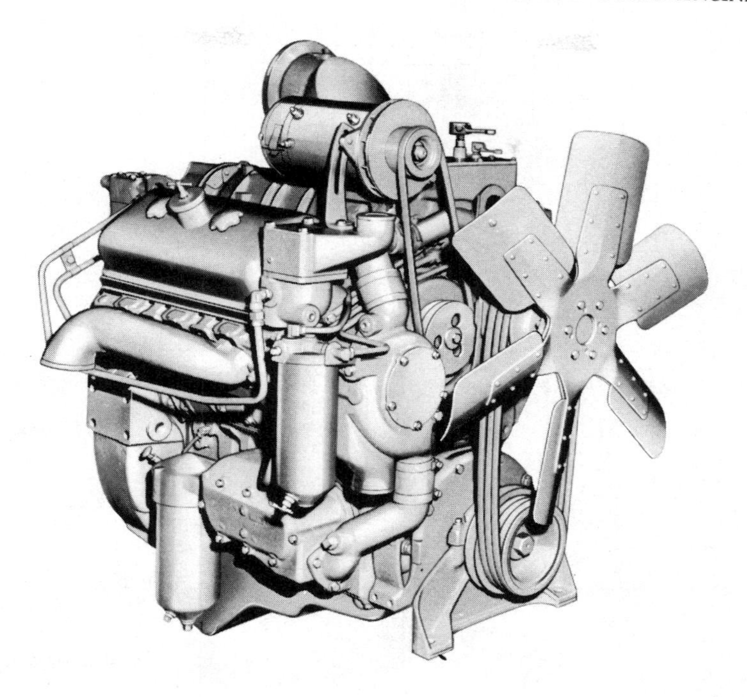

Courtesy Detroit Diesel Engine Division, General Motors

Fig. 26-2. Series 6V-71 diesel engine used in General Motors trucks. This is a vee-type six-cylinder engine.

Answer: The crankshaft is a drop forging of high-carbon steel heat-treated to ensure durability and strength. All main bearing surfaces and connecting-rod bearing surfaces are hardened by the Tocco process (Fig. 26-6).

Each crankshaft main bearing journal is 4 ½ inches in diameter and each connecting-rod journal is 3 inches in diameter.

Main bearing shells are used at each journal. They are of the precision type and are readily replaceable without machining. The shells consist of an upper shell, which is seated in the cylinder block main bearing support, and a lower shell seated in the main bearing cap. A tang at the parting line at one end of each shell prevents endwise or radial movement.

The bearing caps are numbered 1, 2, 3, etc., indicating their

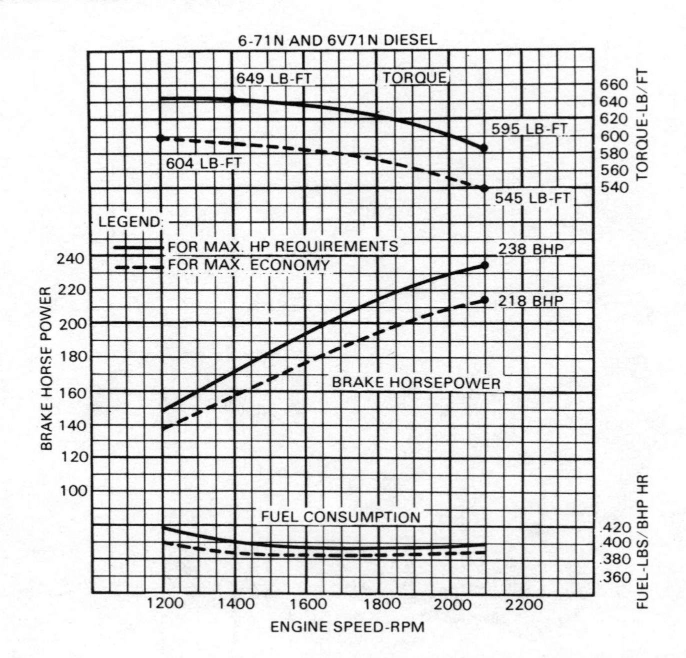

Courtesy Detroit Diesel Engine Division, General Motors

Fig. 26-3. Horsepower, torque, and fuel consumption curves for General Motors series 6-71N and 6V-71N diesel engines.

respective positions, and must always be reinstalled in their original positions after being removed.

Each upper shell has an oil hole in the groove, midway between the parting lines, registering with a vertical oil passage in the cylinder block. Lubricating oil, under pressure, passes from the cylinder block by way of the bearing shells to the drilled passages in the crankshaft and then to the connecting rods.

As the lower main bearing shells have no oil holes, they must not be interchanged with the upper shells. The tangs on the upper bearing shells are centered to aid in correct installation of the main bearing shells.

Describe the connecting-rod assembly of General Motors diesel engines.

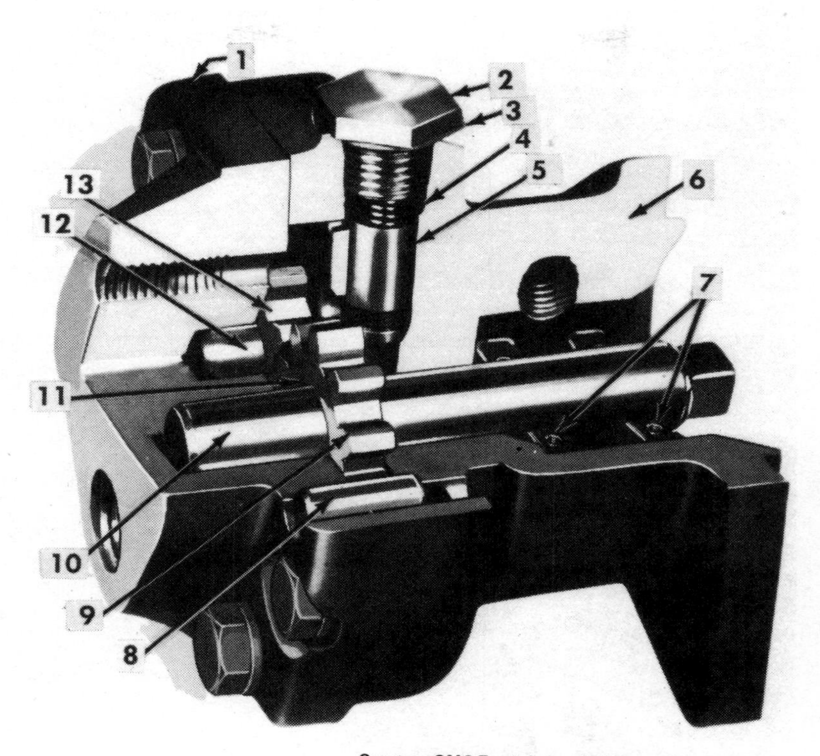

Courtesy GMC Truck & Coach Division, General Motors

1. Pump cover.
2. Valve plug.
3. Gasket.
4. Valve spring.
5. Relief valve.
6. Pump body.
7. Oil seals.
8. Dowel pin.
9. Drive gear.
10. Drive shaft.
11. Retaining ball.
12. Driven shaft.
13. Driven gear.

Fig. 26-4. General Motors diesel engine fuel oil pump assembly. The fuel pump is of positive displacement, gear type. Regardless of the rotation of the engine crankshaft, the fuel pump is always a left-hand rotating pump.

Answer: The connecting rod (Fig. 26-7) is made of forged steel in "I" section. The connecting rod has a closed hub at the upper end and a cap at the lower end. The rod is drilled at the upper end to provide lubrication to the piston pin and is equipped with an oil

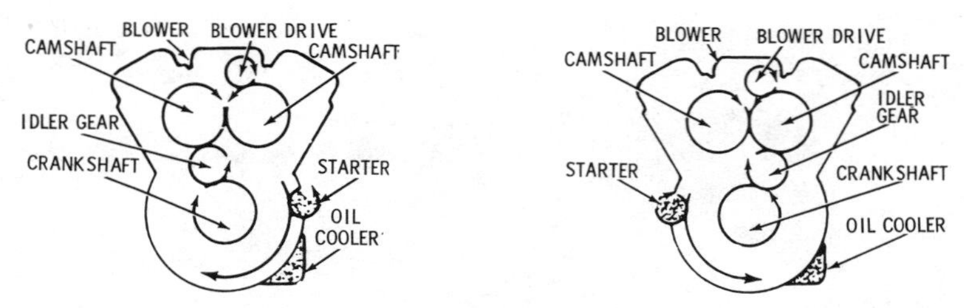

Courtesy GMC Truck & Coach Division, General Motors

(A) Model LD (coach) engine.

(B) Model RC (truck) engine.

Fig. 26-5. Engine rotation and accessory location as viewed from the rear of the engine.

Courtesy GMC Truck & Coach Division, General Motors

Fig. 26-6. Crankcase of a General Motors diesel engine. Illustrated are (1) main bearing cap bolts, (2) main bearing cap, (3) lower main bearing, (4) rear main bearing thrust washer, (5) upper main bearing, and (6) crankcase.

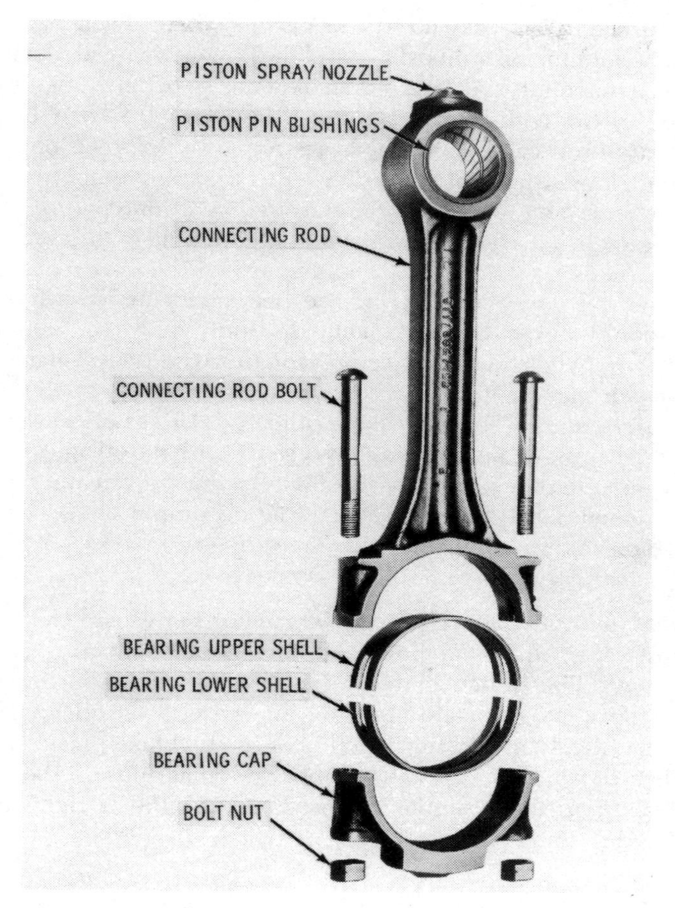

Courtesy GMC Truck & Coach Division, General Motors

Fig. 26-7. Connecting-rod components of a General Motors diesel engine.

spray nozzle for cooling the underside of the piston head as well as lubrication for the piston pin.

A helically grooved bushing is pressed into each side of the connecting rod at the upper end. An approximately 1/8-inch cavity between the inner ends of these bearings registers with the drilled oil passage in the connecting rod to provide a duct around the piston pin for lubricating the piston-pin bearing and allows oil to be forced to the spray nozzle for piston cooling. The piston pin floats in both the piston bushing and the connecting-rod bushing.

The connecting-rod bearing shells consist of an upper shell carried in the connecting rod and a lower shell seated in the connecting-rod cap. These shells are replaceable precision type without shim adjustments. Endwise or radial movement is prevented by a tang located at the parting line at one end of each bearing shell. This correctly positions the bearing in the connecting rod shells.

Upper and lower connecting-rod bearing shells are different and should never be interchanged. Both shells are notched midway between the bearing edges for approximately 3/4 inch in from each parting line. The front side of the lower shell has a circumferential oil groove that terminates at the notched ends. The notches maintain a continuous registry with the oil holes in the crankpins and they provide a constant supply of lubricating oil to the connecting-rod bearings and to the piston-pin bushings and spray nozzle through the hollow connecting rod.

At what points in the cycle do the intake ports open and close, the exhaust valves open and close, and fuel injection occur?

Answer: The timing diagram (Fig. 26-8) shows the cycle in which the opening and closing of the intake ports, the operation of the exhaust valves, and the injection of the fuel take place.

When the engine is properly timed, the markings on the gear train will appear as shown in Fig. 26-9 for either a right- or a left-hand rotating engine.

Give the timing procedure for the two-stroke-cycle General Motors diesel engine.

Answer: In order to properly control fuel injection and the opening and closing of the exhaust valves, the correct relationship between the two camshafts and the crankshaft must be maintained for a smooth-running engine.

The two camshaft gears can be mounted in only one position due to the location of the keyway in each camshaft relative to the cams. The crankshaft timing gear can also be mounted in only one position because one attaching bolt is offset.

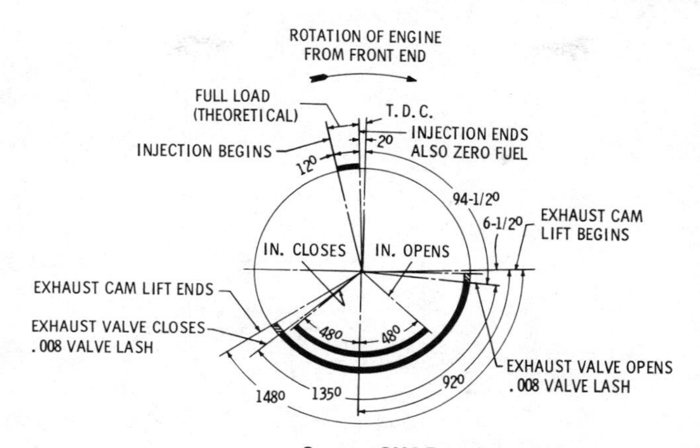

Courtesy GMC Truck & Coach Division, General Motors

Fig. 26-8. Timing diagram for General Motors Model 71 diesel engine.

Preignition, uneven running, and a loss of power result when an engine is "out of time."

A quick check can be made without removal of the transmission, flywheel, and flywheel housing when an engine is suspected of being out of time. The procedure for checking timing is outlined. It is necessary to obtain access to the crankshaft pulley, to mark top-dead-center position of the selected piston, and to the front end of the crankshaft (or to the flywheel), for turning the crankshaft. Then proceed as follows:

1. Remove one valve rocker cover.
2. Any cylinder may be selected, but a cylinder adjacent to a rocker-cover stud is preferred because the stud may be used to mount a dial instructor.
3. Remove the injector at the selected cylinder.
4. Slide a 12-inch rod through the injector tube until the end of the rod rests on top of the piston. Place the throttle in "No Fuel" position and turn the crankshaft slowly in the direction of engine rotation. When the piston reaches the end of its upward travel, as indicated by the rod, stop, remove the rod, and turn the crankshaft opposite the direction of rotation between 1/16 and 1/8 turn.
5. Select a dial indicator with .001-inch graduations and at least

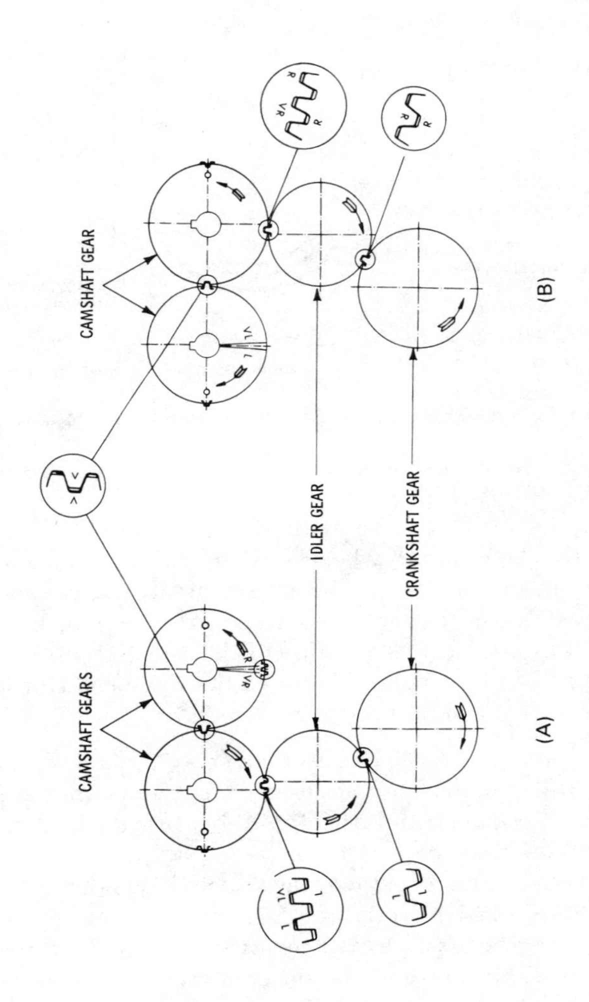

Courtesy GMC Truck & Coach Division, General Motors

(A) Left-hand rotation engines. (B) Right-hand rotation engines.

Fig. 26-9. Gear train and timing marks for General Motors diesel engine.

1-inch spindle movement. An extension long enough to reach the piston just before it reaches the end of its upward stroke must be provided for the indicator spindle.

6. The indicator mounting may be threaded into the rocker-cover stud so that it is over the injector tube; or, the stud may be removed from the cylinder head and the mounting threaded into the tapped hole. Make sure that the indicator spindle is free in the injector tube and is free to travel at least one inch.
7. A suitable pointer should be attached to the front of the crankshaft cover. The outer end of the pointer should extend over the top of the crankshaft pulley.
8. Turn the crankshaft slowly in the direction of engine rotation until the indicator hand ceases to move. Continue rotating the crankshaft until the hand starts to move again.
9. Reset the indicator hand to zero and turn the crankshaft until the indicator reading is .010 inch.
10. Scribe a line on the crankshaft pulley in line with the end of the pointer.
11. Turn the crankshaft slowly, opposite the direction of engine rotation, until the indicator hand ceases to move. Continue to turn until the hand starts to move again.
12. Reset the dial to zero; then, turn until the indicator dial reads .010 inch.
13. Scribe the second line on the crankshaft pulley in line with the end of the pointer.
14. Scribe a third line halfway between the two lines. This line indicates top dead center. If the crankshaft pulley retaining bolt has loosened, tighten it to 290-310 foot-pounds of torque on the torque wrench.
15. Remove the dial indicator.
16. Install the injector assembly, and tighten the injector clamp bolt to 20-25 foot-pounds of torque. Swing the rocker arms into position and install the rocker-arm brackets and bolts. Tighten the bolts to 90-100 foot-pounds of torque.
17. Adjust the valve clearance and time the injector according to "engine tune-up."
18. Turn the crankshaft, in the direction of engine rotation, until the exhaust valves in the cylinder are completely open.

19. Reinstall the dial indicator so that the indicator spindle rests on top of the injector follower. Set the indicator dial to zero, turn the crankshaft slowly in direction of engine rotation until the center mark on the pulley is in line with the pointer.
20. Note the reading on the dial indicator and compare as follows:
    a. If the indicator reading is .225-.235 inch, engine is in tune.
    b. If the indicator reading is .192-.202 inch, timing is retarded by one tooth.
    c. If the indicator reading is .257-.267 inch, timing is advanced by one tooth.
21. After completion of the timing check, remove the dial indicator, remove the shipping caps from the fuel injector fittings, and install the fuel pipes.
22. Remove the pointer from the front cover of the crankshaft.
23. Install the valve rocker cover.

How are the rotating parts of the engine balanced?

Answer: The rotating masses of the engine are completely balanced by the counterweights on the crankshaft. The crank arrangement and firing order selected counteract unbalanced reciprocating forces.

The unbalanced reciprocating couple, due to the end pistons moving in opposite directions, is balanced by a unique arrangement of rotating counterweights. Two weights, rotating in opposite directions with relationship to each other, are used at each end of the engine (Fig. 26-10).

Describe the valve operating mechanism.

Answer: Three rocker arms are provided for each cylinder, the two outer rocker arms operating the exhaust valves and the center one the fuel injector as shown in Fig. 26-11.

Each rocker arm assembly operates on a separate shaft supported by two cast-iron brackets. A single bolt fastens each bracket securely to the top deck of the cylinder head. Consequently, the simple removal of these two bolts permits folding back the complete rocker arm assembly and gives easy access to the fuel injector and the valve springs (Fig. 26-12).

Courtesy GMC Truck & Coach Division, General Motors

Fig. 26-10. Front balance-weight mounting on General Motors diesel engine.

Why is valve lash adjustment important?

Answer: Correct valve lash adjustment is important because of the high compression pressure employed in a diesel engine.

1. Too little clearance causes a loss of compression, misfiring cylinder, and eventual burning of both the valves and the valve seats.
2. Too much clearance between the valve stem and the valve rocker arm results in noisy operation of the engine, especially in the idling range.
3. Specifications for a given engine should be followed for both hot and cold dimensions when resetting valve lash.
4. The adjusting of the valve lash can be changed by means of

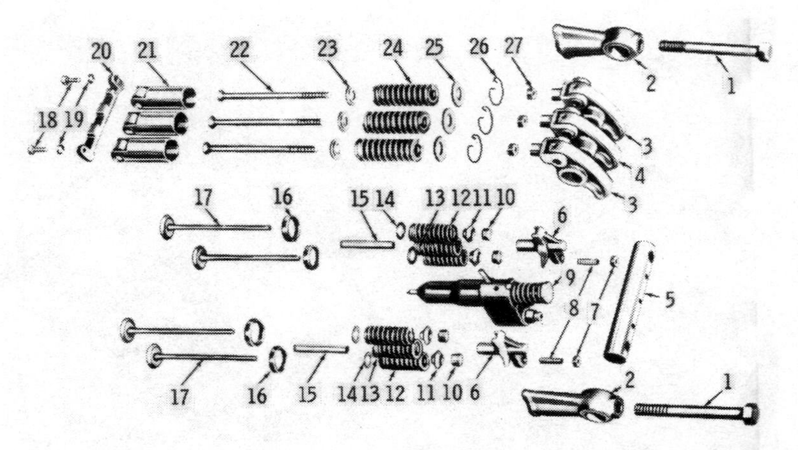

Courtesy GMC Truck & Coach Division, General Motors

1. Bracket bolt.
2. Rocker shaft bracket.
3. Valve rocker arm.
4. Injector rocker arm.
5. Rocker shaft.
6. Valve bridge (4-valve head).
7. Screw nut (4-valve head).
8. Adjusting screw (4-valve head).
9. Injector assembly.
10. Valve spring lock.
11. Valve spring cap.
12. Valve spring.
13. Bridge spring (4-valve head) (early models).
14. Valve spring seat.
15. Bridge guide (4-valve head) (pressed type).
16. Valve seat insert.
17. Exhaust valve.
18. Guide bolt.
19. Lock washer
20. Cam follower guide.
21. Cam follower.
22. Push rod.
23. Spring seat—lower.
24. Push rod spring.
25. Spring seat—upper.
26. Spring seat retainer.
27. Lock nut.

Fig. 26-11. Operating mechanism for valves and fuel injector of General Motors diesel engine.

Courtesy GMC Truck & Coach Division, General Motors

Fig. 26-12. Compressing the exhaust-valve spring on General Motors diesel four-valve cylinder head.

the threaded upper part of the push rod, which is screwed into the push-rod clevis and locked by a lock nut. No special tools are required to make this adjustment.

How does the two-stroke-cycle diesel engine receive air into the cylinder for combustion?

Answer: In the scavenging process employed in two-stroke-cycle engines, a charge of air which is forced into the cylinders by the blower, thoroughly sweeps out all the burned gases through the exhaust valve ports, and also helps to cool the internal engine parts, particularly the exhaust valves, as shown in Fig. 26-13.

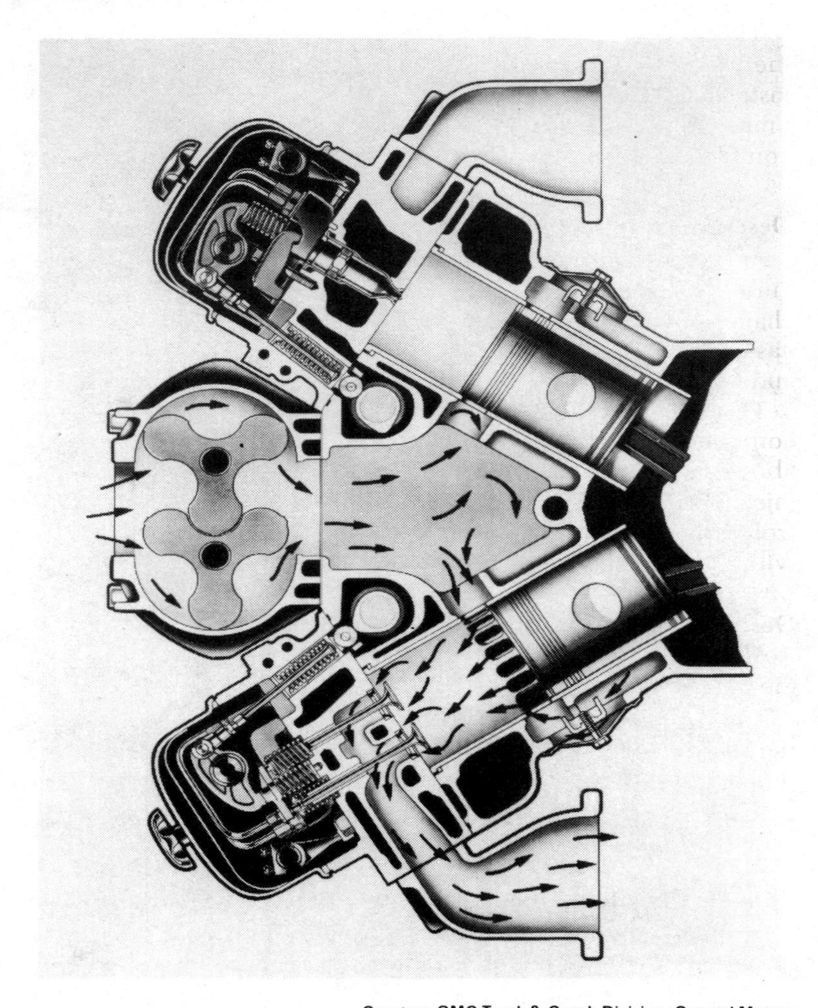

Courtesy GMC Truck & Coach Division, General Motors

Fig. 26-13. Air-intake system through the blower and engine of a General Motors diesel engine. Air is forced into the cylinder by the blower, which sweeps out all the burned gases through the exhaust valve ports and helps cool internal parts of the engine. The blower housing is bolted to the top deck of the cylinder block, between the two banks of cylinders, and supplies the fresh air needed for combustion and scavenging.

The large volume of air circulated through the cylinders makes the use of a high-grade air cleaner essential. The air cleaner is installed at the air intake of the blower. When it seems advisable to limit the intake noise of the engine to a minimum, the use of a combined air cleaner and intake silencer is recommended.

Describe the operation of the fuel injector.

Answer: Before fuel injection can be effected, the fuel pressure must be higher than that of the air charge in the combustion chamber. Consequently, the fuel injector as shown in Fig. 26-14 has three functions: it meters the fuel, it creates high fuel pressure, and it atomizes the fuel.

The unit fuel injector developed and used in these engines combines in a single unit all the parts necessary to perform the above functions, thus providing a complete and independent injection system for each engine cylinder (Fig. 26-15). Load control from full load to no load is produced by rotating the plunger with the control rack (Fig. 26-16).

Describe the procedure for timing the fuel injectors.

Answer: Time of ignition of fuel in each cylinder is governed by the time of injection of the fuel into the combustion chamber.

The injector follower guide (Fig. 26-18) must be adjusted to a definite height for the injector to be timed properly. Injector timing is as follows:

1. Set the governor control lever in "No Fuel" position.
2. Turn the crankshaft manually or by means of the starter until the exhaust valve rocker arms are fully depressed, for the selected cylinder.
3. The fuel injector is identified by a colored tab stamped onto the injector body (Fig. 26-19). The injector is timed with a 1.460-inch (J-1853) gauge.
4. Place the injector timing gauge in the hole provided on top of the injector body with one of the "flats" toward the injector.
5. Adjust the injector rocker arm (Fig. 26-21) by loosening the lock nut and turning the push rod with an end wrench, until the bottom of the timing gauge will just pass over (drag lightly) the top of the injector follower guide.

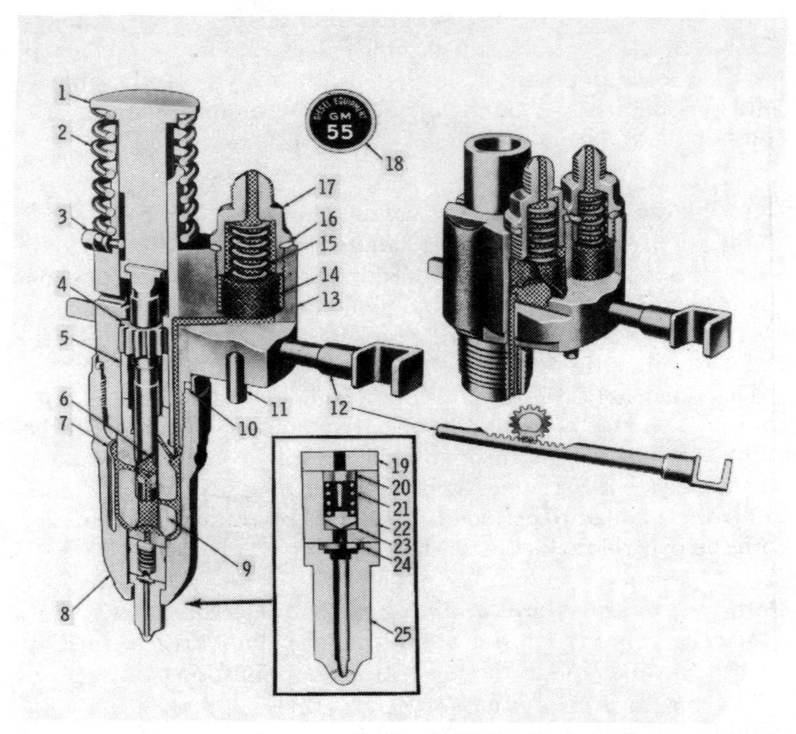

Courtesy GMC Truck & Coach Division, General Motors

1. Follower.
2. Follower spring.
3. Follower stop pin.
4. Gear.
5. Gear retainer.
6. Plunger.
7. Spill deflector.
8. Injector nut.
9. Plunger bushing.
10. Seal ring.
11. Dowell.
12. Rack.
13. Injector body.
14. Filter element.
15. Filter spring.
16. Gasket.
17. Filter cap.
18. Identification tag.
19. Valve seat.
20. Valve.
21. Valve spring.
22. Valve cage.
23. Valve stop.
24. Check valve.
25. Spray tip.

Fig. 26-14. General Motors diesel fuel injector assembly showing the fuel flow in "crown-valve" type of fuel injector.

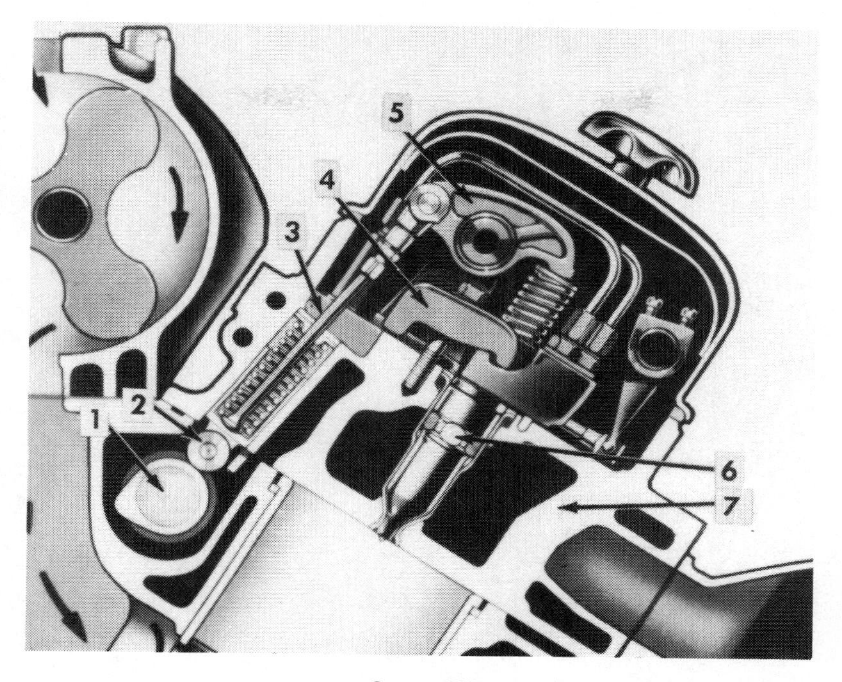

Courtesy GMC Truck & Coach Division, General Motors

1. Camshaft.
2. Cam-follower assembly.
3. Push rod.
4. Hold-down clamp.
5. Rocker arm.
6. Injector assembly.
7. Cylinder head.

Fig. 26-15. General Motors diesel fuel injector installed. Injectors are mounted in the cylinder heads with their spray-nozzle tops projecting slightly below the top of the inside surface of the combustion chamber. A dowel pin in the injector body registers with a hole in the cylinder head for accurately locating the injector assembly. A clamp, bolted to the cylinder head and fitting into a machined recess on each side of the injector body, holds the injector in place in a water-cooled tube that passes through the cylinder head.

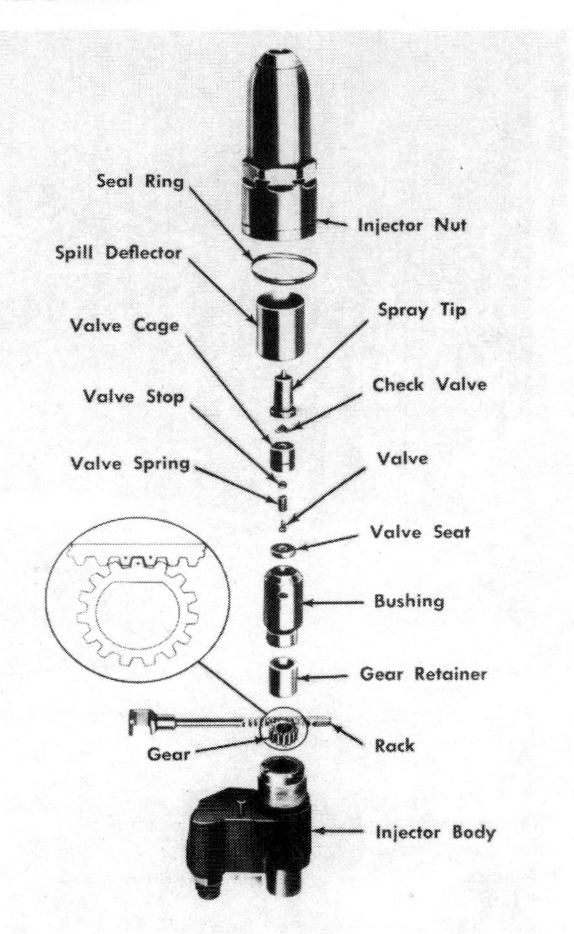

Courtesy GMC Truck & Coach Division, General Motors

Fig. 26-16. General Motors diesel fuel injector assembly.

6. Prevent the push rod turning, and tighten the lock nut. Recheck the adjustment with the injector timing gauge and readjust if necessary (Fig. 26-21).

What factor determines the amount of fuel injected into the cylinder?

Courtesy GMC Truck & Coach Division, General Motors

Fig. 26-17. Removal of General Motors diesel fuel injector. Immediately after the removal of the fuel lines from an injector, the two fuel feed fittings should be protected with a shipping cap to prevent dirt from entering the injector.

Answer: The position of the injector control rack in relationship to the governor determines the amount of fuel injected into each cylinder on General Motors diesel engines (Fig. 26-22).

Cylinders are numbered starting at the front of the engine. The letters R or L indicate the injector location in the right or left cylinder bank when viewed from the rear of the engine when mounted in proper position.

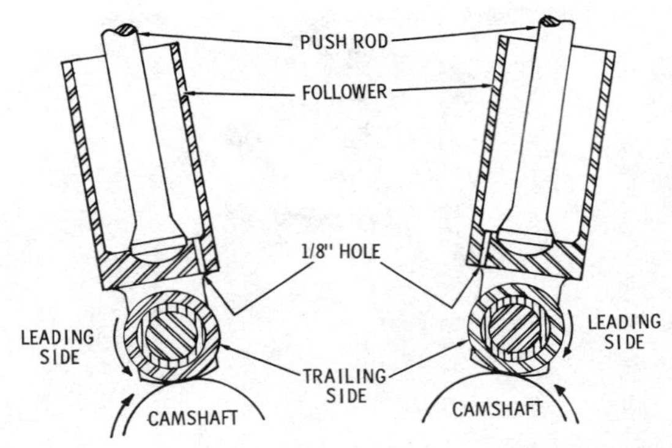

Courtesy GMC Truck & Coach Division, General Motors

Fig. 26-18. Installation of cam followers on a General Motors diesel engine. The oil hole in the bottom of the follower points away from the valves so that the hole is not covered by the follower guide.

What is the procedure for positioning the injector control racks?

Answer: Injector control racks are positioned as follows:

1. Disconnect any linkage to the governor throttle control lever.
2. Loosen the idle adjusting screw until 1/2 inch of the threads projects from the lock nut.
3. Remove right and left rocker arm covers.
4. Loosen all inner and outer injector control rack adjusting screws in both cylinder banks. Be sure all rack control levers are free on the injector control tubes.
5. Move the governor throttle control lever to "Full Fuel" position (Fig. 26-22) and hold with light finger pressure. Turn the inner adjusting screw of #1L injector rack control lever until the clevis pin binds. Tighten the outer adjusting screw.
6. Repeat the procedure for the left bank.
7. Set #2 and #3 injector control racks to the #1 rack for the respective banks of cylinders.
8. When the injector rack adjustment is completed, recheck the clevis pins at the left and right fuel rods to be certain they are snug, while the governor throttle lever is in "Full Fuel" position and the inner and outer adjusting screws are tight.
9. Turn the idle speed adjusting screw in until it projects 1/16 inch from the lock nut, to permit starting the engine.

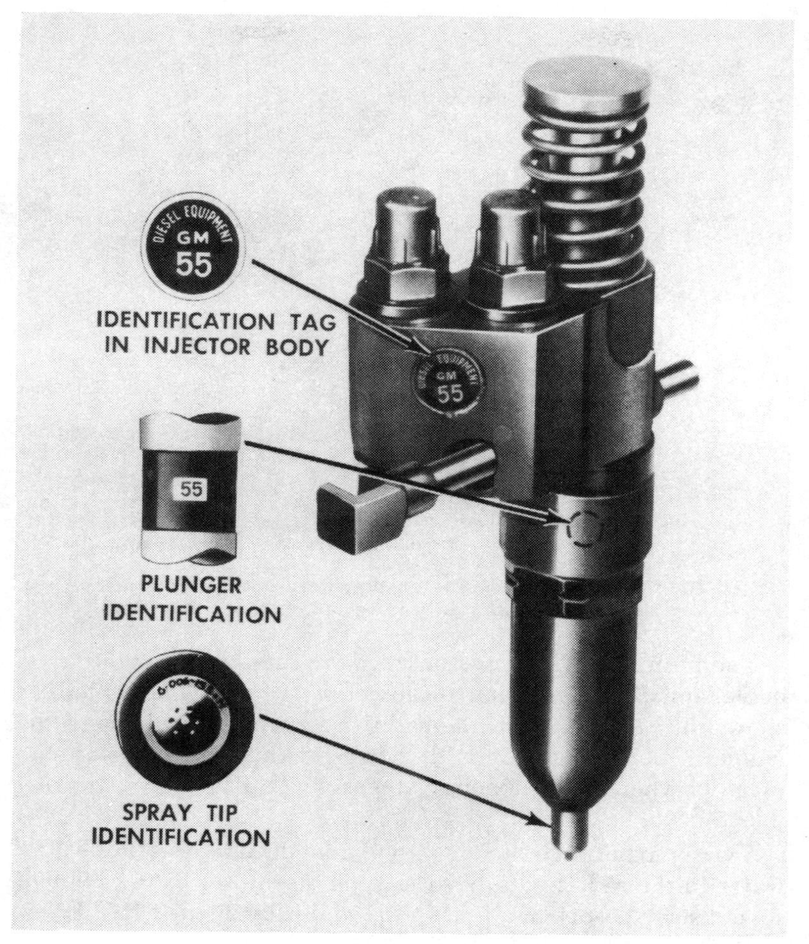

Courtesy GMC Truck & Coach Division, General Motors

Fig. 26-19. Injector identification on General Motors fuel injectors.

Describe the cooling water system of a two-stroke-cycle diesel engine.

Answer: Water is circulated through the oil cooler, cylinder block, and cylinder head by a centrifugal pump mounted on the front end of the blower which is driven by the lower blower rotor shaft by means of a coupling.

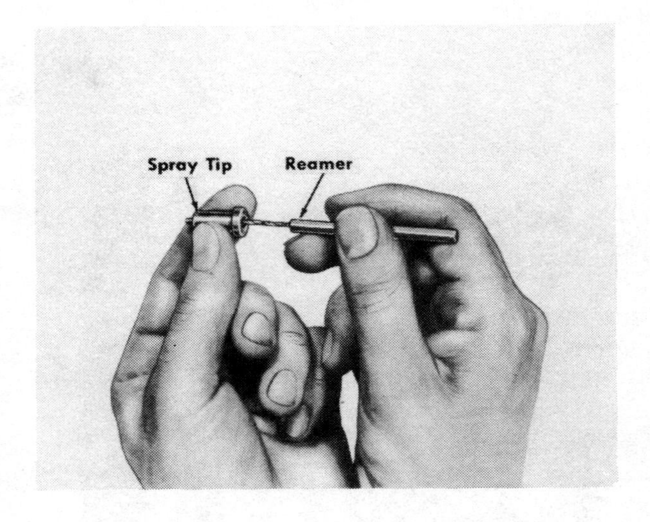

Courtesy GMC Truck & Coach Division, General Motors

Fig. 26-20. Reaming of a fuel injector spray-nozzle tip. Turn the reamer in a clockwise direction to remove carbon deposits.

The pump draws the water from the radiator through the oil cooler and discharges it into the lower part of the cylinder block as shown in Fig. 26-23. Openings in the water jacket around the cylinder bores connect with corresponding openings in the cylinder where the water circulates around the valves and the fuel injectors.

A water manifold bolted to the cylinder head returns the cooling water to the radiator. A by-pass type thermostat is used which should start to open at about 158°F. and fully open at 185° F.

What is the purpose of the governor?

Answer: The governor (Fig. 26-24) serves two purposes:

1. It limits the maximum speed of the engine.
2. It maintains a uniform idling speed regardless of engine friction and temperature.

These purposes necessitate a variation in the amount of fuel injected into the cylinders. To accomplish this, a centrifugal type

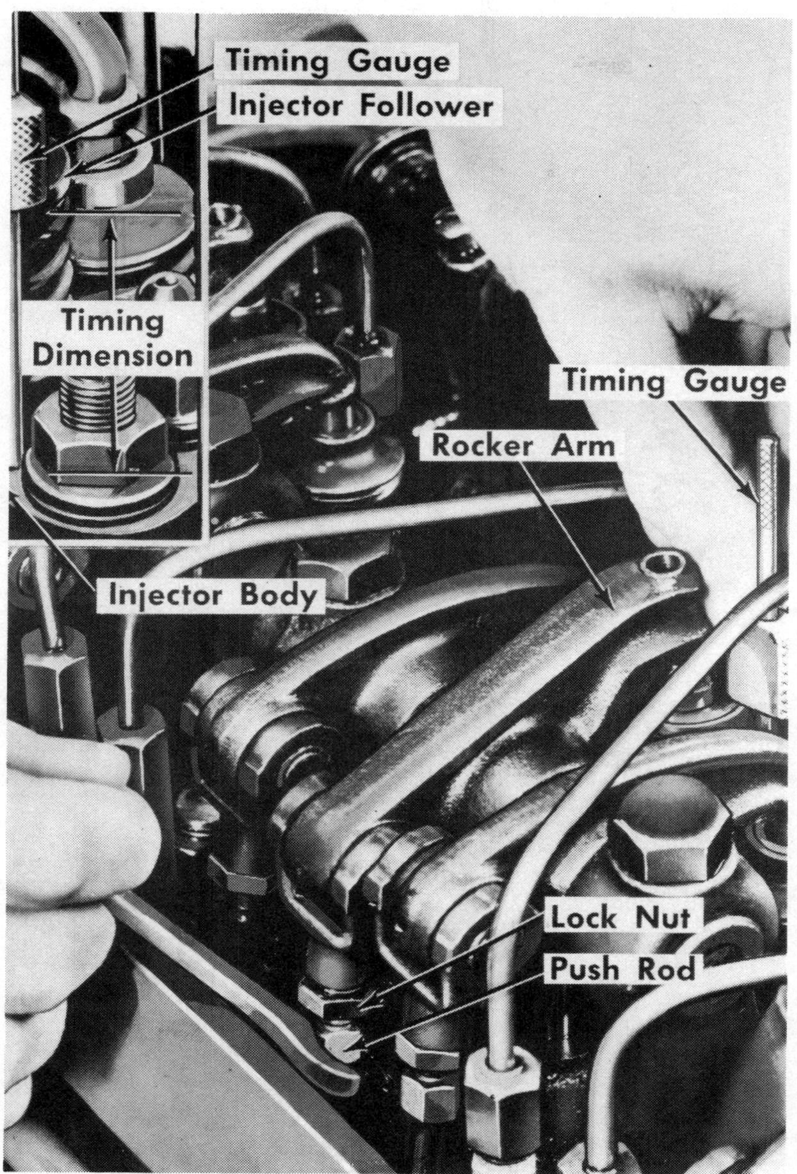

Fig. 26-21. Timing the General Motors fuel injector.

Courtesy GMC Truck & Coach Division, General Motors

Fig. 26-22. Positioning the No. 1 injector control rack on a General Motors diesel engine. The injector control rack must be properly adjusted in relation to the governor.

Courtesy GMC Truck & Coach Division, General Motors

Fig. 26-23. General Motors diesel engine cooling-system circulation for a coach. The centrifugal water pump circulates the cool water through the oil cooler, cylinder block, cylinder heads, and radiator.

governor is used to govern the position of the injector control shaft, independent of the accelerator linkage.

Describe briefly the governor adjustment of General Motors two-stroke-cycle diesel engines.

Answer: The governor has one adjustment, for idling speed only. The high-speed setting is made at the factory and is not adjustable in service (Fig. 26-25).

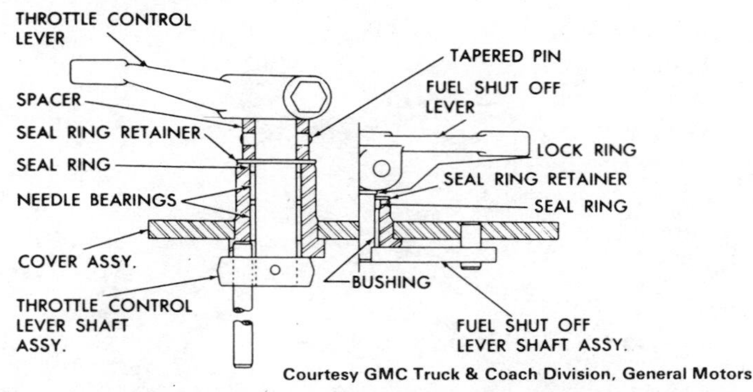

Courtesy GMC Truck & Coach Division, General Motors

Fig. 26-24. Sectional view of governor cover from a General Motors diesel engine.

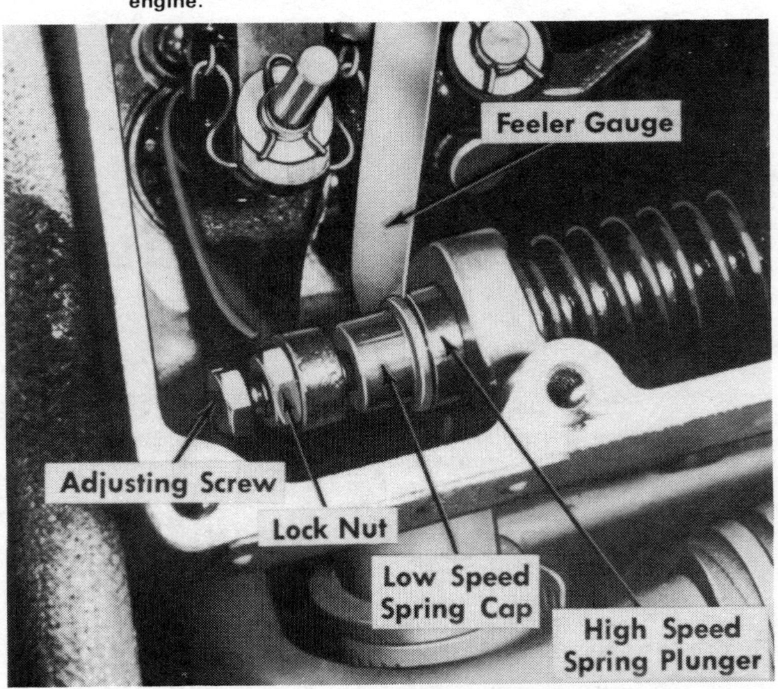

Courtesy GMC Truck & Coach Division, General Motors

Fig. 26-25. Adjusting the governor low-speed gap on a General Motors diesel engine. The gap should be between .001 and .002 inch. Turn the adjusting screw to obtain the desired gap. Recheck the gap after tightening the lock nut.

The spring controlling the low-speed action of the governor has been set to the correct initial load at the factory. This initial setting should not be disturbed while in service unless absolutely necessary (Fig. 26-26).

The idling speed of the engine cannot be influenced by an adjustment of the injector control racks because the governor will automatically counteract such an adjustment.

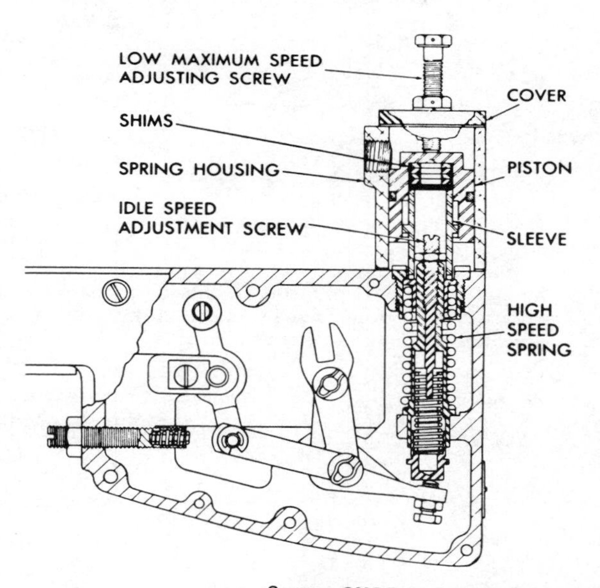

Courtesy GMC Truck & Coach Division, General Motors

Fig. 26-26. Sectional view of late-type, dual-range governor springs on a General Motors diesel engine.

Describe briefly the lubrication system of two-stroke-cycle General Motors diesel engines.

Answer: This consists of an oil pump, oil filter, and oil cooler (Fig. 26-27) with a suitable relief valve in the oil pump and a by-pass valve between the oil pump and the oil filter to ensure positive engine lubrication at all times.

The oil circulation is effected by a gear-type lubricating oil pump (Fig. 26-28) which delivers the hot oil to an oil filter, after which the oil passes through an oil cooler to the cylinder block, and

Courtesy GMC Truck & Coach Division, General Motors

Fig. 26-27. Oil-cooler adapter and by-pass valve on a General Motors diesel engine. A by-pass valve is located in the oil-cooler adapter to assure engine lubrication in the event the oil cooler becomes plugged. If the oil pressure at the inlet side exceeds the oil pressure at the outlet side by 40 psi, the by-pass valve will open and allow the oil to by-pass the oil cooler.

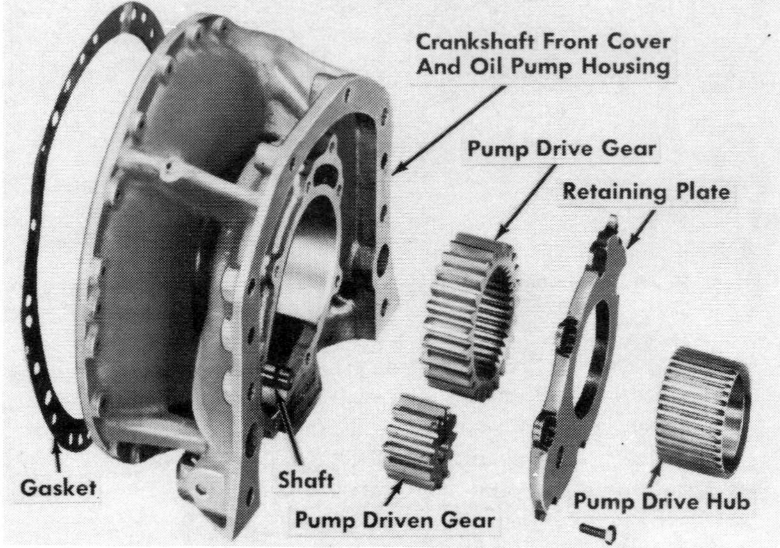

Fig. 26-28. General Motors diesel engine lubricating oil pump details. This pump consists of two spur gears that mesh together and ride in a cavity inside the crankcase cover.

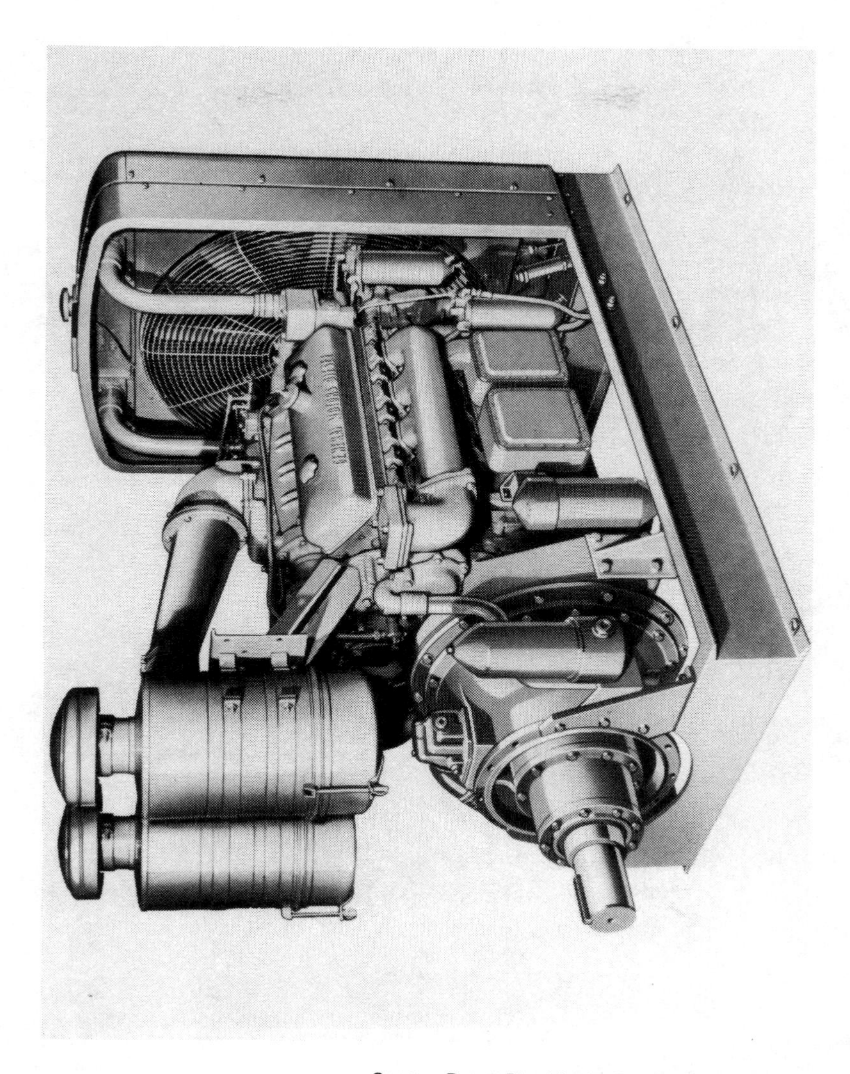

Courtesy Detroit Diesel Engine Division, General Motors

Fig. 26-29. Series 8V-71 General Motors industrial diesel power unit with Torqmatic converter.

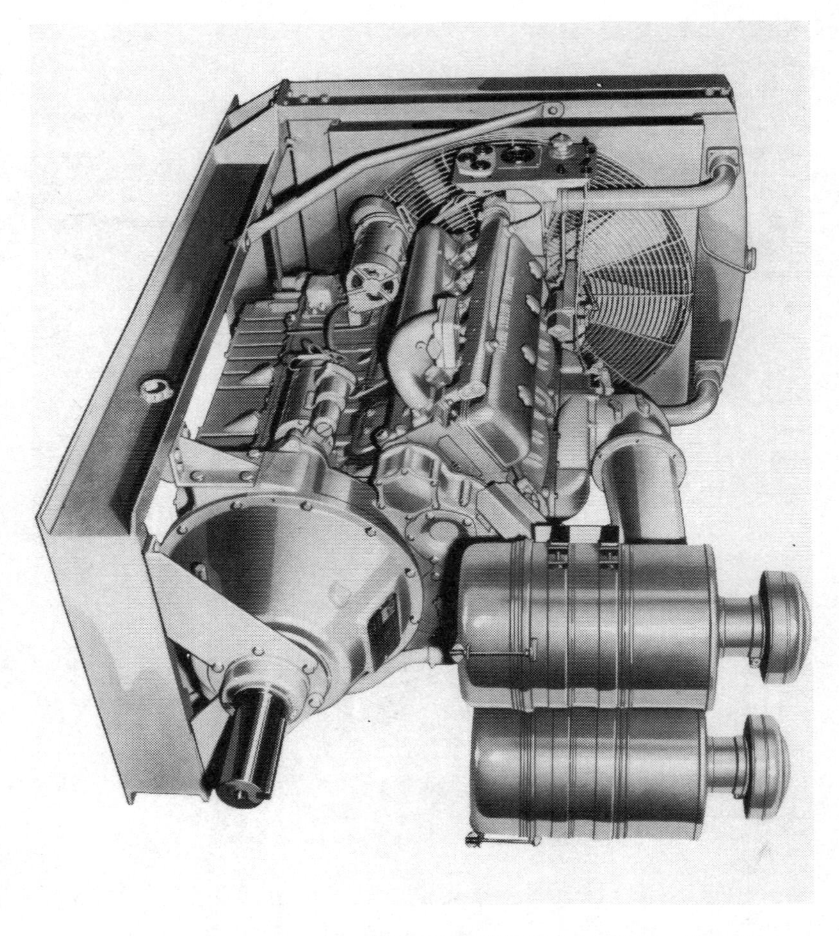

Courtesy Detroit Diesel Engine Division, General Motors

Fig. 26-30. Series 12V-71 General Motors diesel radiator-cooled power unit with PTO. This engine is used on heavy construction, excavating, and road-building equipment.

then to the camshaft, balance shaft, connecting rods, crankshaft, and main bearings.

Are two-stroke-cycle diesel engines used as construction and industrial engines?

Answer: Yes. Industries such as road building, mining, and lumbering are using two-stroke-cycle diesel engines widely either as stationary or as portable power plants (Fig. 26-29).

How are industrial engines cooled?

Answer: When radiator cooling systems (Fig. 26-30) are impractical, heat exchangers are used to transfer heat through copper elements to cold, raw water. The engine water and the raw water are in completely separated, closed systems and they never intermix. The heat exchanger may have an oil cooler for power units having a hydraulic transmission.

CHAPTER 27

Automotive Diesel Engine

What are some of the problems of diesel engine installation in coaches and trucks?

Answer: The problems to be considered in any installation of this type (Fig. 27-1) are as follows:

1. Mounting the power plant.
2. The fuel system.
3. The exhaust system.
4. The cooling system.
5. The electrical system.
6. Engine controls.
7. Transmission systems.
8. Engine starting.

What accessories are included in the power plant?

Answer: The bare engine as delivered by the manufacturer is considered as complete with all engine accessories including the scavenging-air blower, the fuel feed pump, water pump, governor, etc. (Fig. 27-2 and 27-3).

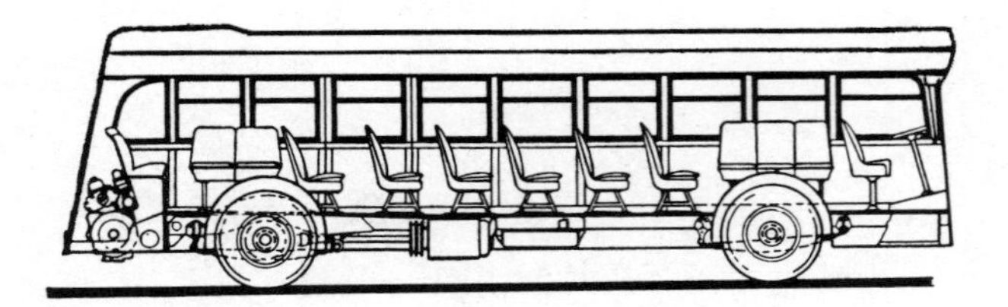

Fig. 27-1. Coach powered by a rear-mounted diesel engine.

Courtesy Detroit Diesel Engine Division, General Motors

Fig. 27-2. Model 8V-71 General Motors diesel engine used on large interstate and cross-country trucks. The standard 8V-71 engine produces 290 horsepower and weighs 2,345 pounds.

Why is engine suspension important in the mounting of diesel engines?

Answer: High compression causes the torque variation of a diesel engine to be somewhat greater than that of a gasoline engine of equal size. At full load the difference is not very serious, and the

Courtesy Detroit Diesel Engine Division, General Motors

Fig. 27-3. Model 12V-71 General Motors diesel engine. This engine is rated at 432 horsepower, weighs 3,300 pounds and is used on large turnpike cruisers.

uneven torque of the diesel engine is noticeable only at the lowest speed. At idling speed, however, the roughness of the diesel can be quite objectionable, and it is necessary to suspend the engine in such a way that its torque impulses cannot be transmitted freely to the chassis.

A considerable amount of research has been made by the various manufacturers to obtain the most satisfactory engine mounting, but in spite of this the design of a satisfactory engine suspension calls for considerable thought and experience. Insulation of the power plant from the chassis is not enough. It is necessary, therefore, to design a suspension system that provides for a definite rocking movement so that the variations in torque may be absorbed as kinetic energy in the whole power plant mass.

When installing diesel engines in trucks the same engine

Courtesy Caterpillar Tractor Co.

Fig. 27-4. Caterpillar Model 1673 truck engine with cutaway view of air-system lines.

suspension problem prevails. In this instance, however, the rear support is placed above the transmission and the power plant is carried at this point by a single large block. As the propeller shaft torque reaction may be very large and practically coincides with the principal axis, the resistance of the suspension must increase very rapidly once the working range is exceeded.

For this reason, separate, adjustable snubbing blocks are placed on each side of the flywheel housing. When the engine is idling, these blocks have clearance and have no restraint on the engine movement but, under load, the power plant rocks over until these blocks make firm contact with their rubber buffers to prevent further movement. As the load is applied only at comparatively high engine speeds, the vibration transmitted through the snubbing blocks is not objectionable.

In truck installations, it is necessary to make sure that the smoothness of the clutch engagement is not going to be upset by the engine movement. There should be no servo action or fore-and-aft movement where clutch control linkage is shortened, causing slipping of the clutch, which produces clutch chatter and is, of course, objectionable.

How is noise prevented in coaches powered by diesel engines?

Answer: Because of its high maximum pressures and the rapid increase in the cylinder pressure just as combustion starts, the diesel engine is inclined to be considerably noisier than a gasoline engine of equal power output. With careful design, it has proved quite possible to keep all objectionable engine noise out of the body and to obtain a noise level actually slightly lower than that of the same coach powered by a gasoline engine.

Noises may be transmitted mechanically or they may be air borne, either across the engine compartment or by pulsations in the air-supply system. Mechanical noises are prevented, of course, by the rubber-mounted suspension system, but it is very necessary to make sure that parts attached to the power plant—exhaust, tail pipes, compressed-air pipes, and so on—are completely insulated from the frame. Air-borne noises are confined to the engine compartment by the usual methods of insulation.

Briefly, the treatment consists of insulating the engine compartment with a one-inch layer of rock wool held in place by light steel panels, making sure that no openings are left in the rear of the body. Air-duct noises are usually dealt with by lining the ducts with absorption material and by fitting a resonant chamber on the air pipe close to the blower intake. The noise level outside the coach is reduced best by closing all possible openings in the engine compartment.

The present standard for coaches is that the noise level, measured in the center of the rear seat at the height of the passenger's car, should not exceed 87 decibels at 40 mph. It is generally found that exhaust noise is more troublesome to deal with than either engine or fan noises, partly because its level is high (87 decibels measured 10 ft. behind the coach), and partly because it finds its way into the coach through the windows.

Describe the fuel system of automotive diesel engine.

Answer: The real difference between a gasoline engine and a diesel engine is in the fuel system. The fuel system of a diesel engine can be compared closely with the electrical system in the gasoline engine. In the electrical system, the spark is timed to the spark plug. In the diesel engine, the fuel is timed into the injector and then into the cylinder after pure air has been compressed to a high pressure, ignition being spontaneous because of the heat of compression. The fuel system of a typical truck diesel engine is shown in Fig. 27-5.

The complete fuel system starts with the fuel line running from the fuel tank through a primary filter, through secondary filters of progressive fineness to the intake of the high-pressure pumps which are assembled in a multiple unit at the side of the crankcase, or they are separately installed in the individual cylinder heads. The high-pressure injection pumps deliver the fuel to the spray nozzles in the individual cylinders by which the fuel oil is sprayed into the combustion chamber.

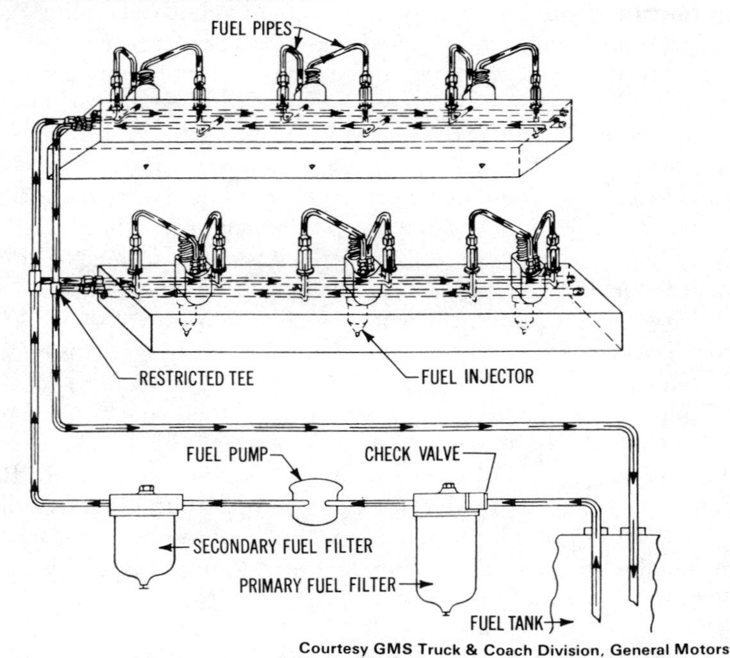

Courtesy GMS Truck & Coach Division, General Motors

Fig. 27-5. Schematic of a typical fuel system of General Motors diesel engines.

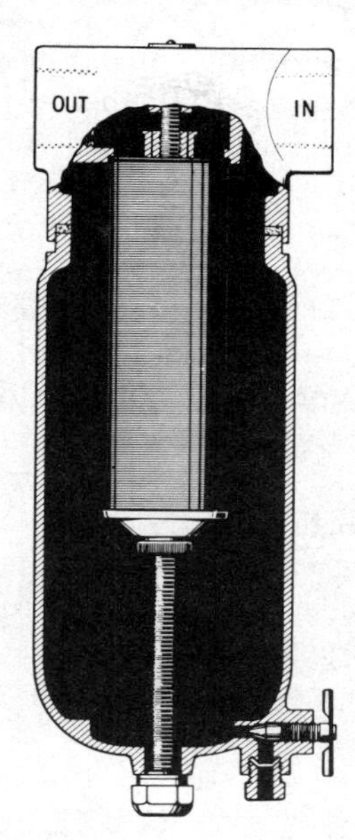

Fig. 27-6. Fuel oil strainer (primary filter) used on Waukesha diesel engines.

Courtesy Waukesha Motor Co.

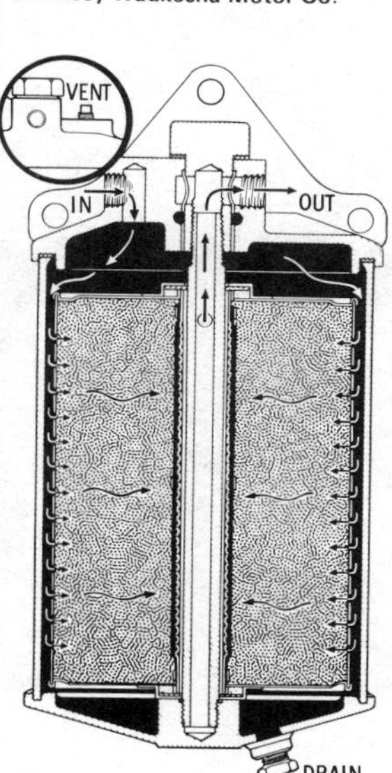

Fig. 27-7. Fuel oil filter used on Waukesha diesel engines.

Courtesy Waukesha Motor Co.

In all diesel fuel injection systems, a surplus of fuel is fed by the transfer pump to the high-pressure pumps in order to maintain a small pressure head on the intake and also to eliminate any possibility of the fuel becoming vaporized from heat and thereby causing a vapor lock, or if not a vapor lock, at least a disturbance of the exceedingly accurate metering which is required of the fuel system in any diesel engine (Figs. 27-8 and 27-9). The spray nozzles, too, are purposely designed so that there is a slight leakage in the overflow from the high-pressure pumps; this drip from the nozzles (Fig. 27-10) is conducted through return tubes back to the fuel tank.

When installing the fuel system, every precaution should be taken to make sure that the fuel tank and fuel pipe lines are absolutely clean when installed. Steel pipe should be used throughout because it has been difficult to obtain leak-proof joints with copper tubing.

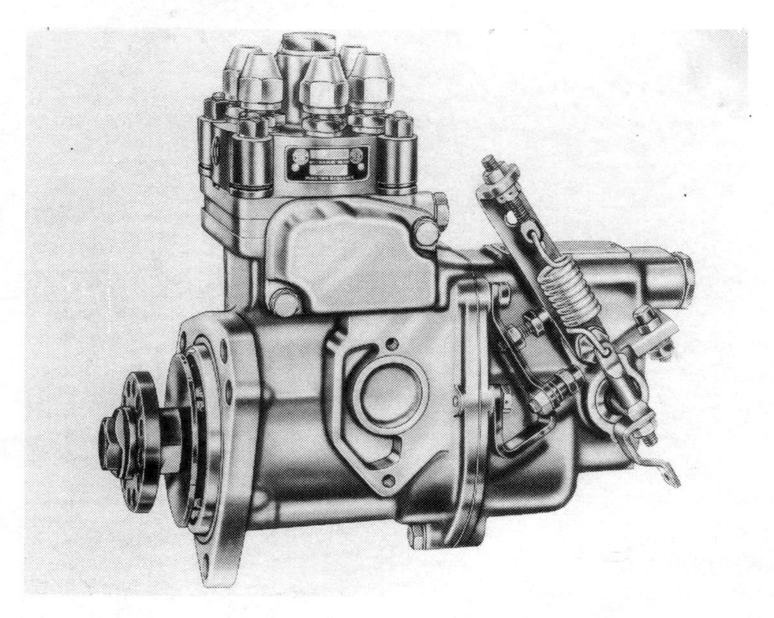

Courtesy American Bosch Arma Corp.

Fig. 27-8. American Bosch PSB fuel injection pump with external-spring governor used on diesel engines.

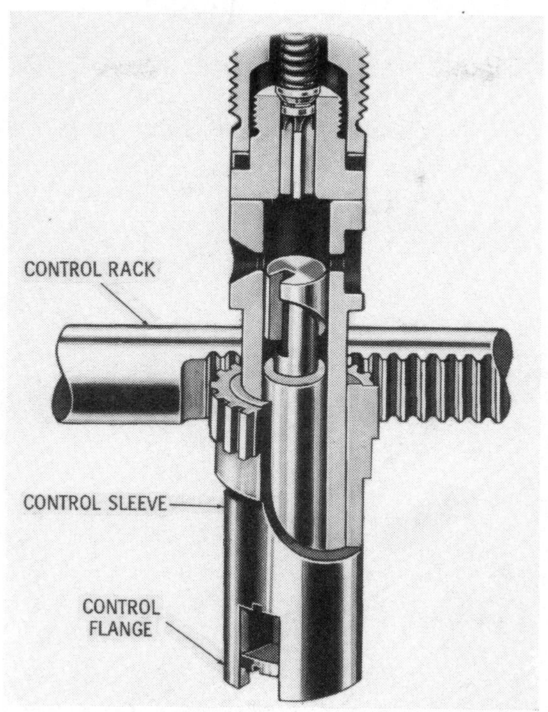

Courtesy American Bosch Arma Corp.

Fig. 27-9. Sectional view of fuel injection pump element. The plunger is rotated by its flanges, which engage slots in the control sleeve. The control-sleeve gear is actuated be the control rack.

How is air supplied for the automotive diesel engine?

Answer: All air taken into the engine must first pass through the air filter, which functions to remove all dirt or dust from the air before it passes into the inlet manifold. In truck installations the air filters are mounted on the blower intake or, in cab-over-engine trucks, on the dashboard. A resonant chamber is built into the base of the air filter.

Illustration of a typical coach air system is shown in Fig. 27-11. An air system of this sort must provide up to 685 cubic feet per minute of clean air, with a total maximum pressure of two inches of mercury. Of this quantity, a loss of about one-third is accounted for in the roof and pillar ducts, one-third in the air-filter casing and

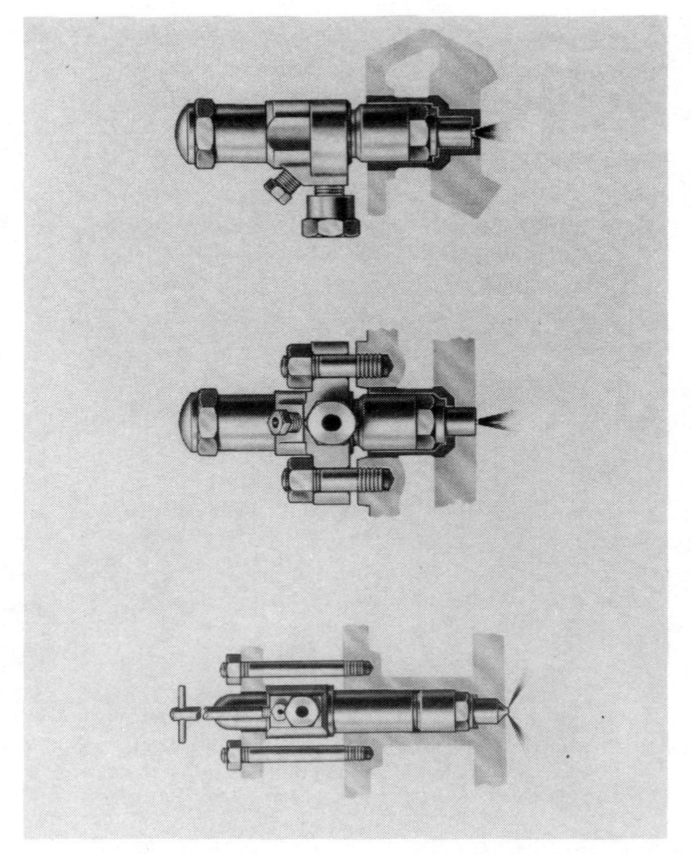

Courtesy American Bosch Arma Corp.

Fig. 27-10. American Bosch nozzle-holder assembly. The assembly should be carefully inserted so that the nozzle tip does not strike against the recess wall.

blower inlet pipe, and the remaining one-third is lost in the air filters themselves.

Air filters are always of the oil-bath type. To ensure maximum mileage between cleanings, the air supply in coaches is drawn from a point well forward on the roof. In designing the air inlet, special precautions should be taken to avoid drawing in rain.

The air-inlet system in addition provides clean air for the air compressor and is connected to the cylinder head of the engine by a breather pipe so as to provide a slight depression (from 1 ½ to 3 inches of water) in the crankcase.

Describe the exhaust system of automotive diesel engines.

Answer: The exhaust system, starting at the attachment of the manifold to the engine and ending with the muffler tail pipe, does not differ in any important respect from that required for the gasoline engine.

One special requirement is that, on account of the rocking tendencies of the power plant, the whole exhaust system should be carried on the engine.

This arrangement does away with the need for special insulated supports and ensures durability, while making it easier to remove the power plant from the frame. In truck installations, of course, some sort of outer support must be used, but fortunately this support need only provide for rotation. The alternative, to make use of flexible tubing, has not worked out so well, as the tubing has, at best, a very limited life.

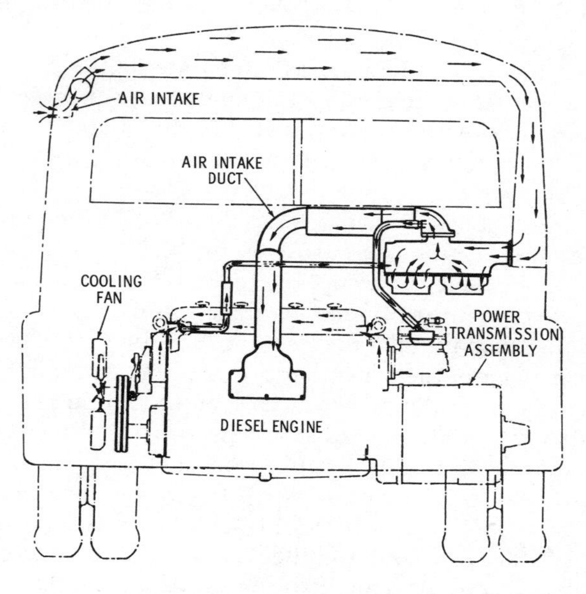

Fig. 27-11. Layout of the air system in a coach powered by a diesel engine.

It has been found that, on the six-cylinder engines of coaches, a double exhaust system with a separate exhaust manifold, exhaust pipe, and muffler for each set of three cylinders has certain advantages in preventing cross interference during the scavenging phase of the cycle. Apart from this consideration, the engine is not particularly sensitive to back pressure although a pressure of seven inches of mercury has been set arbitrarily as the maximum.

In addition to back-pressure test, each new installation is subjected to noise tests, as described already. Mufflers are always of the resistance-plus-resonance type and have special provision for cleaning out the perforations should this expedient prove necessary at times.

Describe the typical cooling system of automotive diesel engines.

Answer: The cooling system used with the diesel engine is also almost identical with that used in gasoline engines. The heat rejection of the diesel engine is only about 87 Btu per minute, per horsepower at full load; thus the radiating capacity need be only about 70 percent as large as that required for a similar gasoline engine.

The radiator is mounted as a separate unit at the front of the engine and is connected to the engine with rubber hoses. The temperature of the cooling water in the engine water jacket is regulated by thermostats located in a housing at the front of the cylinder head.

When the engine is cold these thermostats remain closed and cause engine water to circulate through the water jackets only, and when the engine warms up, the thermostats open and permit the water to circulate through the radiator as illustrated in Fig. 27-12.

Some coach installations have been made with a spring-loaded valve in place of the water thermostat to ensure a full flow of water to the heaters under all conditions, and a system that uses a valve sensitive to both temperature and pressure is under development.

One of the most serious objections that may be leveled against the conventional cooling system is the power absorbed by the fan in moving large quantities of air with low efficiency and a great deal of noise. In cold weather, none, or at any rate as little as possible, of this air is allowed to pass through the radiator.

To overcome this difficulty without facing the complication of

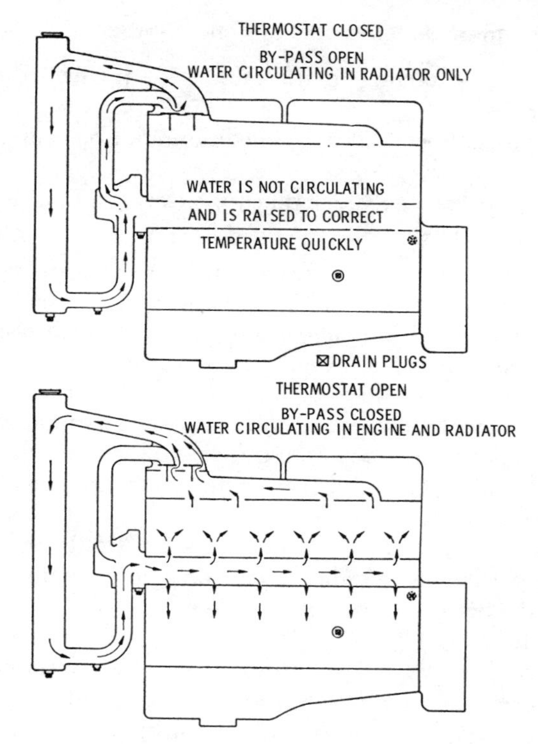

Fig. 27-12. A typical cooling system for a diesel engine. A thermostat is used to control the temperature in the cooling system.

thermostatically controlled clutches in the fan drive or such devices, provision has been made for the interchange of fan-belt pulleys in some instances. By shifting the fan pulley rims, the fan may be driven at either greater than or less than engine speed, according to temperature conditions. Another scheme that has worked well is to use two alternative fans, a large fan during the summer months and a smaller one during the winter.

Describe the electrical system typical of automotive diesel engines.

Answer: The electrical system of automotive diesels does not differ in any important respect from that employed in gasoline driven vehicles.

The function of the electrical system is:

1. To provide starting of the engine by means of a starting motor.
2. To preheat the intake air by means of ignition heaters located in the inlet ports.
3. To provide electrical current for exterior and interior lights and accessories.

The starting motor converts electrical energy into mechanical energy for cranking the engine. The generator is connected to the engine by means of a belt or other form of transmission and converts mechanical energy into electrical energy for supplying current for ignition, lighting, and charging of the battery for starting. The surplus electrical energy is stored as chemical energy in the battery for use when cranking the engine, or whenever the consumption of electrical energy by the connected electrical load exceeds the generator output.

The starting motor and the generator are of similar construction, consisting of three essential elements (armature, field magnets or coils, and the commutator with brushes) although their functions are opposite.

There are three switches in the starting motor circuit: a button switch, a series-parallel switch, and a magnetic switch. The button switch is pressed when it is desired to start the engine and simply operates the series-parallel switch.

The series-parallel switch is of the solenoid type. It has two sets of contacts, the charging contacts and the cranking contacts. The charging contacts can be closed to permit charging of the batteries, and the cranking contacts close the circuits to the starting motor when it is desired to start the engine. The construction of this switch is such that it does not permit both sets of contacts to be open or closed at the same time. One purpose of the series-parallel switch is to connect the two batteries in parallel so as to provide 12 volts for lights and accessories and to permit charging with the 12-volt generator, and the other purpose is to connect the batteries in series to provide 24 volts for the starting motor.

When the button switch closes the circuit, the series-parallel solenoid acts to push the contact plunger, opening the charging contacts and closing the cranking contacts. By this action, the batteries are disconnected from the charging circuit and connected in series to the magnetic switch.

A knob is provided at the bottom of the unit so that it can be manually operated in an emergency, in event the button switch or connection should fail.

The magnetic switch is also of the solenoid type, but it has one set of contacts only, the cranking contacts. When the series-parallel switch brings the series-connected batteries and the magnetic switch into the circuit, the solenoid of the magnetic switch closes its contacts and completes the circuit of the starting motor.

The function of the regulator is to control the output of the generator so as to prevent the output exceeding a safe value at any time and also preventing the discharge of the battery back through the generator when the latter either is running below generating speed or at rest.

The regulator consists of three separate units: a cutout relay, a current regulator, and a voltage regulator. They are mounted together on the same base and located on the dashboard. Each unit is fundamentally a magnetic switch, consisting of a wound core and a hinged armature that opens and closes contact points.

The function of the cutout relay is to close the circuit when the generator speed is sufficiently high to cause a current to flow, and to open the circuit when the generator runs either below generating speed or not at all, so as to prevent discharge of the battery back through the generator.

When the generator is at rest or operating at a low speed, the relay points are open. As the generator speed increases, the voltage increases, and when this reaches the value for which the unit has been set, the magnetic force draws the armature down, closes the contacts, and permits the current to flow to the battery.

The current regulator, which is the center unit on the base, acts when the electrical load is large and the battery is low in order to prevent the generator from exceeding its rated output. Normally the contacts are held closed by the springs on the armature, and the entire output flows through the circuit to the battery. When the current reaches the value for which the regulator has been set, sufficient magnetic force is set up in the regulator core to pull the armature down, open the contact points, and cause the current to flow through a resistance, thereby reducing the generator output.

The voltage regulator is designed to prevent the voltage from exceeding a predetermined value. When the battery is low, the voltage regulator contact points are closed and the unit does not operate. As the battery approaches full charge, the voltage builds up and forces sufficient current through the circuit to pull the armature down and open the contacts, thereby causing the generator field current to flow through a resistance and reduce the generator output.

There are two ammeters provided, main and auxiliary. The main ammeter is connected in series with the generator and the auxiliary ammeter is connected in the ground circuit. The provision of the auxiliary ammeter makes it possible to determine the charge that is being delivered to each pair of batteries. The current indicated by the auxiliary ammeter is half the total output of the generator.

Describe governor operation of automotive diesel engines.

Answer: Gasoline engines are controlled by direct connection between the accelerator pedal and the engine throttle. In a gasoline engine (if used at all), the governor serves no purpose other than to limit the speed.

In diesel engines, however, provision must be made for the comparatively large movement of the engine, which must therefore be governor controlled throughout the range of operation, the manual control being either partially or wholly dependent upon the governor for its effect (Fig. 27-13).

Why is a governor needed?

Answer: The need for a governor may be comprehended easily when it is considered that the medium controlled in gasoline engines is a highly elastic gas, whereas in a diesel engine it is practically an incompressible and inexpansible column of liquid. The result is that rapid movement of the throttle of a gasoline engine results in a somewhat sluggish response on the part of the engine, so that no particular harm is done other than to affect fuel economy to some extent. In the diesel engine, such rapid manipulation of the metering control would result in violent response on the part of the engine which would be destructive not only to the engine but to the driving parts of the chassis.

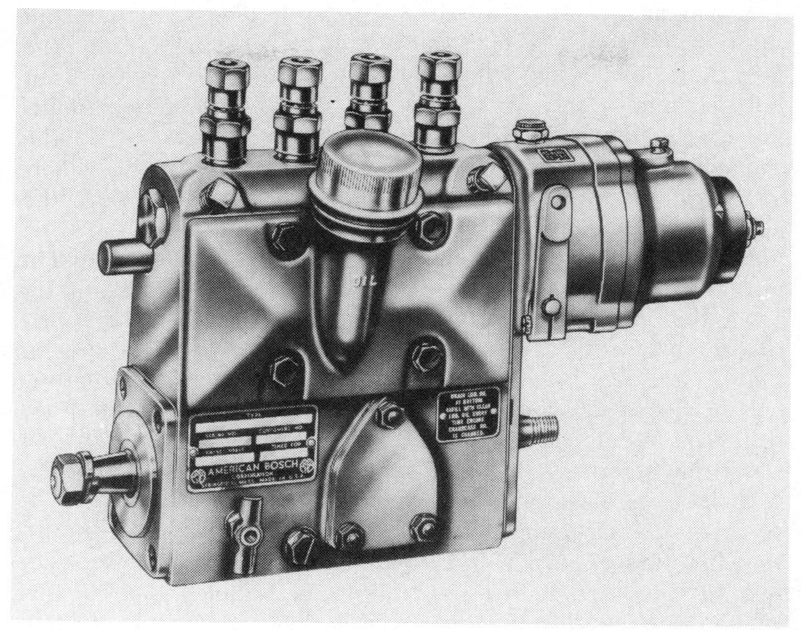

Courtesy American Bosch Arma Corp.

Fig. 27-13. American Bosch four-cylinder APE fuel injection pump with GPA pneumatic governor mounted on the end. The pneumatic governor is operated by air flow into the engine intake manifold.

By operating the controls through the medium of the governor, erratic movement of the accelerator is largely dampened out by the relatively slow response of the governor, and so the engine control is smooth and evened out.

Describe the mechanical governor as a means for control of automotive diesel engines.

Answer: Although various systems of governor action have been extensively tried in the development of automotive diesels, the majority of engines today employ a simple flyball or centrifugal governor operating directly upon the fuel pump metering control rod in such manner that the action of the governor in response to the speed of the engine is to reduce the amount of fuel fed the engine and thus to restrain its speed.

The manual control from the accelerator operates so as to vary the engine speed at which this action will take place. In some governors, this is accomplished by a movable fulcrum on the linkage from the governor to the metering control rod so that the governor becomes in effect a tachometer and takes effect upon the metering rod at the particular speed selected by the control. This arrangement works satisfactorily when properly put together and properly adjusted, except that as the precision of any governor varies with different speeds it sometimes lacks positiveness at a certain given speed.

Another and more simple arrangement is to provide a spring that acts against the governor. The tension is varied by the amount that the accelerator is pressed. Obviously, the greater the tension on this spring, the higher the speed of the governor must be in order to build up sufficient energy to overcome the resistance of the spring and move the metering control rod (Fig. 27-14).

There are also two methods in common use for setting the idling speed of a diesel engine. The most common is the use of a stop on the metering control rod, so that the minimum fuel fed is a quantity just sufficient to keep the engine idling at the selected speed. This method is objectionable, not only in that it fails to hold idling at a fixed speed because of differences in engine temperature, but also because it does not permit the fuel to be entirely shut off when descending grades and thus not only causes a continuous waste of fuel when coasting, but also interferes with engine braking.

A better method is used in connection with the spring-controlled governor just described, in which a second spring with a fixed adjustment is set to overcome the action of the governor at the exact speed at which it is desired to idle the engine.

With this arrangement, when the engine is cold and the fuel oil is stiff, a relatively greater amount of fuel oil will be fed to maintain idling at the selected idling speed. Whereas, as the engine warms up and the oil thins out, the amount of fuel fed for idling will be correspondingly reduced, and at all times the idling speed will be held exactly where it is wanted. In drifting, with the speed of the engine considerably above idling speed, this spring will have no effect, and the release of the accelerator will give the governor unrestricted control, so that the fuel will be shut off entirely until

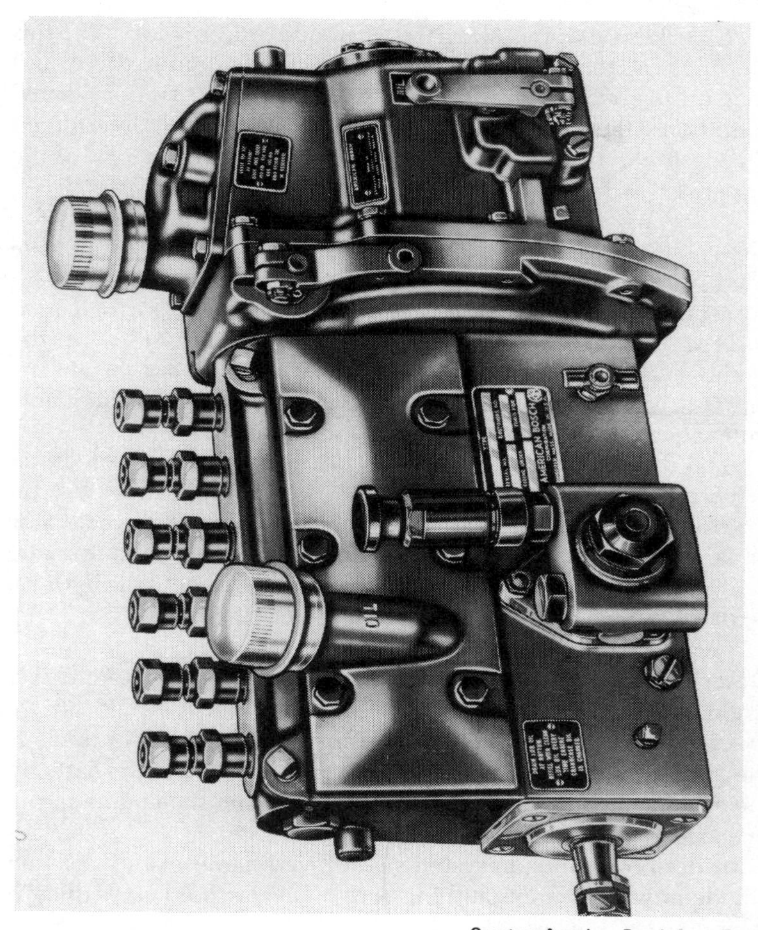

Courtesy American Bosch Arma Corp.

Fig. 27-14. American Bosch six-cylinder APE fuel injection pump with GVA mechanical governor.

either the speed falls to idling range or the accelerator is once more pressed.

Describe the fully automatic system of engine control for use with the hydraulic transmission.

Answer: A very ingenious fully automatic control system has been worked out for use with the hydraulic transmission. With this transmission, the coach is accelerated in the hydraulic drive, and once up to speed, a shift is made to direct drive. Under normal conditions, this shift is made automatically at a predetermined speed, smoothly and silently, without any move on the part of the driver.

To accomplish this shift, a governor driven by the transmission makes contact at the selected road speed. The current first sets the controls to the "No Fuel" position, regardless of the throttle opening set by the driver. Next the hydraulic-drive clutch is disengaged and the direct-drive clutch set ready to engage just as soon as the engine and drive shaft speeds synchronize. The very instant the direct drive is in, the engine throttle is opened again. The whole shift requires about one second to accomplish.

Should the driver wish to prevent the engagement of direct drive or to re-engage the hydraulic drive at any time, he can overrule the automatic controls merely by pushing his accelerator pedal right down. As the vehicle coasts to a stop, the automatic controls will disengage the direct drive and go over to the hydraulic drive (which has a free wheel) at about 12 mph.

Describe the transmission systems used with automotive diesel engines in trucks and coaches.

Answer: By definition, the transmission system of a vehicle is the gear, including the change gear and the propeller shaft, by means of which the power is transmitted from the engine to the live axle.

Modern transmission systems employed in diesel and gasoline engine powered trucks and buses may be classified according to their construction as:

1. Semiautomatic transmissions.
2. Hydraulic transmissions with both manual and automatic controls.
3. Synchromesh transmissions.

A semiautomatic transmission system is basically a three-speed and reverse planetary unit in which the usual friction bands and friction clutches have been replaced by centrifugally operated jaw clutches. It is semiautomatic in that the normal sequence of forward speed changes is effected by the operation of the accelerator pedal and is, therefore, under the driver's control. The shifts to lower ratios occur automatically provided the torque transmitted from the engine to the wheels or vice versa is small.

The semiautomatic transmission consists of a two-step planet with first and reverse sun-gear drive shaft and concentric second-speed quill drive shaft. In first and second speeds, the planet reaction is taken by means of a roller clutch. In direct drive, the entire plane sytem is locked and revolves as a whole, the planet roller clutch freewheeling. Reverse is obtained through the same planet system by disconnecting the planet roller clutch and locking the second-speed quill shaft. The planet then revolves in a reverse direction.

To permit the overrun of the first-speed drive shaft when the transmission is operated in second gear, another roller clutch is interposed between the first-speed shaft and the main clutch of the vehicle.

Second-speed and direct-drive engagement are effected by means of separate centrifugal dog-type clutches that can engage only when the speeds of the meshing parts are synchronized.

A forward–neutral–reverse lever is provided on the dash and along with the accelerator pedal, they are the only normal controls required. For emergencies, a foot button is provided which permits an instantaneous shift from direct drive to second; this shift is accomplished by an air-operated cylinder which forces the direct-speed clutch out of engagement.

When the vehicle is stationary, assuming that the control lever is in the forward position, with the engine idling, the second, direct, and main clutches are disengaged and the transmission is in low gear.

To start the vehicle, the accelerator is depressed; the centrifugal main clutch then engages softly at about 600 rpm. As the vehicle gains momentum, the second-speed clutch is actuated and the faces of the engaging teeth are brought lightly into contact.

To shift from first to second gear, the driver releases the accelerator, allowing the engine speed to decrease until the mating teeth of the second-speed clutch synchronize, at which time engagement takes place.

To shift from second to direct-drive gear, the driver releases the accelerator pedal, eliminating the reaction torque which has held the planets and carrier stationary. The planets, driven by the momentum of the vehicle, revolve in the same direction as the engine actuating the direct-drive clutch and when the planet and drive shaft speeds synchronize, the direct-drive clutch engages.

To shift from direct drive to second gear, the driver releases the accelerator pedal and when the vehicle speed has decreased to a predetermined point, the direct-speed clutch disengages.

To shift from second gear to first gear, exactly the same procedure is used as in shifting from direct drive to second gear.

Describe the operation of hydraulic transmissions in trucks or buses.

Answer: A hydraulic transmission system has been developed by the Yellow Coach and Spicer Organizations. The mechanism comprises:

1. The friction clutch, which connects or disconnects the engine from the turbine.
2. The turbine, including the pump wheel, the rotor, and the turbine housing with its two sets of reaction blades.
3. The reverse gear.
4. The roller clutch, which disconnects the turbine from the angle gear output shaft when in direct drive.
5. The direct-drive lockup or clutch, which connects the through shaft to the angle gearing.
6. The angle gearing.

The turbine (Fig. 27-15) consists of the following parts:

1. Centrifugal-pump wheel, driven by the engine.
2. A three-bladed, three-stage rotor wheel, connected to the propeller shaft.
3. Casing, with stationary blades interposed between the rotor blades.

Torque multiplication reaches a maximum of 5:1 when the bus is started. This ratio is gradually reduced as the vehicle gains speed.

The operation of the turbine can be likened to the operation of a planetary transmission in which the pump wheel is represented by

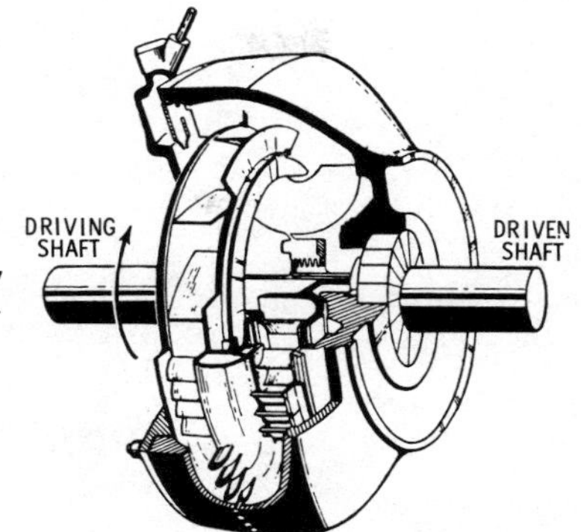

Fig. 27-15. Turbine section of Yellow Coach hydraulic transmission.

the sun gear; the turbine, by the planet gears; the reaction vanes, by the stationary ring gear; and the gear teeth, by a fluid. To complete the analogy, the gears must be imagined as being infinitely flexible and capable of automatically changing their relative size in order to provide the required speed and torque ratios.

Fig. 27-16 shows the liquid flow schematically. As a simplification, the engine-driven pump wheel has been replaced with a conventional centrifugal pump. The fluid, after having been drawn from the reservoir by the pump, is forced against the rotor blades (1) exerting a thrust tending to rotate the moving element. The fluid then passes to the stationary blades (2), which direct it against the rotor blades (3); the stationary blades (4) again change the flow of the liquid directing it against the rotor blades (5). The fluid is then picked up by the pump and the cycle is repeated.

Fig. 27-17 shows the construction of the friction clutch, which engages the turbine element. There are no coil springs, levers, or fulcrum pins. A simple conical spring plate functions in their place. No force is required to hold the clutch in the released position. There are no levers or pins to wear; there are no adjustments. The pressure is distributed evenly over the entire frictional area, thus eliminating distortion. Engagement is smooth and the release is complete.

Essentially, the clutch mechanism comprises the two friction faces, the fabric-lined driven member, the conical-spring plate, and the release-bearing mechanism. The fabric-lined driven member is connected to the turbine impeller through a splined

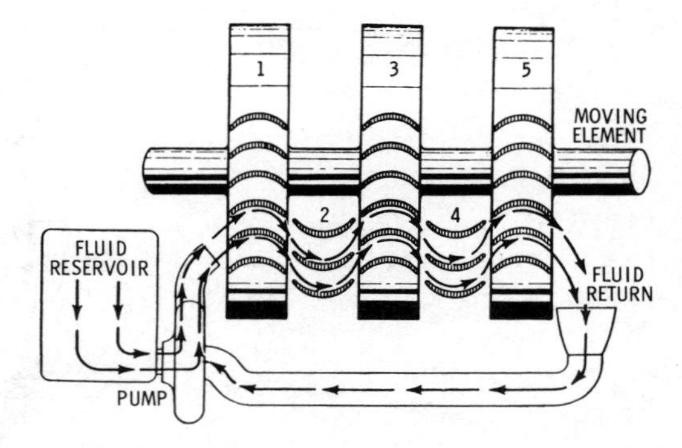

Fig. 27-16. Schematic of Yellow Coach hydraulic transmission.

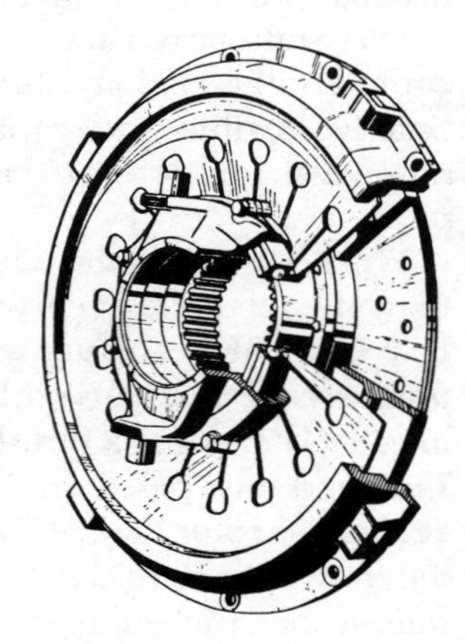

Fig. 27-17. Friction clutch employed in a Yellow Coach transmission.

sleeve, which is mounted on ball bearings. Large annular pockets are provided for the lubrication of the release bearing and also the turbine impeller sleeve bearings.

How does the electropneumatic control system work in connection with the hydraulic transmission?

Answer: A very important development in connection with the hydraulic transmission is the electropneumatic control system. At the correct road speed a simple governor energizes the control mechanism, which automatically performs the sequence of operations necessary to change from hydraulic drive to direct drive or vice versa. There are two important advantages:

1. The driver is relieved of all mental and physical effort.
2. The shift is made at the proper ratio to obtain maximum efficiency.

At a predetermined speed, a governor, by means of a switch located in its housing, provides electrical contact through the direct-drive switch to a solenoid that closes the engine throttle and to a magnet valve that supplies air to release the hydraulic-drive clutch cylinder and to engage the direct-drive clutch cylinder of the vehicle.

The movement of the direct-drive clutch into full engagement breaks the electrical contact through the direct-drive switch, de-energizes the throttle solenoid, and releases the throttle to its original position for direct drive.

Direct drive is maintained until the vehicle slows down to 17 mph, at which time electrical contact in the governor switch is broken, the magnet valve is de-energized and air is released to allow disengagement of the direct-drive clutch and engagement of the hydraulic-drive clutch.

When a vehicle is cruising at a speed above 17 mph in direct drive and rapid acceleration is required, an instantaneous shift to hydraulic drive may be made, thus taking advantage of the torque multiplication which permits higher engine speeds and faster acceleration. To accomplish this shift the driver merely applies a heavier-than-normal foot pressure on the accelerator, which brings the overrule switch into action, de-energizing the magnet valve, which in turn releases the air from the direct-drive and main-clutch cylinders.

How does the syncromesh transmission operate?

Answer: The synchromesh transmission system is a General Motors development and was first incorporated in their Cadillac cars.

Synchromesh, although important from the standpoint of the automobile, is vastly more important from a bus and truck standpoint. The larger the vehicle, the greater are its advantages. Without synchromesh, proper gear shifting is dependent entirely upon the judgment and skill of the driver, but, with a poor shift in a bus or truck, the potential mechanical damage is much greater than with an automobile.

Then too, as compared with the automobile, where a gear shift is improperly made, delayed or missed, the accident hazard is greater. This condition is particularly true when climbing long grades. Synchromesh has contributed much to the safety and easy handling of the bus, and history has repeated itself with the truck.

With large capacity trucks or buses and especially considering truck trains, the power-to-weight ratio is low, while the ratio differences between the transmission speeds are high. Consequently, anything that reduces the time required to shift from one speed to another is advantageous, particularly where vehicles are operated in mountainous country.

Vehicles are sometimes forced to climb long grades at a much lower speed than the capability of the engine, simply because the time required for shifting into the higher gear slows the vehicle to a speed at which it cannot be picked up in the next higher ratio. With synchromesh, a shift of this kind can be made without difficulty, but this procedure, if habitually followed, will affect adversely the life of the synchronizer parts.

The parts involved in the synchronizer element are: main shaft gear, main drive gear, bronze friction cone, synchronizer sleeve, shifting collar, synchronizer-ball spring, synchronizer ball, jaw clutch, and shifting collar pin.

The gear shaft is connected to the jaw clutch by means of splines. The four star-shaped ends of the jaw clutch pass through slots in the synchronizer sleeve, which is located or centralized in the neutral position by the four balls and springs carried in the jaw clutch. The shifting collar is pinned to the jaw clutch; consequently, regardless of the position of the jaw clutch on the main shaft spline, all these parts must rotate or remain stationary with the main shaft.

When the shift fork is moved, an axial force is exerted on the synchronizer sleeve through the four spring-loaded balls, which form a semi-rigid connection between the jaw clutch and the synchronizer sleeve. This force is proportional to the spring load on the balls. The synchronizer cones are then brought into contact. The resultant friction tends to rock the synchronizer sleeve; then the star-shaped ends of the jaw clutch position themselves against either the backward or forward faces of the central sleeve notches. The continued movement of the shift fork causes the four star-shaped ends of the jaw clutch to bear against the inclined surfaces of the synchronizer-sleeve notches. A like force is exerted against the synchronizer cones and this force is proportional to the change in the speed of the synchronizer cones.

When the speeds of the synchronizer cones are equal, there is no load on the jaw clutch to restrain the axial force; hence, the four

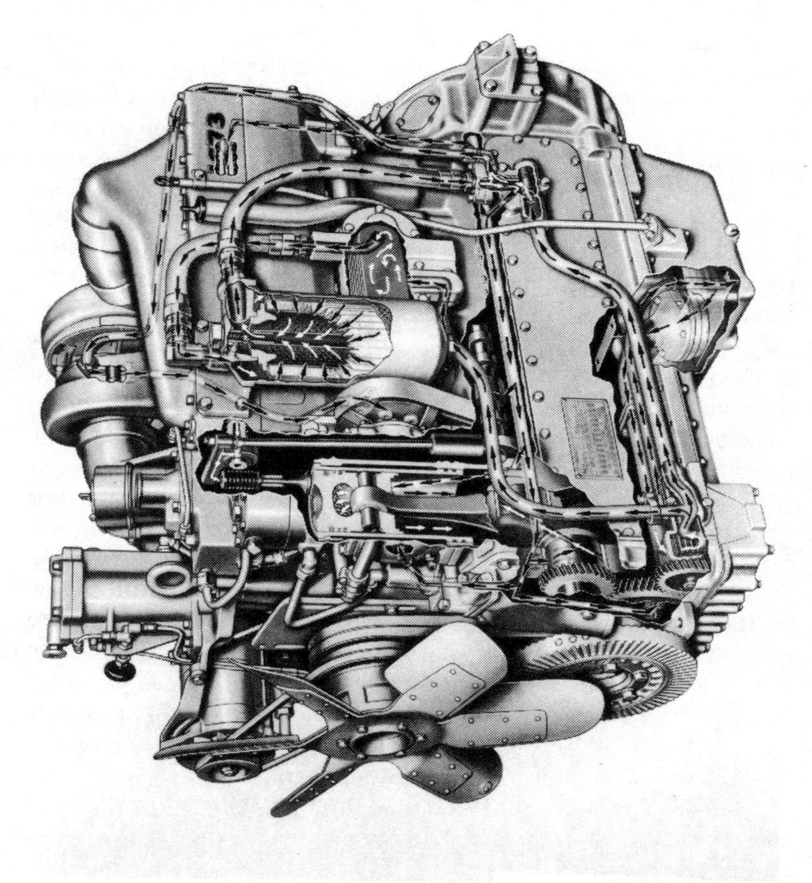

Courtesy Caterpillar Tractor Co.

Fig. 27-18. Caterpillar Model 1673 diesel truck engine with oil-flow cutaway.

star-shaped ends of the jaw clutch no longer bear against the inclined surfaces of the synchronizer-sleeve notches. Consequently, the continued movement of the shift fork causes the four star-shaped ends of the jaw clutch to travel along the synchronizer-sleeve slots until the gear and shaft are coupled mechanically by the meshing of the jaw clutch external teeth with the internal teeth of the gear.

It is impossible to clash the gears in shifting because, while there is any differential in speed between the synchronizing cones, the retarding force preventing engagement is always exactly proportional to the force exerted by the driver on the shift lever in the cab.

There are, of course, other synchronizer systems, some of which have been employed extensively in connection with automobile transmissions, but the inertia-check arrangement is the only method evolved to date that can be applied to heavy-duty vehicles with entirely satisfactory results.

In a typical Yellow Coach synchromesh transmission, three forward speeds and reverse are provided. The torque capacity is 550 foot-pounds. All gearing except reverse is constant-mesh helical type. Splash lubrication is employed. The output shaft is located at the rear of the transmission and engagement is effected through a spiral-bevel pinion gear. With the exception of reverse, all gears are equipped with General Motors inertia-check-type synchronizers.

In a typical truck synchromesh transmission, five forward speeds and reverse are provided. The torque capacity is 400 foot-pounds. All gearing except low and reverse is constant-mesh helical type. Lubrication is by pressure feed from an oil pump driven by the countershaft. The system is basically similar to the method employed in connection with a four-speed constant-mesh coach transmission. With the exception of low and reverse, all gears are equipped with General Motors inertia-check-type synchronizers.

Discuss the problem of starting diesel engines.

Answer: The starting of diesel engines presents no problem at summer temperatures, and when in proper condition, the diesel engine starts with far greater reliability than the average gasoline engine. This is because in a diesel engine the fuel reaches the cylinders on the first turn of the crankshaft, whereas in the gasoline

Courtesy Caterpillar Tractor Co.

Fig. 27-19. Caterpillar Model D330 automative diesel engine.

engine, many turns are often required before a sufficiently rich mixture is available for combustion.

During cold weather, however, their positions are somewhat reversed. Although gasoline engines present all too well-known difficulties in starting, starting is nearly always certain if the choke is used. In the diesel engine, on the other hand, starting demands that the air in the combustion chamber reach ignition temperature before a start is possible.

Variable injection timing has proved highly effective in improving the starting characteristics of diesel engines, even when not of a type that normally starts well. The reason for this is that the temperature of the air at injection depends upon the timing of injection. With fixed injection, the timing is necessarily early, so that it occurs before the piston has fully compressed the air, while the air is still relatively cool. With variable injection, the timing is later,

Courtesy Caterpillar Tractor Co.

Fig. 27-20. Caterpillar Model 1693 diesel truck engine.

at starting, so that it occurs well toward the end of the compression stroke when the air is at a maximum pressure.

What methods of starting are used on automotive diesel engines?

Answer: Various auxiliary methods of assisting cold-weather starting are used on diesel engines, but in automotive diesel engines only two methods are regularly employed: glow plugs and intake-air heaters. Both of these devices are electrical, drawing current from the battery.

Glow plugs serve to heat the air in the combustion chamber directly and are exposed to the destructive effort of the flame at all times. They are a necessity on all separate combustion-chamber types of engines, but are so unsatisfactory that strenuous efforts are being made to dispense with them. Not only is the necessity of their operating during cranking a serious handicap, due to the extra load upon the battery, but their exposure to the combustion results in serious curtailment of their life.

Electric manifold heaters are of two principal types, both of which serve to heat the incoming air before it is introduced into the combustion chamber. They are not satisfactory for separate combustion-chamber types of diesels, but work well with open combustion-chamber, air-cell, and Lanova-type engines.

In one type, a single element of large capacity heats the entire contents of the manifold, as well as warming the manifold itself. In the other type, separate heaters are disposed in each of the intake ports, so that much higher heating efficiency is provided along with more uniform heating of the charges for all cylinders. In both types, all current consumption occurs before the starter comes into action, so that full cranking power is available.

A recently perfected timing switch automatically times the duration of heating, according to the atmospheric temperature, and by controlling the starter-relay circuit prevents simultaneous heating and starting while insuring the earliest start feasible.

CHAPTER 28

Marine Diesel Engine

What types of diesel engines are used for marine purposes?

Answer: Marine diesel engines do not differ in any important respects from those used for stationary purposes. They are built in various sizes from a few horsepower up to thousands of horsepower, as required, to furnish propulsion for small boats and oceangoing liners (Fig. 28-1).

Diesel engines employed in towing and dredging services may be rated from 100 horsepower to 1,000 horsepower or more, depending upon service requirements.

Give the advantages and disadvantages of two- and four-stroke-cycle diesel engines for marine purposes.

Answer: Diesel engines for ship propulsion purposes are being manufactured in two general types, designated as either two-stroke-cycle or four-stroke-cycle, according to the cycle of operation. The two-stroke-cycle type is preferred for very large ship installations because of the savings in floor space and in weight per

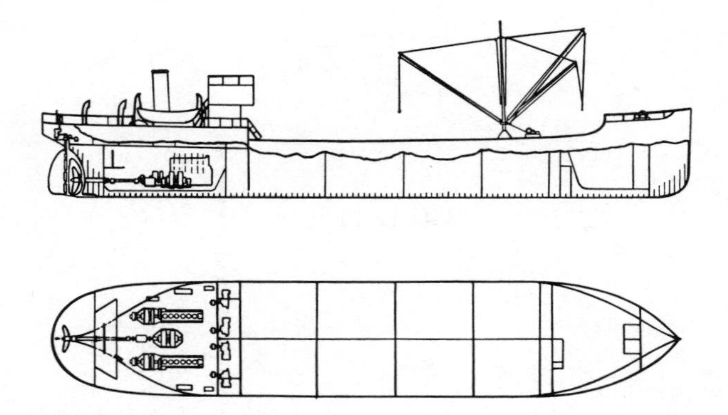

Fig. 28-1. Arrangement of diesel engines in a typical marine installation. The engines are directly connected to electrical generators whose electrical energy is converted to mechanical energy by means of interconnected electric motors, which in turn drive the propeller shaft.

horsepower, with a corresponding low cost of manufacture.

The four-stroke-cycle engine, on the other hand, is slightly more efficient in operation, as the two-stroke-cycle engine requires an auxiliary air pump for scavenging and refilling the cylinders with air (Fig. 28-2).

This air pump outside the cylinder requires considerable power for its operation, as it must supply a volume of air greater than the volume of the cylinder for thorough scavenging and this reduces the mechanical efficiency of the engine.

The two- and four-stroke-cycle engines may be further classified according to their particular construction as:

1. Double acting.
2. Single acting.
3. Opposed piston.

It should be observed that the terms two-stroke-cycle and four-stroke-cycle have no relation to the heat cycle upon which the operation is based, as either one or both may utilize the same heat cycle in its operation.

Courtesy Burmeister & Wain American Corp.

Fig. 28-2. B&WO-Lathrop Model D-45-V, three-cylinder, four-stroke-cycle marine diesel engine has fresh-water cooling and a 12-volt, 40-ampere alternator as standard equipment.

What is the principal advantage of the double-acting marine diesel engine?

Answer: This type of engine is found only in two-stroke-cycle installations. Principally, the double-acting engine has fuel oil injected into the cylinder by two separate fuel valves, one located in the cylinder head and the other located in the cylinder wall below the lowest position of the piston.

In the design of diesel engines it was early recognized that if both ends of the piston cylinder were closed and if both ends of the piston were utilized as pressure surfaces, the amount of power obtainable from a given size of cylinder could be approximately doubled.

Because the cylinders of double-acting engines are closed at both ends, however, difficulties are experienced in keeping them lubricated and in cooling the pistons. Also the packing of the piston rods is subjected to both heat and pressure and must be renewed frequently to keep it from leaking.

The scavenging air required for driving out the remnants of the exhaust gases and for filling the cylinders with a change of fresh air is usually provided by electrically driven compressors.

While the double-acting engine has been developed to the point where it is entirely reliable and is widely used for marine propulsion, it has not supplanted the single-acting engine. It has found its application where a high concentration of power on a single shaft is required or where fore and aft room is restricted.

Describe the operation of the opposed-piston diesel engine.

Answer: An opposed-piston engine may be either the two- or the four-stroke-cycle type and it is equipped with two pistons for each cylinder. The lower piston is attached to the crankshaft in the usual way, whereas the upper piston is attached by means of a rod to a yoke above the upper end of the cylinder. The ends of this yoke are connected by side rods to connecting rods extending downward to two crank journals, one on each side of the crankshaft that is driven by the lower piston (Fig. 28-3).

In an engine of this type, the two pistons move inward on the compression stroke, compressing the air between them, and the space between the pistons forms a combustion chamber into which the fuel is injected. The combustion pressures then force the two pistons apart and as they travel in opposite directions to the ends of the cylinder, each acts on its own connection to the crankshaft, the center crank throw pushing downward and the two side crank throws pulling upward.

When reaching the end of their strokes, the upper piston uncovers the exhaust ports and the lower piston uncovers the scavenger ports and the gases are blown out of the cylinder.

Describe two systems for starting diesel engines.

Answer: As previously noted there are two methods of starting diesel engines, namely:

1. Compressed air.
2. Electricity.

The latter method of starting is used for small marine engines only

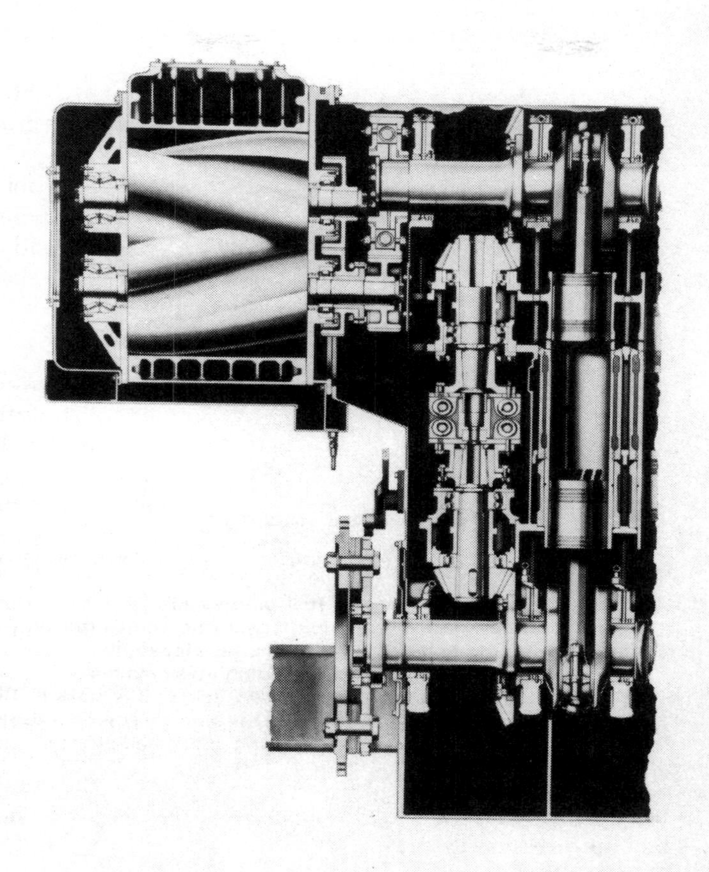

Courtesy Fairbanks, Morse & Co.

Fig. 28-3. Longitudinal section of Fairbanks-Morse Model 38D8 ⅛ opposed-piston diesel engine.

and usually consists of an electric motor receiving its source from an unfailing supply of power, usually a storage battery. It is similar in construction to that used in automotive diesels (Fig. 28-4).

The compressed-air system is universally employed for starting medium- and large-sized marine diesels. There are two principal types of compressed-air starting systems. They are:

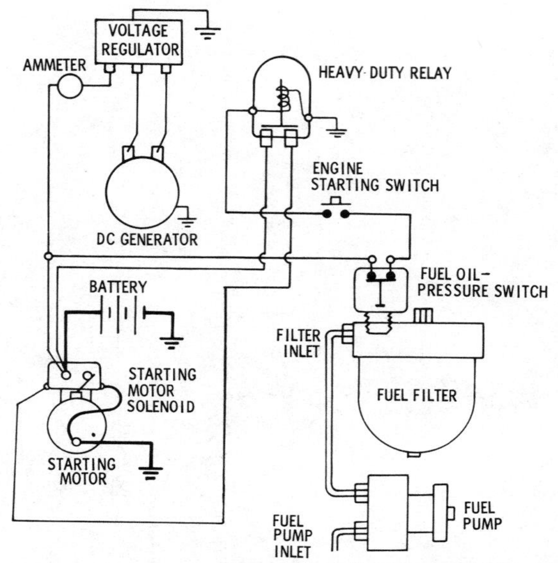

Courtesy Detroit Diesel Engine Division, General Motors

Fig. 28-4. Diagram of General Motores fuel oil-pressure switch, starting motor, and cutout-switch wiring. The automatic starter output device is recommended where the engine is located at a distance from the operator or where surrounding noise makes it difficult for the operator to determine when the engine has started. The fuel oil-pressure switch opens to break the starter solenoid circuit at a speed between maximum cranking speed and engine-idling speed.

1. The low-pressure system.
2. The high-pressure system.

In the low-pressure system, the air is stored at the starting pressure, while in the high-pressure system the air is stored at a pressure higher than that used for starting.

The low-pressure system (Fig. 28-5) consists of one or more air compressors (usually electrically operated) whose function it is to supply compressed air to the receivers at the required pressure. The compressed air is stored in receivers furnished with suitable gauges and valves as indicated.

The high-pressure system (Fig. 28-6) differs from the low-pressure type mainly in that the air is compressed to a higher

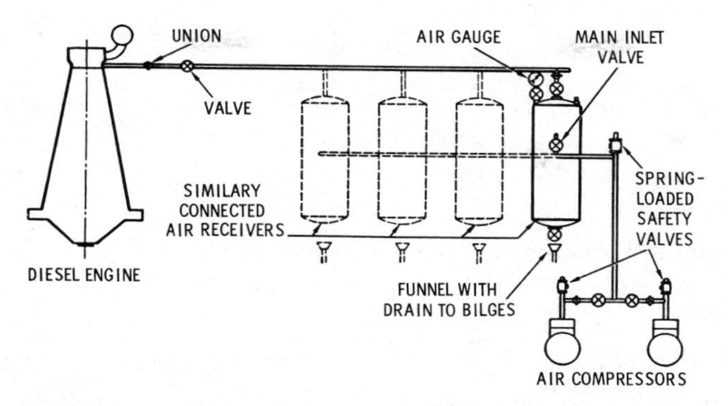

Fig. 28-5. Basic arrangement of low-pressure, compressed-air starting system.

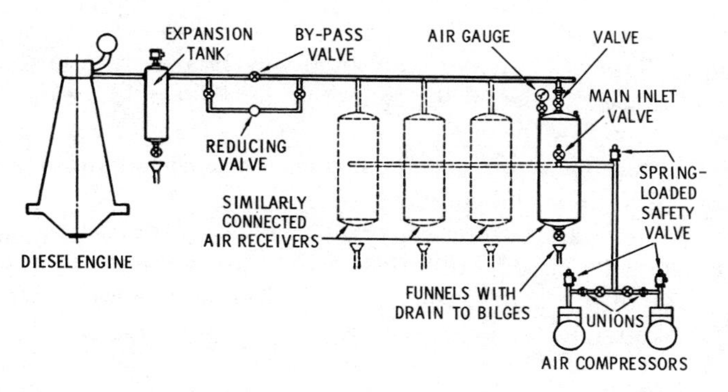

Fig. 28-6. Basic arrangement of high-pressure, compressed-air starting system.

pressure in addition to having a reducing valve installed between the air receivers and the engine.

Describe the cooling system for marine diesel engines.

Answer: Water is universally used as a medium for cooling the cylinders of diesel engines. Water was formerly used for cooling the pistons also, but the difficulties associated with leaking of piston-cooling water have resulted in the wide use of oil for piston cooling.

The method of circulation of water is by means of centrifugal or rotary pumps. For small- and medium-sized diesel engines, these pumps are directly connected to the engine, but for large diesel engines the pumps are driven independently, usually by electric motors.

Where the pumps are driven by the engine, it is customary to provide additional pumps, independently driven, so that oil and water service can be started before the engine is and continued after it is stopped.

What methods of engine cooling are used in marine diesel engines?

Answer: There are two methods employed in diesel engine cooling

1. Direct cooling of the engine or engines by the water surrounding the vessel.
2. Cooling the engine or engines by fresh water that is in turn passed through coolers after leaving the water jackets.

The first method involves pumping the water from the sea through the water jackets and then overboard. Although this method is the simpler of the two, it is seldom used, because the water surrounding the vessel is often charged with salt and impurities, resulting in clogging and scale deposits in the cylinder heads and on the cylinder liners.

Fresh-water cooling consists of a closed circulation system, and it is extensively used in all modern diesel installations. It has the advantage of preventing the accumulation of mud and scale deposits resulting from the direct use of salt water (Fig. 28-7).

Describe the cooling water pumping system.

Answer: The equipment comprising a modern system consists of a circulating pump of the centrifugal type usually in duplicate, a heat exchanger, a storage tank for fresh water, and a raw-water pump of the centrifugal type, to circulate sea water through the heat exchanger.

The raw-water pump may also serve to pass cooling water through the lubricating oil cooler. In some installations the pump that circulates the fresh water through the water jacket takes its suction directly from the return line from the jacket system, instead of from the storage tank.

When this arrangement is used, it is desirable to include in the system an expansion tank, placed in an elevated position that ensures that the system is full at all times.

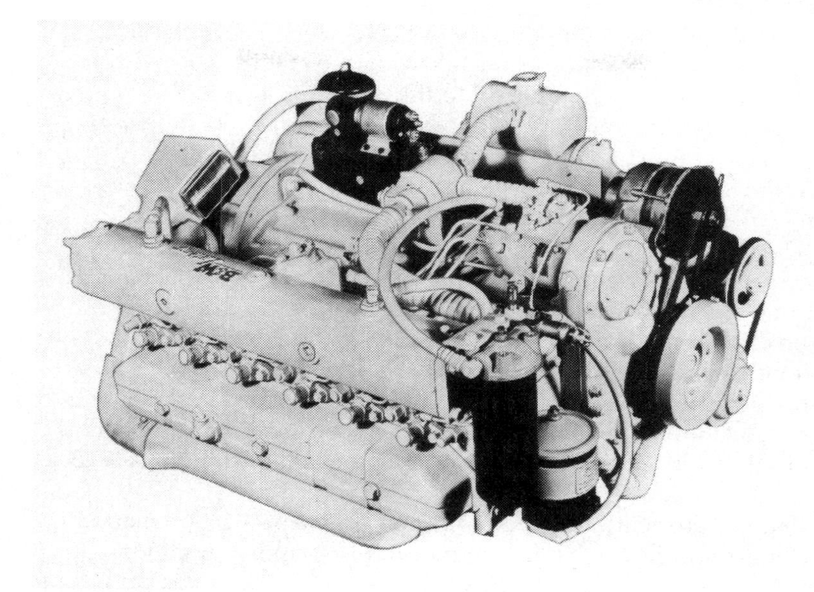

Courtesy Burmeister & Wain American Corp.

Fig. 28-7. B&W Lathrop Model D-110-H six-cylinder, four-stroke-cycle marine diesel engine has fresh-water cooling and a 12-volt, 40-ampere alternator and opposition rotation pairs as standard equipment. This is a low-profile horizontal engine measuring 21½ inches from the bottom of the oil pan to its highest point—12½ inches above the crankshaft center line—which permits installation in limited overhead space.

In multiple-engine installations, it is customary to cross-connect the cooling system and to provide duplicate equipment so that either system can be used on one or all engines.

In smaller installations, it is common practice to use attached centrifugal pumps of the reversible type, whose functioning is independent of the reversal of the engine.

It should be observed, however, that when attached pumps are used, auxiliary pumps should be provided that are independently driven. These are usually combination water and lubricating oil pumps that can be used to circulate both water and lubricating oil before the engine is started and also after it is shut down to prevent overheating.

Describe the lubricating oil system for marine diesel engines.

Answer: The function of the lubrication system is to provide an unfailing supply of suitable oil to the cylinders and the bearings. Cylinder lubrication has a twofold purpose: to maintain the required oil film to separate the liner surface from the faces of the piston rings as they slide up and down, and to act as a seal to prevent gases escaping past the rings.

The customary method of applying lubricating oil to the cylinder walls is by means of small pumps known as mechanical lubricators, with a common suction reservoir, and a discharge tube from each pump leading to a connection in the side of the cylinder through which the oil passes into a hole in the liner. The oil is supplied in measured quantities; the object is to supply only a sufficient amount to maintain a film of lubrication and continually replace the oil that is burned or escapes past the piston rings (Fig. 28-8).

Bearing lubrication is accomplished by means of a circulatory system through which a large amount of oil under suitable pres-

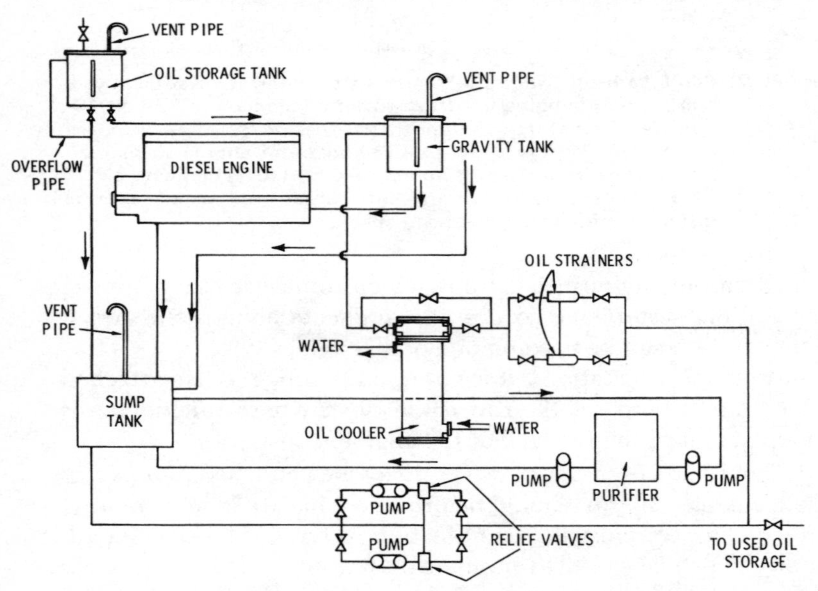

Fig. 28-8. Piping diagram of lubricating oil system of marine deisel engine having a gravity tank feed.

sure is forced through the bearings, then cooled, cleaned, and returned to the bearing.

In general, the lubrication system is made up of a sump or reservoir, from which a circulating pump draws the oil and discharges it through the coolers and filters to a manifold that has a branch leading to each main bearing.

Some of the lubricating oil flows out the ends of the main bearings, while the remainder of it passes through an axial hole in the crankshaft to the crank throw bearings. Here more oil is lost out the end of the bearings, and the rest passes up through axial holes in the connecting rods to the wrist-pin or crosshead bearings. From here, all the oil may be discharged from the bearings into the crankcase, or a portion may be passed through pipes to the interior of the pistons and thence to the crankcase or oil sump. From the crankcase or sump the oil is returned to the pump for recirculation (Fig. 28-9).

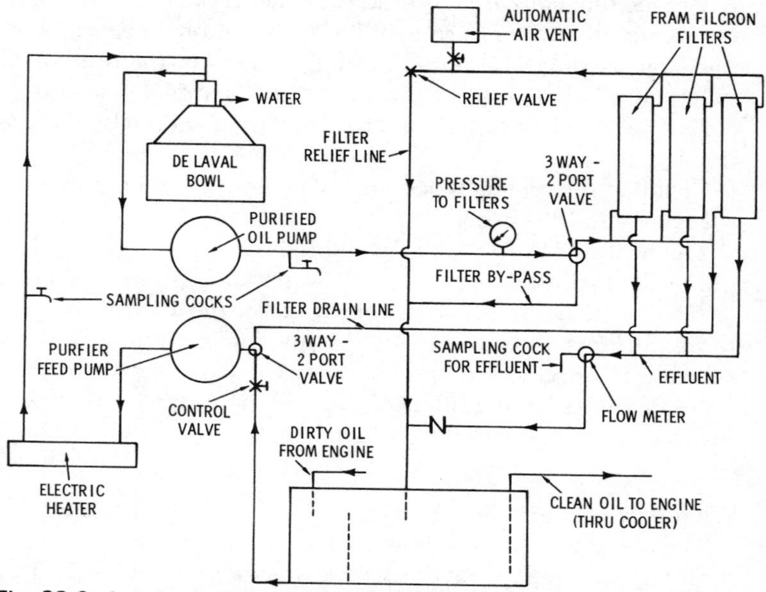

Fig. 28-9. Lubricating oil piping diagram for the DeLavel-Fram purifier system. This diagram shows a complete piping layout as well as pumps, valves, heater, filters, and centrifuge for a typical marine diesel installation.

Describe the systems of supplying fuel oil for marine diesel engines.

Answer: The fuel oil system as required for diesel operated vessels includes any or all equipment used to move the fuel oil from its storage points to the combustion chambers in the cylinder of the engines.

In practice such systems differ greatly, depending upon the size of the vessel and type of diesel engine employed. Small- and medium-sized power plants generally use a higher gravity fuel oil of better quality, whereas large oceangoing vessels use a lower grade of fuel oil, which usually requires special treatment in the form of heating before being injected into the cylinders of large diesel engines.

How is the fuel oil heated for large marine diesel engines?

Answer: This oil heating treatment begins at the fuel bunkers, where heating coils are required to reduce its viscosity for pumping under all temperature conditions. Additional heating is supplied by passing the fuel oil through heaters of the tube or film type, supplied with steam from an auxiliary boiler in order to reduce its viscosity sufficiently for good atomization by the injection valves.

Cleaning the fuel oil is accomplished by a combination of centrifuging and filtering.

The heating treatment of fuel oil reduces its viscosity and simplifies the process of separation of water and impurities from the oil in the centrifuges. After centrifuging, the fuel oil is passed through filters in its passage to the fuel injection pumps. A final filtering is provided at the injection valves, some widely used types of valves having filters of the fine-mesh strainer type incorporated in the valve body.

What drive systems are used in marine diesel installations?

The drive system as used in marine diesel applications may be defined as a method of power transmission between the prime mover and the propeller shaft.

The various systems employed to transmit power from the diesel engine to the propellers are:

1. Direct drive.
2. Reduction-gear drive.
3. Electrical drive.

Answer: In the direct-drive system, the engine may be directly connected to the propeller shaft, or as is more common with a sailing clutch or other type of coupling, between the engine and the propeller shaft.

Describe the reduction-gear drive used for ship propulsion.

Answer: Reduction gears employed for ship propulsion consist essentially of a pinion or pinions connected between the engine and the propeller shaft. This arrangement lends itself to greater flexibility, as one to four engines may be used to drive a single propeller shaft.

With single-engine drive the engine may be located on either side of the center line and the drive pinion located on one side of the gear on the horizontal center line, or the engine may be located on the center line of the ship sufficiently high to connect with a pinion on the vertical center line of the gear.

With an arrangement of two engines per shaft, both engines are symmetrically positioned on each side of the center line, each connected to a separate pinion on opposite sides of the gear (Fig. 28-10). The usual arrangement is to locate the reduction gear aft of the engines, but in some installations the reduction gear is placed forward of the engine, with the line shaft extending aft between the two engines. For an arrangement of four engines per shaft, two engines are placed forward of the reduction gear and the remaining two aft of it (Fig. 28-11).

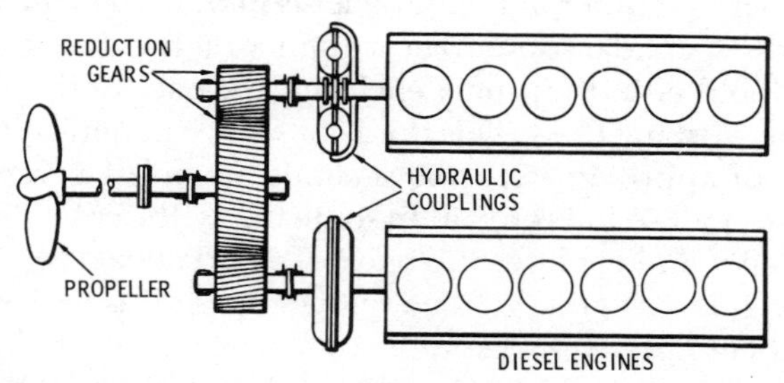

Fig. 28-10. Twin-engine drive arrangement of a ship with reduction-gear drive.

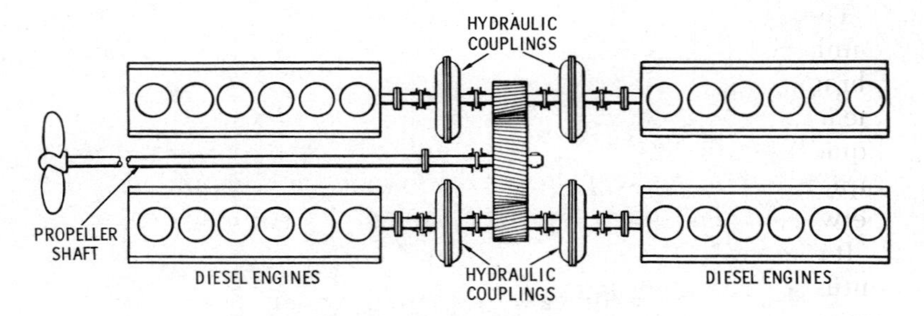

Fig. 28-11. Quadruple-engine drive arrangement of a ship with reduction-gear drive. Engines are placed so that each occupies the corner of the rectangle with the center of the propeller gear located in the intersection of the rectangle diagonals. A pinion on each side of the main gear has a shaft extending forward and aft, with a single engine coupled to each end.

The lubrication system of the reduction gears is usually from the engine if the gears are of the built-in type, but it is independently supplied when the gears are of the open type. This lubrication is usually under pressure through nozzles, which direct the oil against both sides of the gear teeth.

Describe the hydraulic coupling used in ship propulsion.

Answer: The hydraulic coupling used in diesel propulsion is principally the same as that adopted for power transmission in automobile engines. By definition, a hydraulic drive coupling is one in which power is delivered from the engine to the propeller, through the medium of a liquid without any mechanical connection.

The primary member of the impeller is mounted on the driving shaft and the secondary member or runner on the driven shaft.

A cover is bolted to the runner enclosing the back of the impeller to retain the working fluid. This fluid consists of mineral oil having a viscosity of approximately 190 seconds Saybolt Universal at a temperature of 130° Fahrenheit. In some types, the cover is bolted to the impeller and encloses the back of the runner.

The hydraulic coupling is filled with oil by means of a pump or an overhead gravity tank. A stationary housing encloses the entire coupling and serves to catch leakage and return the oil to a container when the coupling is disengaged.

The impeller and the runner are provided with an unequal number of straight radial vanes and two circular concentric diaphragms which form a series of concave recesses or cups. When the impeller is rotating, and assuming the couplings to be full, the liquid in the passages between the vanes of the impeller flows rapidly outward under centrifugal force and passes over the gap between the impeller and the runner.

It then flows radially inward between the vanes of the runner until it reaches the inner diameter of the working circuit. There it returns across the gap to the impeller, and the cycle is repeated again.

The rotating vortex ring of liquid follows a circular spiral. Its path may be likened to a long coil spring bent until its ends join and form a ring, the pitch of the spiral representing the relative speed between the driving and driven members of the coupling.

Power is transmitted by the release of kinetic energy, as the vortex ring of liquid flows radially inward between the vanes of the runner. This kinetic energy is generated during the outward flow of the liquid between the vanes of the impeller.

For a given horsepower and speed, the coupling is selected so that there will be a difference of about three percent in the speed between the driving and the driven members. Thus, if the engine—that is, the driving member—operates at a speed of 500 rpm, the driven member will operate at approximately 485 rpm. It is the difference in centrifugal force resulting from the difference in speed that causes the liquid to circulate between the two members of the coupling.

To obtain rapid clutching and declutching there is a valve in the oil supply to the coupling, and a series of ports in the periphery of the rotor housing.

These ports are covered by a ring valve that has a small amount of axial movement and is operated through the claws and linkage by an outside lever.

If the ring valve is moved to uncover the ports, the oil is thrown out by centrifugal force and the runner ceases to revolve. The valve in the supply line is connected to the ring-valve operating gear so that when the ring valve is opened to empty the coupling, the supply valve is closed, and when the ring valve is closed, the supply valve is open.

Since, as already pointed out, there is no mechanical connection between the driving and driven elements in the coupling, a great amount of flexibility is obtained by using this type of drive. Thus, by a simple procedure of dumping the oil from the coupling of one engine, it may be disconnected from the gears without interrupting the running of the others.

What is the principle of the electromagnetic coupling used in ship propulsion?

Answer: A coupling of this kind depends upon electromagnetic lines of force between the two rotating members for its functioning. One of the rotating members is attached to the engine shaft and the other to the pinion shaft.

It acts upon the principles of a squirrel-cage induction motor, and has collector rings and brushes through which the energizing current is supplied. The current entering the collector rings induces a current in the squirrel-cage winding of the inner member. The lines of force set up across the air gap oppose a separation of the two members, in a similar manner to that of a magnet and its plunger.

The amount of slip between the two rotating members amounts to approximately 2.5 percent. Thus, if the engine rotates at 250 rpm, the pinion shaft will rotate at a speed of approximately 244 rpm.

The connection between the two members is interrupted by a disconnection of the two wires leading to the collector rings. The similarity between this coupling and the previously considered hydraulic coupling lies in the fact that in both types the power is transferred from one shaft to the other without any mechanical connection. In the hydraulic-coupling method, the power is transferred by the action of a liquid interposing between a set of vanes mounted upon the engine and the pinion shaft respectively.

Describe the electromagnetic coupling used in ship propulsion.

Answer: In the electromagnetic coupling, vanes are substituted for electric windings and the power is transferred from one shaft to another by means of the magnetic action between them.

Since the connection between the engine and its pinion will be instantaneously interrupted by the simple manipulation of a switch, this feature can be used to advantage with a multiengine installation. Thus, for example, half the engines may be run astern, the other half ahead, and the ship maneuvered by energizing either the ahead or the astern couplings as desired, connecting the propeller either to the ahead or to the astern engines. When maneuvering in this manner, however, only half the power is available in either direction.

What are the advantages of electromagnetic coupling over other types of couplings for diesel-propelled ships?

Answer: Electromagnetic coupling has certain advantages over other types, which should be given consideration in the planning of geared diesel propulsion plants. Among these may be mentioned:

1. The torque that the coupling is capable of transmitting is independent of the speed.
2. The coupling makes very rapid maneuvering possible because of its torque capacity and simplicity, and rapidity of engaging and disengaging which consists in operating only a simple switch applying the excitation current or interrupting its supply.
3. There are minimum fire hazards since there is no oil involved and the insulating material used is strictly of the Class B or fireproof type.
4. The electric coupling is not subject to mechanical wear.
5. The only auxiliary equipment necessary with the electric coupling, namely the control panel, may be located anywhere convenient, as it is connected to the coupling by means of cables.
6. The electric coupling has considerable flexibility in design in providing the desired flywheel effect on the engine.

By simple control means, the losses in the electric coupling at cruising speeds may be reduced to compare favorably with those of the hydraulic coupling, particularly on applications with severe space and weight limitations, where the percentage loss in the hydraulic coupling at half power may be double the percentage loss at full power.

The flexibility of operation introduced by the use of either the hydraulic or the electromagnetic coupling is very similar and the choice between one or the other is one of expediency.

The hydraulic coupling requires for its performance a constant supply of proper oil, whereas the electromagnetic coupling must be supplied with an unfailing source of electricity.

Describe the electrical drive for diesel marine engines.

Answer: The electrical drive has certain advantages over the previously described reduction-gear drives in that it protects the engines from the shocks of the propeller and makes it possible to operate the ship at a reduced speed with one or more engines not operating.

Its disadvantages as compared to reduction-gear drive are higher first cost, increased weight, higher fuel consumption and maintenance, resulting in lower overall efficiency.

The basic components of the diesel electrical drive are a motor to drive the propeller shaft and one or more diesel-driven generators to produce current for operating the motor with the required control equipment.

Maximum flexibility in arrangement of machinery results from the fact that with the exception of the propulsion motor, which necessarily must be attached to the propeller shaft, there is no restriction as to where the remainder of the equipment may be installed.

What type of current is used for diesel electrically driven installations on ships?

Answer: With a few exceptions most diesel electrically driven installations are of the direct-current type. This is due mainly to the inherent speed characteristics of diesel engines.

If alternating current were used, it would be necessary to operate the generators in parallel. To operate AC generators in parallel, the very closest speed regulation and exact angular velocities of all prime movers is necessary.

Since this is difficult to obtain with even the most perfect governors, alternating current is not suitable for diesel electrical propulsion. Direct current not only overcomes these difficulties, but possesses many advantages in the way of operation control and reverse power.

What control methods are used on diesel electrically driven ships?

Answer: The control methods used in the diesel electrical drive is usually the Ward Leonard or variable-voltage control. This type of control, being general in its field of application, controls the speed of propulsion motors by varying the generated voltage of the main generators. The generated voltage is controlled by both the field strength of the generators, and the speed at which they are driven by the diesel engines.

The diesel engines, as prime movers, have a speed range from idling (which is usually 50 percent of full speed) to maximum rate

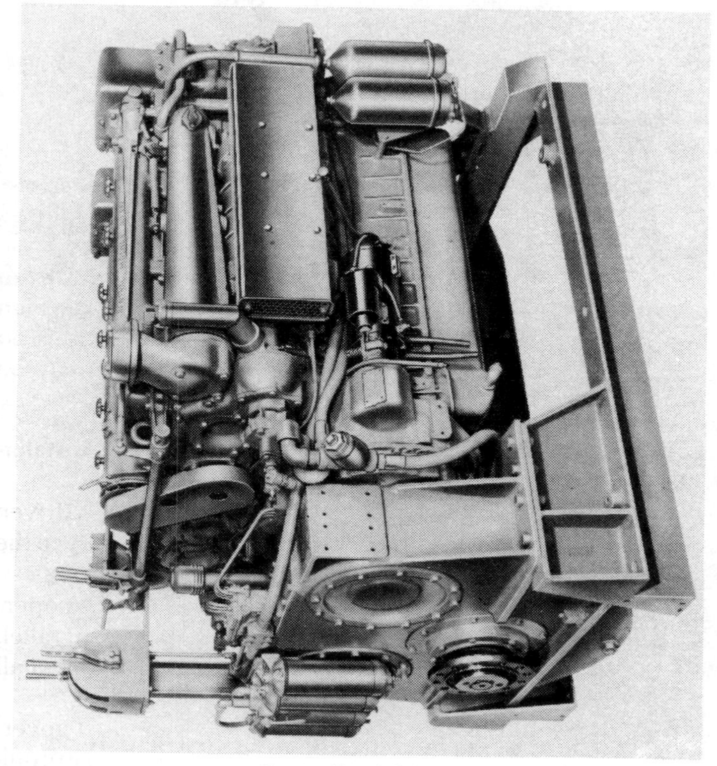

Courtesy Detroit Diesel Engine Division, General Motors Corp.

Fig. 28-12. General Motors Series 6-71 Twin Model 12005A marine diesel engines coupled to a single output shaft through a heavy-duty gearbox. Either engine may be shut down and the other kept running if only partial power output is required or if adjustment is necessary on an engine.

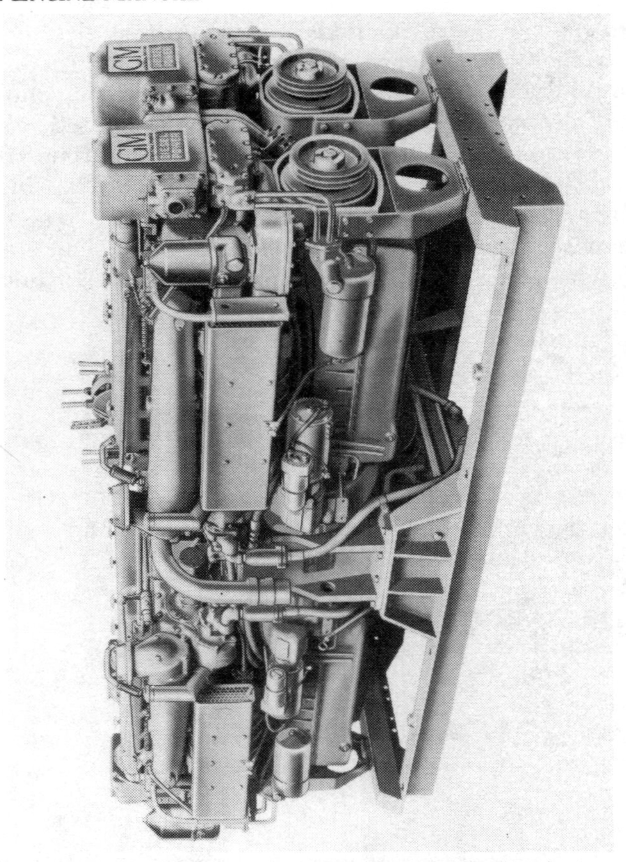

Courtesy Detroit Diesel Engine Division, General Motors Corp.

Fig. 28-13. General Motors Series 6-71 Quad marine diesel engine consisting of four Series 6-71 inline engines driving a common output shaft. Each engine is a complete, self-contained unit and may be shut down for reduced power demand or for maintenance.

speed. This speed is regulated by governor control of the engines through a suitable remote control system.

As the field current is of small value, the losses in the system are negligible and the voltage or power generated can be controlled so that it is just sufficient to give the propulsion motors the speed desired.

The system requires excitation and control power from a constant potential source. This is generally a small auxiliary generator that is either directly connected or belted to the main generator shaft. When used with a variable speed engine, its voltage is held constant by a voltage regulator. When the output from the main generators is transferred from the propulsion circuit to an auxiliary bus, the generators can be operated as self-contained units having their own voltage and speed control.

When the generators are being used on the propulsion circuit, variation of the motor torque at any given voltage is accomplished by adjusting the strength of the motor field. This type of control is very flexible as to speed graduations, and the ship can be maneuvered either from the bridge or from the engine room of the large vessels.

As the controlling circuit handles only a small value of current, and the main line switches are opened and closed only at zero voltage, the entire equipment is comparatively simple both electrically and mechanically.

What types of generators and motors are used in diesel electrically driven ships?

Answer: The generators and motors are usually shunt-wound and are separately excited from small exciter generators, usually mounted on the same shaft as the main generators.

The main generator armature as well as that of the propulsion motor is usually connected in series, the unit for each propeller shaft forming a single series circuit.

In a system of this kind, the motor field excitation remains constant, while the generator field excitation is varied from zero to maximum, regardless of the number of generators used to regulate the propeller speed.

The diesel engine rotates at constant speed under governor control similar to that of a land installation, but because of the series arrangement of the generators, it is not necessary for the engines to run at the same speed.

The fact that the propeller speed is controlled by a generator field rheostat makes the bridge control feasible.

When there are multiple-engine installations with an exciter generator on each main generator shaft, all of the exciters may not

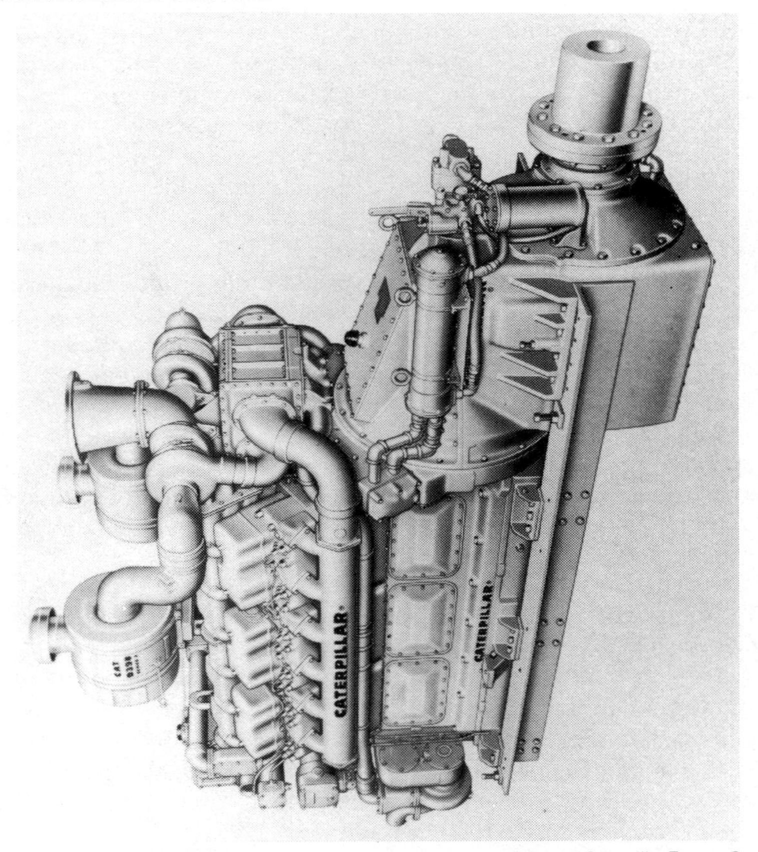

Courtesy Caterpillar Tractor Co.

Fig. 28-14. Caterpillar Model D398 marine diesel engine.

be needed for field excitation; therefore, one or more exciters may be used for miscellaneous motor-driven auxiliaries.

When a large amount of current is used, as when loading or unloading in port, a main generator may be utilized for the electrical supply of winch motors.

Describe supercharging of marine diesel engines.

Answer: Supercharging, as applied to internal-combustion engines, is the process of supplying air for combustion at a pressure greater than that attained by natural or atmospheric induction.

The purpose of supercharging is to increase the output of an

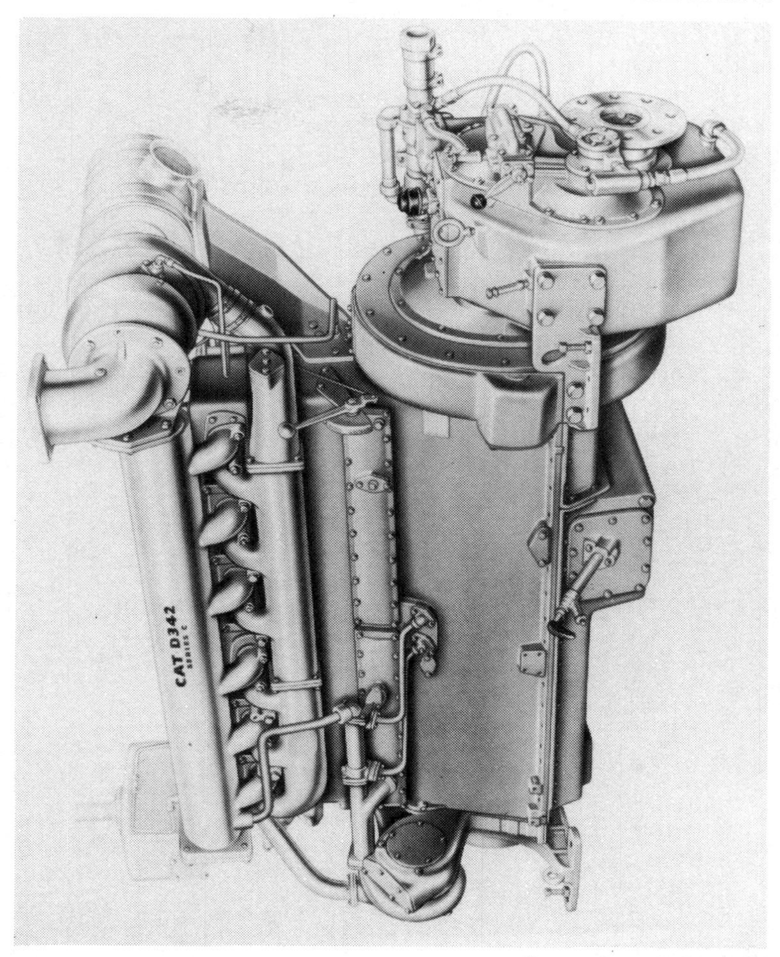

Courtesy Caterpillar Tractor Co.

Fig. 28-15. Caterpillar Model D342 marine diesel engine.

engine, the combustion of greater quantity of fuel per cycle being made possible by the greater charge of air.

The first types of superchargers to receive attention were the exhaust turbocharger and the gear-driven rotary blower. The former consisted of a gas turbine on the extended shaft of a centrifugal blower and derived its energy from the exhaust gases of the engine.

The second type, the rotary blower, was mechanically driven from the engine by means of a light high-speed gear train. Both types were designed to maintain sea-level pressure in the induction manifold at various altitudes and thereby maintain substantially sea-level output.

Since the early development in the field of aviation, supercharging has found useful application to many installations of diversified nature, such as racing boats and automobiles, diesel locomotives, cargo, passenger, and naval vessels, power plants for peak load conditions, and in recent years to a limited extent for passenger automobiles.

CHAPTER 29

Diesel Electrical Power Plant

What are the two classes of diesel-powered power generating plants?

Answer: Diesel power plants for generation of electrical power and light may be divided into two classes depending upon methods of installation:

1. Stationary.
2. Portable.

Stationary power plants are built to supply power on a permanent basis with one or several diesel engines usually directly connected to its individual generator (Fig. 29-1).

At the present time there are millions of diesel horsepower used in stationary power plants. These engines range in various sizes up to several thousands of horsepower in one unit (Fig. 29-2).

The use of diesel power in comparatively small municipal power plants has expanded greatly during the past several years, particularly in plants using units of horsepower ratings ranging

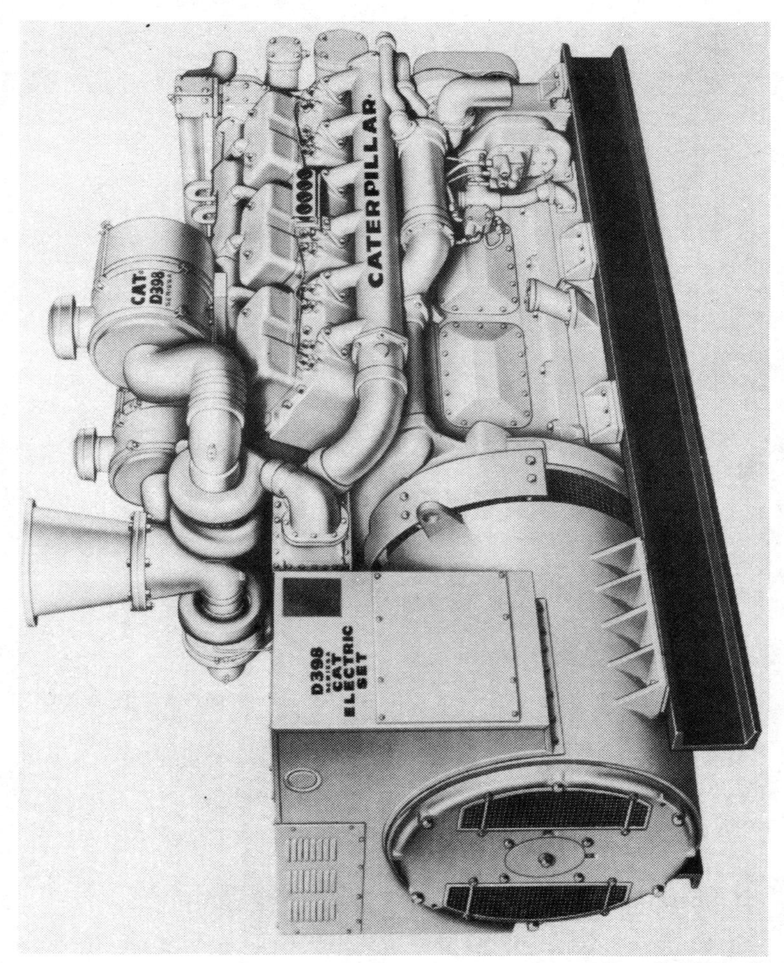

Courtesy Caterpillar Tractor Co.

Fig. 29-1. Caterpillar Model D398 diesel engine with generator unit.

from 500 to 2,000 horsepower. Such plants are usually built with similar engines. Rural electrification plants have used diesel engines in view of their exceptional economy under variable load conditions (Fig. 29-3).

Many utility companies also use diesel engines in their power systems. In some instances the demand for additional power at the

Courtesy Worthington Corp.

Fig. 29-2. Worthington SEH diesel engines used in large electric-power-generating plant.

end of the transmission line may require increasing the size of the line. The diesel engine installed at the end of the line supplements the power line so that the required voltage will be maintained or it will act as a standby for peak load service.

Diesel engines are also used quite extensively in large buildings, including department stores, hotels, and hospitals, and often sup-

Courtesy Worthington Corp.

Fig. 29-3. Worthington SEHTP 8 turbocharged engine in R.E.A. plant.

plement power furnished by steam generating plants installed in such buildings (Fig. 29-4).

The success of diesel power in stationary plants of various types (Fig. 29-5) is dependent upon the economics involved. In addition, other conditions may prevail such as availability of space, quality and quantity of water, and load characteristics (Fig. 29-6).

Courtesy Worthington Corp.

Fig. 29-4. Worthington SW four-stroke-cycle turbocharge engines.

Some industries also desire to have their source of power within their own plant so as to avoid the possibility of failure of power from outside sources due to storms or other conditions beyond control.

In what ways are portable generating plants used?

Answer: Utility companies are also exploring the use of porta-

Courtesy Worthington Corp.

Fig. 29-5. Worthington SW six-cylinder, inline diesel engine driving one of main pumps in a Florida municipal pumping plant.

ble generating plants; however, comparatively few of those plants have been built so far.

Such generating plants, including diesel engines, generators, and all auxiliary equipment, are mounted on railways cars, truck trailers, or on skids for semiportable units.

Courtesy Worthington Corp.

Fig. 29-6. Worthington SEH diesel engine. This engine is completely assembled and tested at the plant before shipment.

Each portable plant is complete in itself and can be transported from one location to another ready for immediate production of power, as soon as cable connections can be made to the system requiring the service.

In the event there is an unusual peak power requirement in a particular season, the unit can supplement the permanent generating facilities. Also, water conditions sometimes require temporary

Courtesy Worthington Corp.

Fig. 29-7. Worthington SW diesel engine is a four-stroke-cycle engine designed for high-pressure turbo-supercharging.

additional capacity to supplement hydroelectric plants. Such units can also be used in expediting work on a transmission line between the point of generation and the community being served by placing the portable plant at or near the community, making it unnecessary to encounter the hazards of accidents with a hot line. The portable power plant is also used in the event of power failures, particularly in severe weather areas.

What fundamental requirements should be considered for location of a diesel electrical power plant?

Answer: In determining the location of a diesel electrical power plant, it is necessary to observe certain fundamental requirements such as:

1. Distance from plant to load center.
2. Cost of fuel delivered to plant.
3. Cost of land on which plant is to be built.

Other factors that have a definite bearing upon the economy of the plant are such items as:

1. Suitability of soil in order to obtain a reliable foundation for building and machinery.
2. Availability of an unfailing supply of suitable cooling water in close proximity to the plant in order to keep the pumping cost of cooling water at a minimum.
3. Consideration must also be given to possible additional power units with necessary space allotment for increased fuel storage, cooling, and auxiliary equipment (Fig. 29-8).

What considerations should be given the building layout for a diesel engine installation for a generating plant?

Answer: When planning a diesel engine installation for generation of electrical power and light, close attention should be given to the most economical space utilization, bearing in mind the overall appearance and maintenance requirements.

A survey should be made to determine possible future power requirement, and provision should be made for possible future addition of similar power units.

Consideration should also be given to location of switchboards, office space, toolrooms, and toilet facilities for operating personnel.

Additional space requirement consists of lubricating oil storage, auxiliary equipment such as purification apparatus for lubrication and fuel oil, water, and cooling apparatus.

It is recommended that building plans be worked out in close cooperation with the manufacturer of equipment, and all details and dimensions of machinery and accessories be obtained prior to the erection process.

Courtesy Gardener-Denver Co.

Fig. 29-8. Gardener-Denver Model RLA compressor has a capacity of 108-1592 cfm to 150 psi.

All equipment should be installed in accordance with the requirements of the National Board of Fire Underwriters in addition to any local laws and regulations in force at the particular location of installation.

What other power plant equipment is required?

Answer: The piping diagram for a typical diesel electrical power plant is shown in Figs. 29-9 and 29-10. There are two methods of engine jacket cooling employed, depending upon the availability and purity of the water at the plant location. They are:

1. The open cooling system in which the water, under pressure, flows through the engine jackets, and is wasted or recirculated through a cooling tower.
2. The closed cooling system consists of two complete systems

of circulating water, one for raw water and the other for water jacket cooling.

The open cooling system is generally unpopular because of the fact that the recooling of water in a cooling tower is dependent upon the evaporation process, and an accumulation of impurities leaves a scale deposit in the engine water jackets which requires frequent cleaning to remove the impurities.

In the closed cooling system, the engine jacket cooling water is pumped through the system by means of a pump which circulates the water through the engine and then to a heat exchanger where the heat is dissipated through the raw water. When an oil cooler is utilized, it may be placed in either the raw-water system or the jacet water system depending upon the engine manufacturer's recommendation.

It is economically advantageous, particularly in large plants, to be equipped for oil storage for several months so that fuel oil may be purchased to take advantage of market conditions.

The equipment required in a diesel electrical power plant depends upon the kilowatt capacity involved. In a typical average-sized plant the following major equipment is usually involved:

1. Diesel engines, governors, and generators.
2. Switchboards and accompanying wiring.
3. Air-intake silencer.
4. Exhaust silencer.
5. Air filters.
6. Starting-air compressor.
7. Jacket water pumps.
8. Water heat exchanger.
9. Cooling tower.
10. Fuel oil tanks.
11. Lubrication oil system.
12. Gauges, thermometers, meters, and controls.

What is the function of air-intake and exhaust systems in generating plants?

Answer: For successful operation of diesel engines, an adequate and properly designed air-intake and exhaust system is necessary.

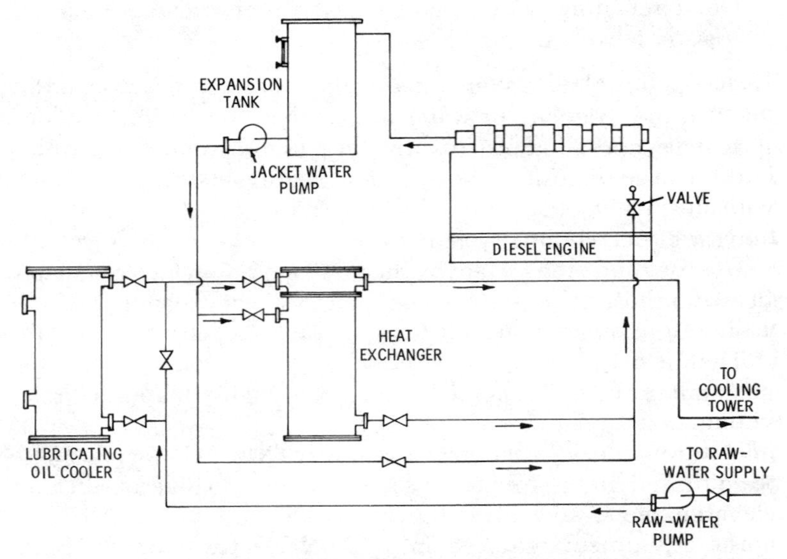

Fig. 29-9. Piping diagram of a typical closed water-cooling system with shell and tube heat exchanger.

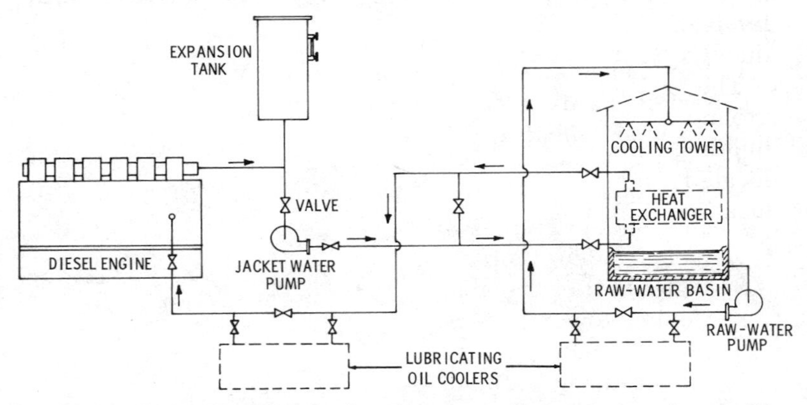

Fig. 29-10. Piping diagram of a closed water-cooling system equipped with a cooling tower.

Intake air free from foreign material must be supplied to the engine in adequate quantity for proper combustion of the fuel. The amount of air required for proper combustion and scavenging depends upon the type and design of the engine.

The function of the exhaust system is that of handling the discharge gases with a minimum of both pressure drop and noise.

Describe the air-intake system for a diesel-powered generating plant.

Answer: These consist of a duct for leading the combustion air into the engine manifolds from the air-induction point together with air filters and properly designed air-intake muffling apparatus as may be required (Fig. 29-11).

The function of the air-filter system is to prevent foreign particles from entering into the engine, causing damage to cylinder walls and to piston rings.

The most satisfactory method used to eliminate objectionable air-intake noises is by means of a suitable intake-air muffler. Providing a large plenum chamber between the air filters and the air inlet pipe to the engine will also tend to reduce sound pulsations at the air filter. Supplying the air into the filter housing from a high elevation has been found effective in eliminating objectionable noise.

The location of the air intake for the engine is determined partly by the design of the power plant structure and partly by the limitation imposed by the length and diameter of the air-intake duct required.

The most common method of air-intake locations is near the ground adjacent to the building, in the building wall, or on the building roof, although in some instances air intakes have been located inside the power plant building (Fig. 29-12). In this latter location, however, the air requirement for fuel combustion will

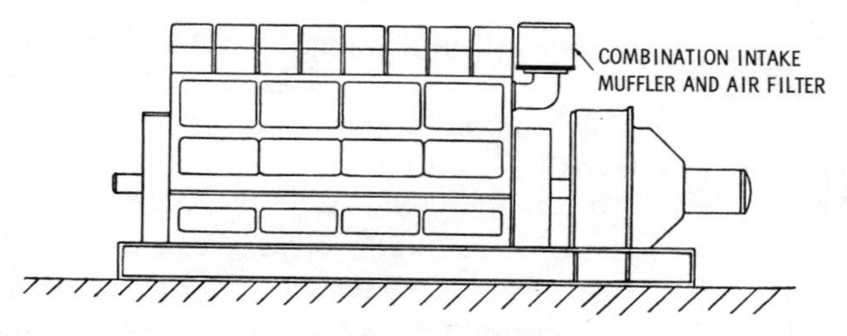

Fig. 29-11. Installation of intake muffler and air filter for diesel engines in a generating plant.

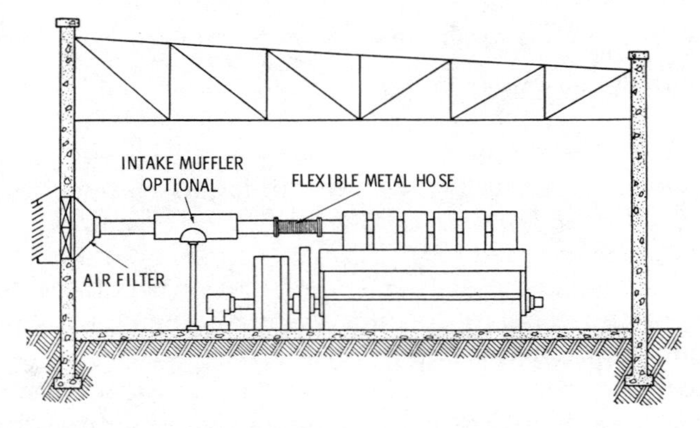

Fig. 29-12. An alternative method of installing an intake muffler and air filter in a diesel-powered generating plant.

require a greater amount of ventilation, thus making it difficult to adequately heat the building during cold weather.

Describe the air-exhaust systems for diesel engines in generating plants.

Answer: The proper operation of the exhaust system of a diesel engine requires three independent separate functions. They are:

1. The system must carry the exhaust gases from the engine to the point where its rejection will not be objectionable.
2. It must release the exhaust gases with a minimum of noise.
3. The system must impose a minimum of back pressure on the engine exhaust.

To obtain the best possible exhaust systems, a great deal of experimentation has been performed both by the engine manufacturers and by the suppliers of engine auxiliaries.

The need for scientifically designed silencers or mufflers will be readily apparent if it is realized that approximately 30 percent of the fuel energy in a diesel engine is released as exhaust, which if improperly dealt with can cause serious trouble in the form of poor engine performance and noise.

If the released engine exhaust gases were given a completely free passage, a very large silencer out of proportion to the size of the engine would be required. Most silencers are a compromise between a low back pressure and excessive overall dimensions of the installations.

The path the gases are forced to take should be as gently curved as possible. Sharp turns must be avoided. Baffle plates that deflect the gases in a reverse direction are likely to set up undue back pressure.

Any space provided for the passage of the gases within the silencer must not be less than the area of the main exhaust duct or manifold. The volume of the silencer should be slightly greater than the capacity of the engine cylinders.

Most silencers are fabricated by welded steel plate although some are made of cast iron. Those most generally adopted consist fundamentally of a large cylindrical vessel with baffle plates mounted on the inside. The exhaust conduit from the engine is led into one end of this vessel, and a conduit from the other end leads the gases at a greatly reduced pressure into the atmosphere.

What factors are important in installation of an air-exhaust system in a diesel-powered generating plant?

Answer: The following pointers should be observed when installing an exhaust-air system:

1. The gas-carrying conduit should have a sufficient diameter.
2. The exhaust line should be as short as possible with provision made for expansion and contraction.
3. The number of bends in the conduit should be as few as possible.
4. The conduit between the silencer and the atmosphere should extend above the building roof.

Special spark-attester silencers are employed in locations where dirt and live sparks must be removed for safe and satisfactory performance. Such silencers are equipped with a specially constructed spark trap at the extremity of the system, which effectively eliminates any fire hazards.

What types of current are generated in diesel-powered generating plants?

Answer: The electrical equipment required for a diesel electrical power plant does not differ in any important respect from that required in any other stationary power generating plant (Figs. 29-13 and 29-14).

There are two methods used in power distribution with respect to type of current generated. They are:

Courtesy Worthington Corp.

Fig. 29-13. Worthington SEH 6 turbocharged diesel engines in Mideast pipeline station.

1. Alternating current.
2. Direct current.

Alternating current is employed extensively in all large plants, particularly where distribution distances are considerable. It is customary to employ certain standard frequencies and voltages.

Courtesy Worthington Corp.

Fig. 29-14. Worthington SEHGO 8 turbocharged engines in municipal power-plant service.

Standard frequencies in the United States are 25, 50, and 60 cycles per second, the 60-cycle system being the most common.

Voltages are also standardized but are usually dependent upon the KVA ratings of the generators. They may be from 120 volts up to 4,160 volts. The standard excitation voltage is 125 volts direct current.

Alternators are nearly always built for three-phase service but require a small independent direct-current generator for supply of excitation. These are called exciters and are usually mounted on a common shaft with the alternator.

Because of the necessity for a standard lamp voltage (usually 110-120) and a standard voltage for motors, it is necessary to employ transformers where the generated voltage is too high for standard lamps and motors.

It should be noted that when transmission distances are great, alternating current of high voltage and power transformers are commonly used.

Direct current is used to some extent, particularly in small plants and where the transmission lines required are relatively short. One particular disadvantage in the use of direct current is that only direct-current or universal-type motors can be employed for power use.

How is the electrical current distributed from the generator?

Answer: Since two-wire direct-current distribution at the conventional low voltage is very uneconomical because of the high weight of copper in transmission lines, the three-wire direct current is universally employed.

The early three-wire systems were operated with two similar generator units connected in series and the neutral connected to the center point.

Because of the high cost of machinery and accessories involved, this system has been largely superseded by the three-wire generator, popularly known as the Dobrovolsky machine.

Briefly, this machine consists of a standard two-wire generator with a coil of high reactance and low resistance connected permanently across diametrically opposite points on the armature. The connection between the coil terminals and the armature is facilitated by two slip rings and accompanying brushes mounted on the machine in the conventional manner.

Instead of one balance coil, two such coils are sometimes used. This reduces the effect of armature resistance on unbalancing the voltages. When two external balance coils are used, four slip rings are required.

When the loads on the two sides of the system differ, the difference between the currents in the outside lines flows in the neutral wire and through the balance coil, which offers only a small resistance to direct current.

Why is a switchboard used in the power station?

Answer: Switchboards are used as means of control of generating machinery and load and therefore are the connecting link between the motor or lamp and the power supply unit.

The instruments, meters, protecting devices, and switches required depend generally upon the size of the system, that is, the generating capacity of the station.

Generally a switchboard consists of slate, ebony, asbestos, or steel panels of certain standard sizes, upon which the control equipment and meters are mounted.

For control of diesel-powered generator units, direct-control switchboards are most generally used. They are designed for control of small- and medium-capacity installations where low cost is of prime importance and a complete installation of only a few panels is required.

In the direct-control switchboard, it is customary to mount the equipment for each generator and feeder on individual panel units. Thus, for example, if a power station consists of say three AC generators operated in parallel and five feeders, there would be three generator panels and five feeder panels required for the installation.

Three-wire, direct-current switchboards are used for combined power and lighting installations. In these boards, a third or neutral bus is generally provided for those feeders requiring the neutral lead. The voltage from positive to negative is usually 250 volts with the voltage from positive to neutral and negative to neutral 125 volts. In some installations of this type, the voltages are 500/250 volts. The generators require an equalizer on each side.

Alternating current switchboards of the direct-control type for voltages of 500 volts or less have many characteristics in common with those for direct current. They are used for 110-, 220-, and 440-volts service. The switching is accomplished through knife switches with either fuses or circuit breakers.

A typical generator and feeder panel may be assembled to form switchboards for low-voltage substations used in industrial plants and in public buildings.

What kinds of automatic voltage regulators are required on AC installations?

Answer: Due to the inherent poor voltage regulation of AC generators, automatic-voltage regulators are generally required on most alternating-current installations.

These take various forms, depending upon the requirement, and may be divided into three general groupings, namely:

1. Rheostatic.
2. Vibrating contact.
3. Electronic.

Almost all the voltage regulators now in use are designed to control the current in either the alternator field or the exciter, or by a combination of the two.

In electronic rectifiers, a part of the alternating current is rectified by electronic tubes to furnish direct current to either the shunt field of the exciter or the field of the alternator. The amount of excitation current is controlled electronically in accordance with the generator voltage.

How does high altitude affect the operation of diesel engines?

Answer: The starting and operation of internal-combustion engines encounters certain difficulties at higher altitudes.

These difficulties are not commonly noticeable until an altitude of about 3,000 feet is reached. Above this altitude, it is usually necessary to make certain changes in the diesel engine to facilitate starting, to increase power output, and to eliminate incomplete combustion.

At an elevation of 3,000 feet, a diesel engine will have lost approximately 10 percent of its sea-level rating, but at 6,000 feet this loss will be approximately 21 percent.

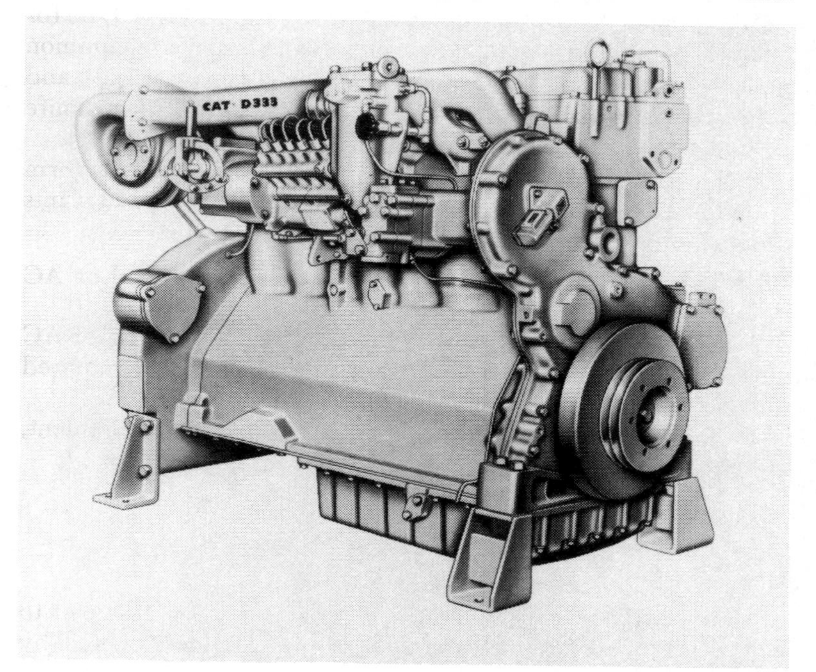

Courtesy Caterpillar Tractor Co.

Fig. 29-15. Caterpillar Model D333 industrial engine.

From these figures, it is readily apparent that some provisions should be made to overcome a part of the power loss and the engine starting difficulties encountered at high altitudes.

A small part of the aforementioned losses may be regained by making various adjustments to the engine.

Since the air at higher altitudes is lighter, a longer period is required to effectively burn the fuel oil. Thus the fuel injection-pump timing should be advanced about one degree per 1,000 feet of elevation over standard timing, unless the compression ratio is increased; then the timing can remain the same as originally set. This will aid in starting and combustion, resulting in a slight reduction of the power losses.

Also, due to the lightness of air at higher altitudes, the engine cylinders do not fill as well as at sea level; therefore, the compression pressures are lower, causing difficult starting and poor combustion. This can be partly alleviated by changing the combustion chamber liners and cylinder head gasket thickness so as to obtain a higher compression of the lighter air entering the cylinders. When a thinner gasket is installed, it is necessary to reseat the valves into the head as much as the reduction of the gasket thickness.

It is also necessary to reduce the amount of fuel entering the cylinder as with the original setting because with the smaller amount of oxygen, the combustion is incomplete and a smoky exhaust results.

CHAPTER 30

The Diesel Locomotive

Give the definition of a diesel locomotive.

Answer: A diesel locomotive may be defined as one in which the prime mover consists of a diesel engine usually directly connected to an electric generator whose energy is converted into mechanical power by means of interconnected traction motors.

The traction motors (usually two to four per locomotive unit) are mounted on the axles and equipped with suitable reduction gears for transfer of power to the driving wheels (Fig. 30-1).

Diesel locomotives are built in various sizes depending upon their application, such as for switching, freight, and passenger service (Fig. 30-2).

What types of fuel injection systems are used on diesel locomotives?

Answer: At present, practically all diesel engines used in railroad service use the solid or direct fuel injection system.

Courtesy Electro-Motive Division, General Motors

Fig. 30-1. General Motors locomotive truck for use with Model DD-35, 5,000-horsepower freight locomotive. This is a four-axle, four-motor Flexicoil truck that weighs 75,000 lbs. and has an overall length of 22 ft. 8 in.

The great advantage of the solid fuel injection system lies in the elimination of the air compressor, which is a very delicate mechanism that requires maintenance and expert attention.

Are railroad diesel engines two- or four-stroke-cycle engines?

Answer: Railroad diesel engines are built on both the four- and the two-stroke-cycle principle. The evident advantage of the two-stroke-cycle engine is that for the same piston and rotative speed the number of working strokes is double that of the four-stroke-cycle engine. Thus with the same mean effective pressure, the power of the engine should be doubled.

Experience, however, proved that the mean effective pressure of a two-stroke-cycle engine can be made equal to that of a four-stroke-cycle engine only when the method of scavenging is sufficiently thorough to permit perfect combustion. With some systems of scavenging this is rather difficult at high speeds, as there is not sufficient time for the scavenging air to expel the gases and fill the cylinders with fresh air.

The best system of scavenging is the straight-through arrangement by which the scavenging air is admitted at one end of the cylinder and permitted to escape at the other end, sweeping out the gases in its path through the cylinder.

Courtesy Electro-Motive Division, General Motors

Fig. 30-2. General Motors GP-35, 2,500 horsepower general-purpose, broad-range domestic freight locomotive.

What is the chief advantage of four-stroke-cycle railroad diesel engines?

Answer: The four-stroke-cycle engines have the advantage in that for the expelling of the burned gases and for the intake of fresh air, two complete strokes are provided. This ensures the thoroughness of the operations. In other words, in a four-stroke-cycle engine, the cylinder is thoroughly cleaned of burned gases and again refilled almost to its full volume with fresh air. This results in good combustion and high mean effective pressure and permits high piston and rotative speeds.

Compare the operation of two- and four-stroke-cycle railroad diesel engines.

Answer: To fully comprehend the working principles of the four-stroke-cycle and the two-stroke-cycle engines the following data will be of assistance.

In a four-stroke-cycle engine, four strokes of the piston are required to complete one cycle of events. They are:

1. The intake stroke.
2. The compression stroke.
3. The power stroke.
4. The exhaust stroke.

The crankshaft will make two revolutions per cylinder for each power stroke. During the intake and exhaust strokes, the piston functions as an air compressor, the operation of which consumes power.

In a two-stroke-cycle engine, only two strokes of the piston are required to complete the cycle of events. Intake and exhaust take place during part of the power and compression strokes. Each downward (power) stroke of the piston delivers a power impulse to the crankshaft. Therefore, a two-stroke-cycle engine has twice as many power impulses as a four-stroke-cycle engine, with the same number of cylinders and operating at the same speed.

As the piston in a two-stroke-cycle engine is not required to function as an air pump, an external means of supplying air must be provided. A specially designed blower which handles a large volume of air at a low pressure is used for this purpose. The blower forces air into the cylinder through ports in the cylinder liner wall, thus expelling the exhaust gases and filling the cylinder with a fresh charge of air for combustion.

Describe the diesel locomotive power unit.

Answer: The heart of any diesel locomotive is its diesel engine. Structurally, diesel locomotives may differ depending upon their assignment and specifications by the prospective user (Fig. 30-3).

Diesel locomotives were first built in the United States about 1925. At present, development of specially designed diesel engines for railway service has been stressed by most builders of engines. In many instances, however, certain series of diesel engines may be used with minor modifications in railroad, stationary, and marine service (Figs. 30-4 and 30-5).

Typical of this development is the General Motors series of diesel engines which have found extensive use in their line of railroad locomotives as well as in the LST vessels of the United States Navy.

This series of locomotive engines are built with 6, 8, 12, and 16 cylinders, with a 45-degree angle between the cylinder banks, as illustrated in Fig. 30-6. All operate on the two-stroke-cycle principle with uniflow scavenging. Scavenging air is supplied by positive displacement helical three-lobed blowers. Air intake is by

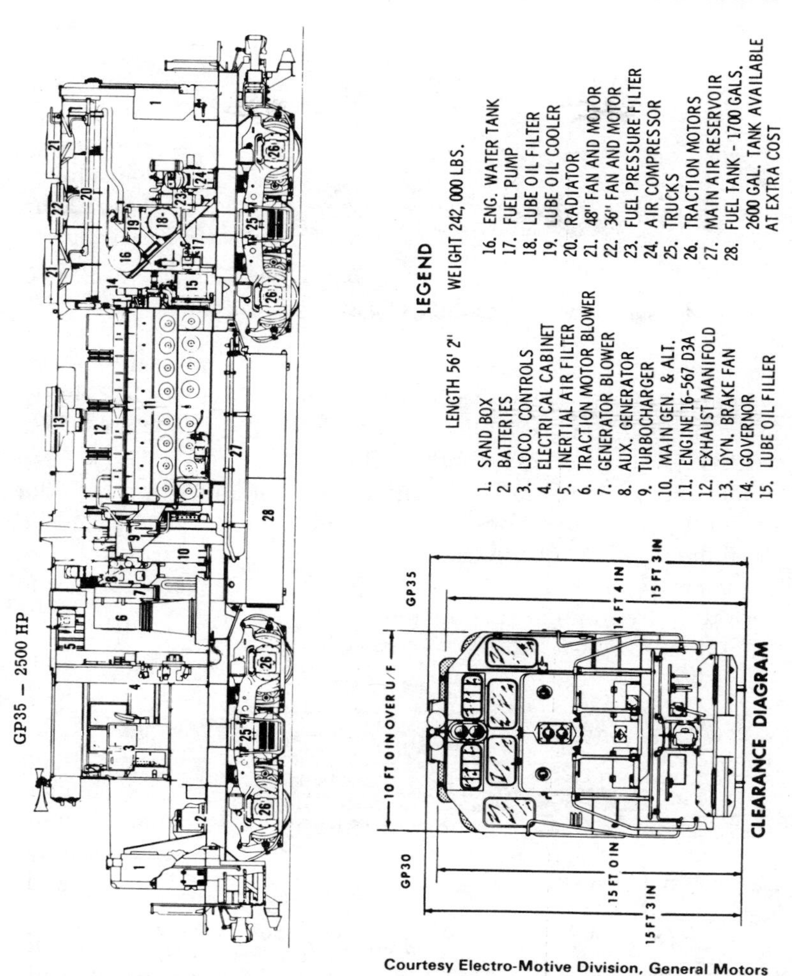

Courtesy Electro-Motive Division, General Motors

Fig. 30-3. Diagram of General Motors GP-35, 2,500 horsepower diesel locomotive.

cylinder ports, and exhaust gases are discharged through four valves in the cylinder head.

Describe the General Motors Model 567 locomotive diesel engine.

Answer: The General Motors Series 567 diesel engine as manufactured by the Electro-Motive Division is a V-type, two-stroke-

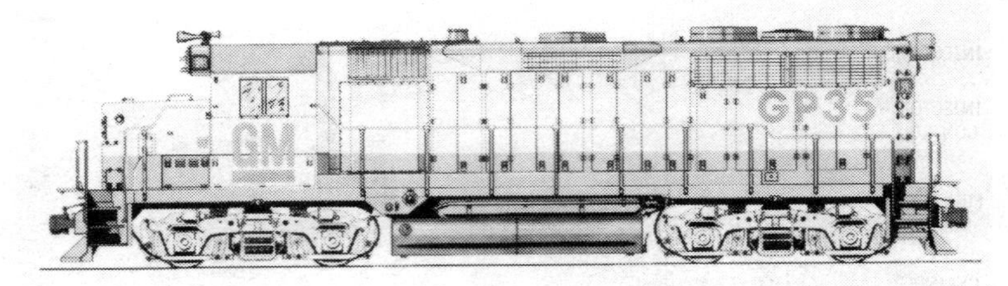

Courtesy Electro-Motive Division, General Motors

Fig. 30-4. General Motors GP-35 general-purpose freight locomotive.

cycle engine that incorporates the advantages of relatively low weight per horsepower, full scavenging-air system, solid fuel injection, high compression, and full horsepower development at relatively low engine speed (Fig. 30-6).

The diagram in Fig. 30-7 will serve to identify the cylinder locations, ends, and sides of the engine. The governor, electro-pneumatic governor control, water pumps, and lubricating oil pumps are mounted on the front. The blowers, oil separator, and generator are mounted on the rear.

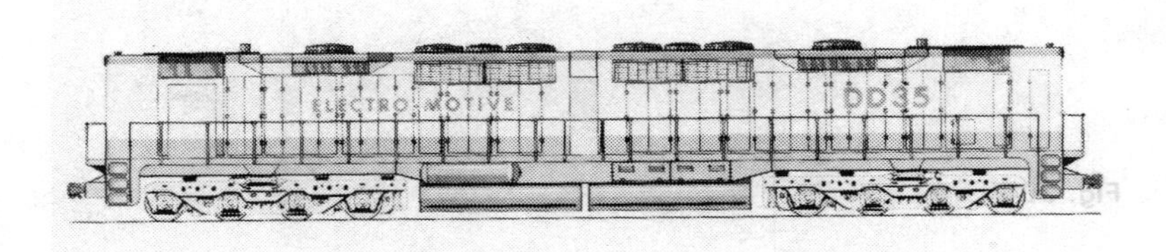

Courtesy Electro-Motive Division, General Motors

Fig. 30-5. General Motors DD-35, 5,000-horsepower freight locomotive.

How are the cylinder heads constructed?

Answer: Cylinder heads (Fig. 30-8) are made of alloy cast iron and are clamped in place in the counterbored, cast-steel, head-retainer castings in the top deck of the crankcase. An indexing plate on the cylinder head fits over a dowel in the top deck to correctly locate the head.

The cylinder head is cast with cored water passages, with openings matching those in the cylinder liner. The engine cooling water

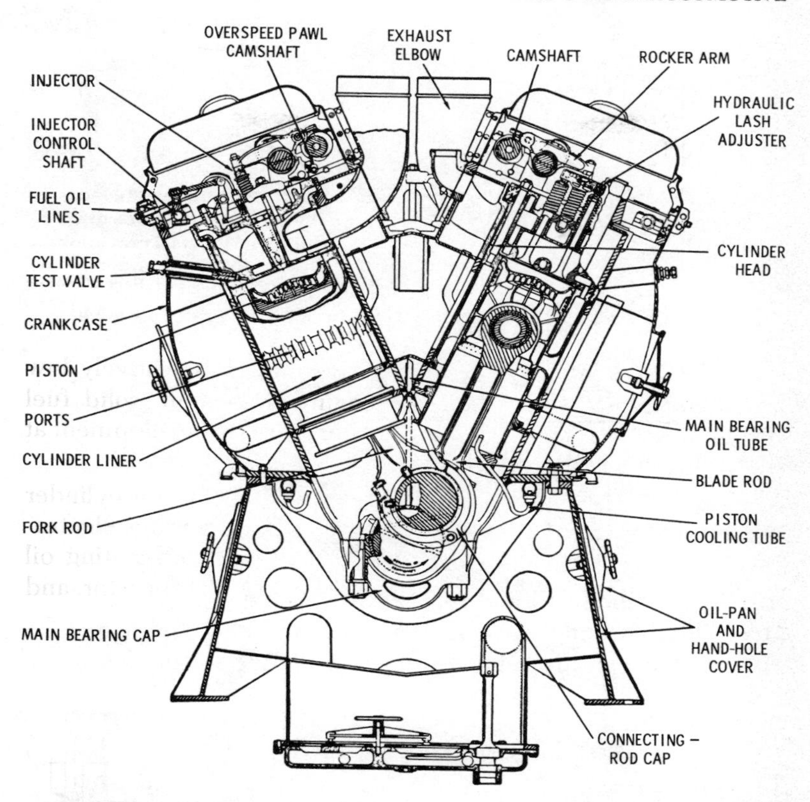

Fig. 30-6. Cross section of General Motors Model 567 locomotive diesel engine.

is circulated through the passages in the liner, up through the head, and is discharged from side ports in the head to outlet ports in the crankcase. Synthetic rubber seal rings, above and below the circumferential outlet passage in the cylinder head, seal the water passages. The exhaust ports in the head line up with a water jacket exhaust port in the crankcase.

Describe the exhaust valve arrangement on the General Motors Model 465 engine.

Answer: The cylinder head is equipped with four exhaust valves, valve guides, and valve springs. The valve guides are

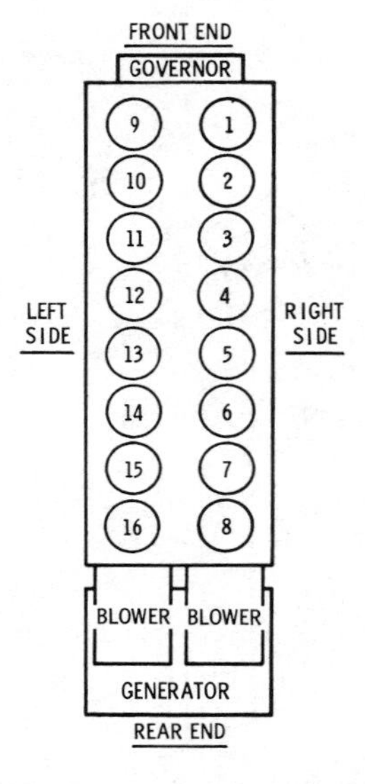

Fig. 30-7. Diagram showing the location and engine auxillaries of a General Motors 16-cylinder locomotive diesel engine.

pressed into the head and reamed for proper valve-stem clearance. The valve and valve-spring assembly is held in place by a tapered valve-spring retainer and two mating locks. Shims are

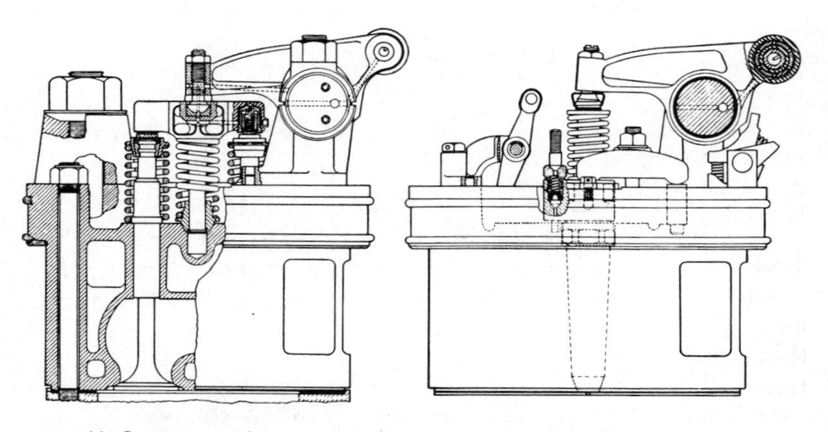

(A) Section at exhaust valve. (B) Section at fuel injector.

Fig. 30-8. Cross section of the cylinder head of General Motors Model 567 locomotive diesel engine.

used on the end of the valve stem, under the valve-stem cap, to compensate for variations in height of valve stems above the cylinder head, which varies with grinding of valve seats and refacing of valves. The valve-stem cap is held in place by a retaining ring.

The valve bridge operates the two exhaust valves under one rocker arm. A return spring and ball seat are held in place on the stem of the bridge by a retaining ring. The ball seat rests in a socket bored in the cylinder head, and the spring applies a pressure against the rocker arm, so that the valve-bridge seat will stay in contact with the rocker arm.

The hydraulic lash adjuster absorbs the lash between the valve stem and the valve bridge. The assembly consists of a cylinder, piston, spring, ball check, and a ball-check guide. A snap ring retains the assembly within the lash-adjuster cylinders. A lash adjuster is pressed into each end of the valve bridge.

Oil flows through a drilled passage in the valve bridge to the top of the lash adjuster, past the ball check, and into the cylinder. When the rocker arm depresses the valve bridge, a slight movement of the piston in the lash adjuster seats the ball check, trapping the oil.

Describe the rocker arm arrangement on the cylinder head.

Answer: Three rocker arms are mounted on the cylinder head. Two rocker arms actuate the four exhaust valves; the third arm operates the injector plunger. The rocker arms are operated directly by the camshaft, through a cam follower mounted at the end of each rocker arm. The opposite end of each rocker arm has an adjusting screw for setting the injector timing, or adjusting the lash adjuster to the valve-stem setting.

Describe connecting-rod construction.

Answer: The connecting rods are of the interlocking, blade, and fork rod construction. The blade rod oscillates on the back of the upper bearing shell and is held in place by a counterbore in the fork rod.

One side of the blade rod bearing is longer than the other and is known as the "long toe." The blade rods are installed in the right bank of the engine, with the long toe toward the center of the engine.

The fork rods are installed in the left bank of the engine. Serrations on the sides of the rod at the bottom match the serrations on the two-piece hinged-bearing basket. Rods and bearing baskets are machine-fitted in sets and are numbered to match. If either the basket or the rod is defective, both parts must be replaced as a complete unit.

Describe the connecting-rod bearings.

Answer: Connecting-rod bearings consist of both upper and lower shells. The upper shell is held in position by dowels in the fork rod. The lower shell is doweled to the bearing basket, which holds the bearing assembly rigidly in place and ensures proper alignment with the crank journal. The outside of the upper shell provides the bearing surface for the blade rod. The inside surface of the bearing shell is lead-tin plated to permit rapid seating on a worn crank journal. If this lead-tin plating has worn off, the bearing shells must be used on the same crank journal, and not changed to other crank journals in the engine.

No adjustment of connecting-rod bearings is provided. When bearing clearances exceed the limit, replace them with new bearings.

The connecting-rod bearings are lubricated by oil from the adjacent main bearing, fed through drilled passages in the crankshaft. A hole drilled through the upper bearing shell conducts oil to the blade rod bearing surface.

Describe piston construction in the locomotive diesel engine.

Answer: The alloy cast-iron pistons used in the engine are of the two-piece or "floating" type. The body of the piston is supported by a piston-pin carrier which is held in position within the piston by 3/16-inch snap ring. A .060-inch copper thrust washer is inserted between the piston-pin carrier and the piston bearing plate.

This type of construction permits more efficient cooling of the piston crown, eliminating "hot spots," better lubricaton of the piston pin and bushing, and more evenly distributed piston wear and load.

The piston cooling oil is directed through a drilled passage in the piston-pin carrier and discharged through another drilled passage diametrically opposite. The piston pin and bushing are lubricated by the constant flow of piston cooling oil.

The pistons are treated by a process which etches the surface, giving it a dark gray color. This treatment, known as the Lubrize process, helps to retain the lubricating oil on the surface of the piston.

Describe crankshaft construction of the Model 567 locomotive diesel engine.

Answer: The crankshaft is of drop-forged carbon steel with electro-hardened (Tocco-hardened) main and crank throw journals. Drilled passages in the crankshaft provide a means of carrying a continuous flow of lubricating oil to the main and connecting-rod bearings.

The main generator armature is, in effect, the flywheel for the engine and is attached to the crankshaft by means of a flexible coupling, as shown in Fig. 30-9. The coupling consists of two disks, one bolted to the crankshaft flange, the other bolted to the armature flange. The outer edges of the disks are bolted together with a wide rim, or spacer, between them. One large bolt locates the disks and spacer in the correct position. A special tool is used to remove the coupling bolts. The rim is drilled for an engine turning bar and marked in degrees to facilitate timing the engine.

Describe main bearing construction.

Answer: The main bearing shells are solid bronze with a lead-tin-plated bore which helps to form a better bearing surface during the break-in period. The bearing shells are precision made, requiring no hand scraping or shims for fitting.

The bearings have tangs which locate them in the correct position and prevent them from turning. All lower bearings have three tangs, except the center bearings on the 16-cylinder engine, which have only two tangs. The upper bearings have one tang, making it possible to rotate them out of position for inspection or replacement.

Lubricating oil for the main bearings is received from the engine lube-oil manifold through a passage in the frame. The oil flows into a groove in the upper bearing, then through a drilled passage in the crankshaft to the adjacent connecting-rod bearing.

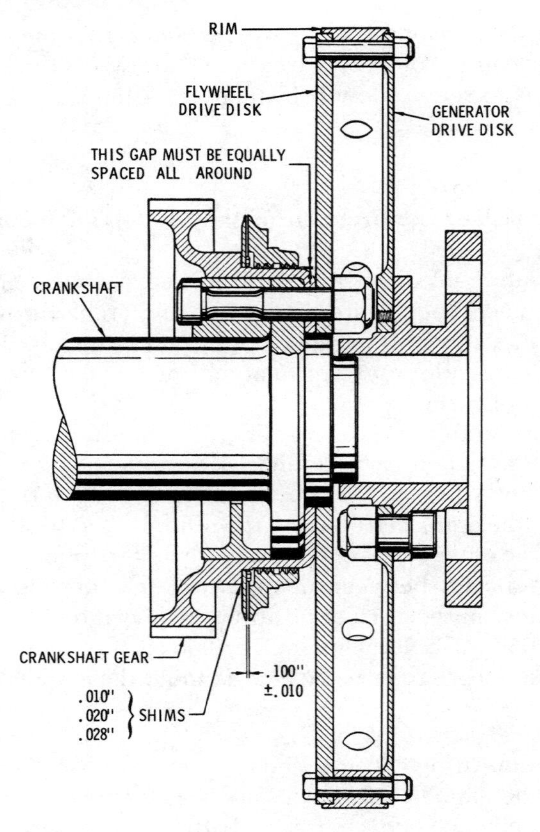

Fig. 30-9. General Motors Model 567 diesel engine showing crankshaft rear-end details and generator coupling.

What is the function of the harmonic-balancer hub in the Model 567 engine?

Answer: The harmonic balancer is located on the front end of the crankshaft and consists of a two-piece hub, laminated springs, and a rim. The function of the harmonic balancer is to dampen torsional vibration inherent in all crankshafts. Each spring pack contains approximately 84 leaves. These packs are depressed 3/32-inch by the pivot pin. The springs receive lubricating oil from the engine through drilled passages in the harmonic-balancer hub.

Describe the camshaft and camshaft drive in the Model 567 engine.

Answer: The camshaft in each bank of the engine is made of flanged segments, one segment for each cylinder. These segments are bolted and doweled together, each segment being marked with the letters "A" and "B" on the flange for correct location. Each segment contains an injector cam, two exhaust-valve cams, and two bearing journals. Flanged stub shafts are bolted to both ends of the camshaft. The camshaft drive gear and counterweight assembly is bolted and doweled to the rear-end stub shaft. A counterweight is also mounted on the front-end stub shaft.

The camshafts are driven by a train of spur gears from the crankshaft gear at the end of the engine. The idler gears turn on stub shafts mounted on the end plate of the crankcase. Oil lines carry engine lubricating oil to each stub shaft floating bushing.

How are the camshaft bearings supported?

Answer: The camshaft bearings are supported by bearing blocks mounted on the top deck plate of the crankcase. These blocks are correctly located by square keys. The bearings consist of upper and lower shells, each equipped with two tangs. These bearings are lubricated by engine lubricating oil that flows through the bore of the camshaft and through radial holes to each bearing. The stub shaft bearings also consist of upper and lower shells which are supported by brackets bolted to the end plate of the crankcase. The thrust of the camshaft is taken by the rear-end stub shaft bearings.

What is the purpose of the over-speed trip mechanism on locomotive diesel engines?

An over-speed trip mechanism is provided as a safety feature on all engines to stop the injection of fuel oil into the combustion chambers should the engine speed become excessive.

How does the over-speed trip mechanism work?

Answer: If the engine speed should increase to approximately 880 rpm, the over-speed trip mechanism will shut down the engine. Fig. 30-10 shows this mechanism in both the latched and tripped positions. In the latched position a notch in the reset lever engages with the trip pawl, so that the pawl camshaft is held away from the

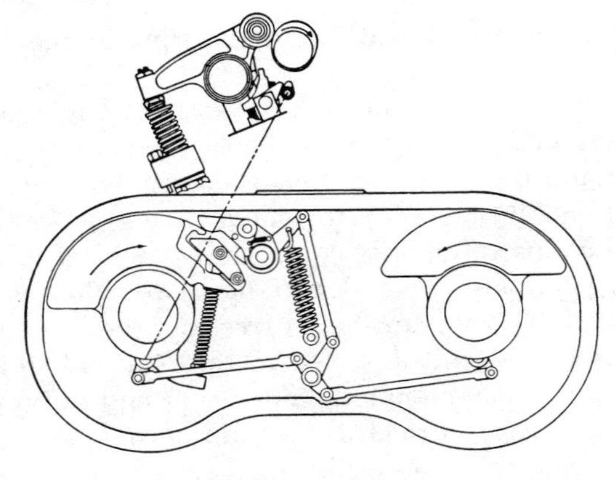

(A) Latched position.

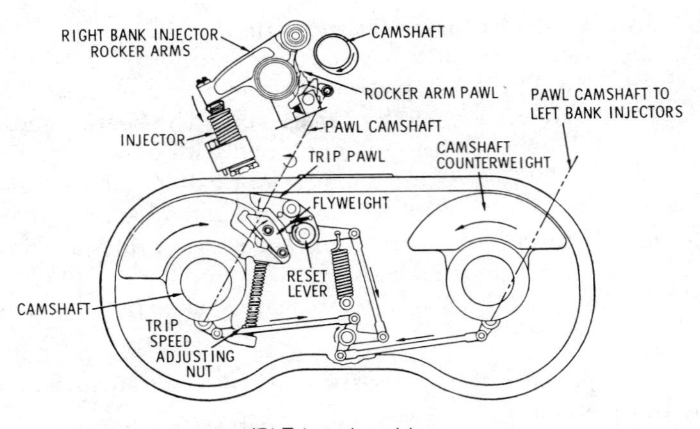

(B) Tripped position.

Fig. 30-10. Over-speed trip mechanism of General Motors Model 567 diesel locomotive engine.

rocker arm. Thus the engine camshaft is free to operate the injector rocker arm in the usual manner.

When the engine speed reaches the setting of the over-speed trip, the centrifugal force of the flyweight (mounted on right-hand camshaft counterweight) overcomes the tension of the spring, and the flyweight moves out and engages the trip pawl. This pushes the shoulder of the trip pawl out of the notch in the reset lever and the spring-loaded linkage rotates the pawl camshaft. The pawl camshaft then engages the injector rocker arm pawl. Then, as the engine camshaft lifts the injector rocker arm, the rocker arm pawl snaps into a notch at the back of the rocker arm, holding the injector in a depressed position, so that the camshaft no longer operates the injector plunger. This will shut down the engine.

To restart the engine, the over-speed trip reset lever must be moved in a counterclockwise direction until the trip pawl snaps into the notch in the reset lever. When the engine is rotated, the engine camshaft will raise the injector rocker arm sufficiently to allow the rocker arm pawls to move away from the notch in the rocker arm. The injector rocker arms will then resume their normal operation.

Describe the lubrication system of the Model 567 engine.

Answer: The engine lubricating oil system (Fig. 30-11) is a pressure system using two positive-displacement and gear-type pumps combined in a single unit and separated by a spacer plate. One pump delivers oil for the pressure lubricating system, the other for piston cooling. The oil supply for these pumps is drawn from the oil cooler tank strainer chamber through a common suction pipe.

A scavenging-oil pump is used to draw oil from the engine oil pan through a strainer, pump it through the lube-oil filter to the cooler core section of the oil cooler tank, and return it to the strainer chamber. Excess oil overflows to the oil pan.

Describe the operation of the oil pumps.

Answer: When the engine is started, the pressure pumps pick up oil from the strainer chamber. Oil pressure may read low on the instrument panel gauges until the initial cycle of circulation is completed.

If the oil is too cold and too heavy or the cooler core is dirty, the oil will flow over the cooler core into the strainer chamber to keep the pressure pumps supplied. A constant oil level is maintained in the cooler core chamber by a baffle plate or "dam." Oil flowing over this dam returns to the engine oil pan. This keeps the scavenging-pump suction pipe under oil.

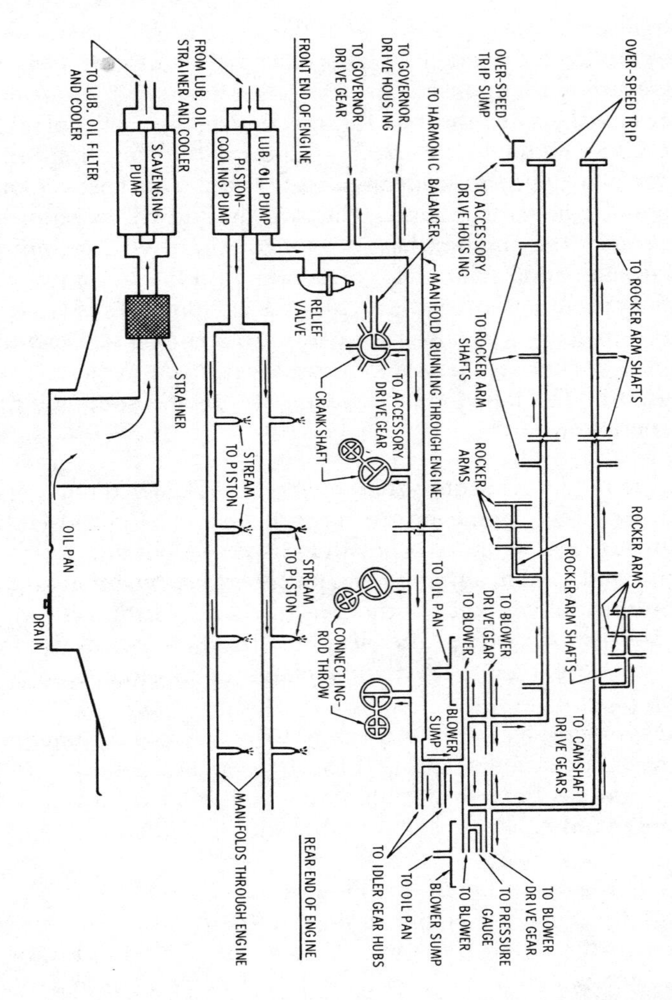

Fig. 30-11. Lubricating oil diagram of General Motors Model 567 diesel locomotive engine.

Where is the proper oil level for the Model 567 diesel engine?

Answer: With the engine running, the oil level should always be between the "low" and the "high" marks on the bayonet gauge in the engine oil pan. The oil level can be checked with the engine running at any speed. Oil is added by removing the filler cover on the oil cooler tank strainer chamber, and pouring the required amount through the strainer basket.

When the engine is stopped, all the oil in the cooler core chamber will flow through the pilot hole (located at the top of the strainer chamber) and into the engine oil pan bayonet-gauge reading to the "system charged" level. This level is below the "system uncharged" level because some oil is trapped in the lube-oil filter, oil lines, and engine.

What is the proper oil pressure reading for the Model 567 diesel engine?

Answer: Lubricating oil pressure must be maintained at all times. When starting a cold engine, allow it to idle for some time and see that pressure starts to build up almost immediately. The pressure should rise on a cold engine to about 40 pounds and it will run between 20 and 30 pounds at 800 rpm with the engine warm. At idling speed, pressure should be at least six pounds.

With normal operation, the lubricating oil pressure should not drop below 20 pounds per square inch with the engine running at 800 rpm. However, if the water temperature cannot be held below 180°F., it will be permissible to operate the engine with lubricating oil pressure, at the rear end of the engine, as low as 15 pounds per square inch. This low limit on pressure can be permitted only when the oil temperature is high as a result of the water temperature being above 180°F. As the piston cooling system has no relief valve, pressure is dependent upon oil temperatures, viscosity, and engine speeds. Average pressure will be 20 to 30 pounds per square inch with the engine running at 800 rpm and 4 to 8 pounds per square inch at idling speed, with a warm engine.

Describe the water-pump operation on the Model 567 diesel engine.

Answer: The water pumps are of the centrifugal type and rotate in a counterclockwise direction. They are mounted on the front of the engine. The 12- and 16-cylinder engines each have only one pump. Pumps are identical for all engines except for the position of the impeller housing. The one-piece cast bronze impeller is keyed to the pump shaft and held in position by a shaft nut.

The shaft rotates in two ball bearings which are mounted in the housing. The inner bearing is held in place by a retainer and snap ring. The outer bearing is fitted with a water slinger that prevents any water leaking past the seal from entering the bearings. The bearings are lubricated by means of an oil cup in the pump housing. The pump shaft is equipped with carbon and Garlock seals. Water leaking past the seals will run out through a hole in the bottom of the casting.

Describe the cooling water system on the Model 567 diesel locomotive engine.

Answer: The cooling system of the engine consists of the radiators, oil cooler, and water-tank assembly in addition to the water pumps, cooling-water manifolds, and passages of the engine. Water is drawn from the water tank by centrifugal water pumps located on the engine. These pumps discharge water to the water-inlet manifold through the cored passages of the cylinder liner and cylinder head, and out through the outlet manifold. From the engine, the water flows to four groups of radiator sections located under the roof of the locomotive body. Here the water is cooled and returned through the oil cooler to the water tank.

How is the blower driven?

Answer: Each blower is driven by a blower drive gear in the camshaft gear train. A flanged quill shaft extends through one of the tubular rotor shafts and is bolted to the rotor gear. The opposite end of the quill shaft is serrated and fits into a hub mounted on the blower drive gear.

Describe the operation of the blower system.

Answer: The blower (Fig. 30-12) consists of a pair of lobed aluminum rotors which revolve in a closely fitted aluminum housing. Each rotor has three helical-shaped lobes. This type of construction produces a large volume of air at low pressure and ensures a positive supply of air to the engine, regardless of engine speed. The 6- and 8-cylinder engines have one blower and the 12-and 16-cylinder engines have two blowers.

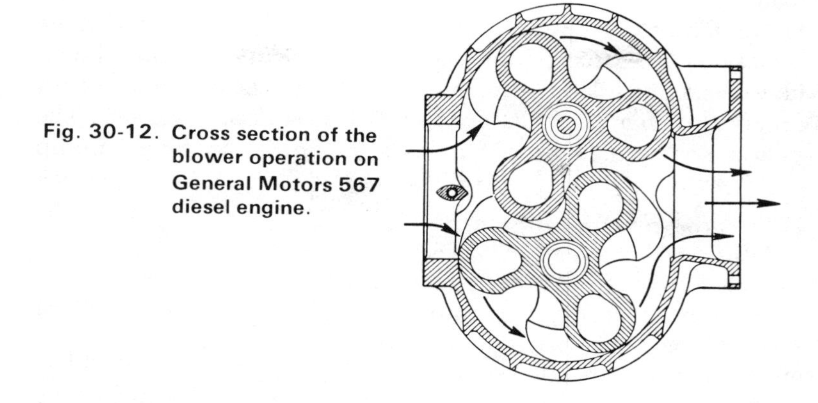

Fig. 30-12. Cross section of the blower operation on General Motors 567 diesel engine.

Each rotor is pressed onto a tubular steel shaft. The engine ends of these shafts form journals which carry the rotors in bearing blocks bolted to the blower end plate. The ends of the shafts which extend through the end plate on the opposite end of the blower are serrated. Flanged hubs, with a serrated bore, are pressed onto the serrated tubular shafts and serve as bearing journals and drive flanges for a matched pair of helical rotor gears. This construction provides a fixed relationship between the rotors and the rotor gears. Bearing blocks on the outer end of the blower are bolted to the end plate and carry the thrust from both directions. All bearing blocks in the blower have sleeve-type Babbitt bearings, which are diamond-bored after application.

The blower bearings are pressure lubricated by the engine lubricating oil which is supplied to the blower housing by a tube at the top of the camshaft drive housing. Lubricating oil for the bearings is supplied through drilled passages in the blower end plates. The oil is conducted from the front to the rear end plate through a seamless tubing which is cast into the blower housing. The rotor gears are lubricated by the gear teeth dipping into a reservoir of oil, the level of which is maintained by a standpipe at the inlet end of the blower drain line. The oil returns to the camshaft drive housing through a drain from the bottom of the blower end-plate cover.

Oil seals are fitted into recesses in the end plates to prevent oil from leaking around the rotor shafts and entering the rotor housing.

Gaskets are not used between the end plates and the blower housing. A fine silk thread around the housing, inside the stud line, together with a very thin coating of nonhardening gasket compound, provides an airtight and oiltight seal.

How is the governor driven?

Answer: The governor is driven through a 90° bevel-gear drive mounted on the front of the engine. The bevel-gear drive is driven from a spur gear in the accessory drive-gear train by means of a serrated shaft. The governor drive assembly is lubricated by the engine lubricating oil through drilled passages in the housing.

The governor is a device used for controlling the speed of the engine at a setting determined by the throttle position. On multiple-unit road locomotives, the speed control arm on the governor is actuated by the electropneumatic governor control. The mechanical throttle linkage on switches is connected directly to the governor control arm.

Describe the operation of the fuel oil system.

Answer: The fuel pump draws fuel oil from the supply sump in the fuel tank through a suction line and suction filter. After leaving the pump, the fuel is pumped through the Ful-Flo filter and the sintered bronze filter to the injectors. The injectors use only a small part of the fuel pumped through them. The surplus fuel oil lubricates and cools the injectors and then returns to the fuel tank through the five-pound relief valve and sight glass. The use of the correct grade of fuel oil is of the utmost importance.

Describe the operation of the fuel injector in the Model 567 diesel locomotive engine.

Answer: Each cylinder in the engine is equipped with a single fuel injector. The injector is seated in a tapered hole in the center of the cylinder head so that the end of the injector is flush with the bottom of the cylinder head. The injector is held in position by a crab secured with a stud and a nut.

The external working parts of the injector are lubricated by engine oil that flows from the end of the rocker arm to the top of the plunger follower guide. The internal mechanism is lubricated and cooled by the fuel oil that flows through the injector.

Fuel oil enters the injector, passes through the sintered bronze filter in the inlet passage and fills the annular supply chamber around the bushing. The surplus fuel flows out of the annular supply chamber through another sintered bronze filter in the outlet passage. The filter in the outlet passages prevents a reverse flow of fuel carrying dirt into the injector when the engine is shut down.

Fuel is injected into the cylinder by the reciprocating motion of the plunger. The plunger is operated with a constant stroke, by a rocker arm from the camshaft. The adjustment screw at the end of the rocker arm is used for timing the plunger stroke.

The quantity of fuel injected into the cylinder is controlled by rotating the plunger. The plunger is rotated by the control rack which is operated by a lever connected to the injector linkage. An adjustable link is provided for setting the control rack.

Upon what factors does the maintenance schedule of diesel electric locomotives depend?

Answer: It is necessary when setting up a maintenance schedule to have a predetermined value by which the proper intervals for certain items to receive maintenance can be specified. Yardsticks that may be used are:

1. Mileage.
2. Time intervals.
3. Fuel consumption.
4. Hours of service.
5. Revolutions performed by the engine.
6. Electrical output of the generator.

None of the foregoing factors, however, will give an absolute indication of the amount of work performed by all parts of the locomotive.

The mileage made by a locomotive in a given period is dependent upon operating and grade conditions and train tonnage, and is not a consistent indication of the amount of work performed by the locomotive power plants.

However, mileage does have a direct bearing on the service rendered by the wheels and their connecting parts.

The use of time intervals such as one month, three months, six months, or twelve months as intervals for doing certain kinds of work on the locomotive is a very simple and convenient method of scheduling the work, due to the fact that the government requires that certain work and inspections be made at the same intervals and records maintained accordingly.

The use of the amount of fuel oil consumed by an engine as a yardstick for scheduling maintenance requires that a method be provided for measuring the fuel oil and maintaining a record of the amount consumed by the locomotive.

The amount of fuel consumed by a locomotive for power indicates quite accurately the amount of work performed, but it does not indicate the load factor while the work was being done, which has a very important bearing on the resulting condition of the engine.

The use of actual hours of service as a yardstick to determine when maintenance is needed requires maintaining an accurate record in order to determine what the actual hours of operation were. This is rather difficult to do, inasmuch as the locomotive may be operating in and out of several terminals and have some out-of-service time at each point, which is not a true indication of the amount of work done. In addition, the load factor and operating conditions causing locomotives to stand idle are not known.

In order to use the number of revolutions performed by the engine as a yardstick for maintenance, it would be necessary that the diesel engines be equipped with a revolution counter. These counters give a true indication of the number of revolutions the engine has made, but they do not indicate how many revolutions were performed with the engine idling and when operating at partial or full load.

The use of generator output as a yardstick for determining maintenance requires that the power plant be equipped with instruments for recording the amount of energy produced. These instruments would give a true indication of the work performed by the generator but not the work performed by the engine, inasmuch as no current is generated while the engine is idling.

A survey of the maintenance schedule in effect on some of the railroads at the present time according to The Railway Mechanical Engineer indicated that the common yardsticks for scheduling maintenance on switch locomotives are time intervals such as the time of the monthly, quarterly, semiannual, or annual inspections as required by government regulations; whereas mileage is used as the yardstick for road locomotives with a tendency to group the work items so that the majority of them will fall due at the same time as the government inspections and thus prevent holding the locomotives out of service unnecessarily.

A few railroads have established classified repairs in their maintenance schedules covering older types of equipment, but the tendency at the present time on more modern equipment is to handle the maintenance currently, and make heavy repairs as inspections indicate the work is necessary.

What are the component parts of the fuel system in a diesel electric locomotive?

Answer: In a typical diesel electric locomotive, each unit constituting the locomotive has its own fuel tank with a capacity of approximately 1,200 gallons. With reference to Fig. 30-13 showing the fundamental fuel system of a diesel electric locomotive, the tank sump and the supply sump are connected by the emergency fuel cutoff valve.

The suction line to the engine enters the supply sump and can draw fuel only if the cutoff valve is open. The emergency fuel cutoff valve is a federal requirement and has to be operative from inside the unit as well as from both sides near the filling plug. It is a spring-loaded valve that is to be used in event of fire to stop the engine when it cannot be stopped in the usual manner. Each fuel tank has two vents, one on each side, with four-inch flame arrestors terminating outside the locomotive. This arrestor voids any possibility of a fuel tank explosion.

Before fuel oil enters the supply pump, it passes through the suction filter. Until recently, these were three-thousandths spaced wire-wound drums. The handle which operates the drum scraper

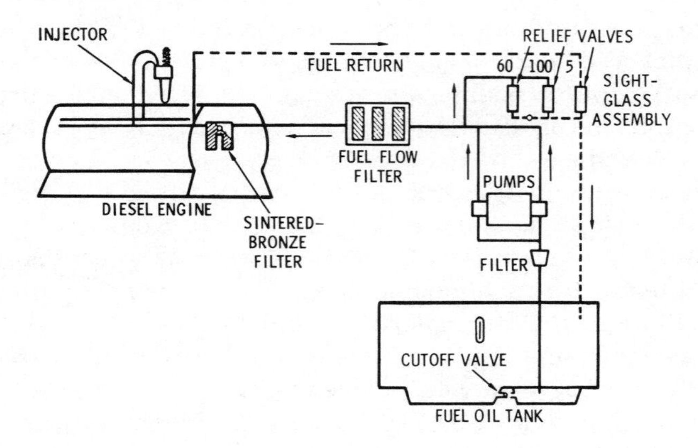

Fig. 30-13. Diagram of fuel system of a diesel locomotive engine.

should be turned every hour while the engine is running. The latest filters, however, are waste-packed and should be changed regularly. The elements are repacked with eight ounces of Wastex.

The fuel pump is either single or dual. The pumping principle is known as the internal-gear principle. An electric motor furnishes the power applied to the rotor, which is transmitted to the idler gear with which it meshes. The space between the outside diameter of the idler and the inside diameter of the rotor is sealed by a crescent-shaped projection fastened to the cover. As the teeth come out of mesh, there is an increase in volume, which creates a partial vacuum. The fuel oil enters the pump to fill this vacuum and stays in the space between the teeth of both the idler and rotor. When the teeth mesh, the fuel oil is forced from this space and out of the pump. The capacity when pumping against 35 pounds pressure is two gallons per minute. However, it is capable of pumping against a pressure of 150 pounds.

The Ful-Flo filter is connected between the fuel pump and the sintered bronze filter. Inside the filter are three elements made of closely wound string. When the sight glass which indicates 60 pounds pressure shows fuel by-passing, turn the handle on the duplex filter to determine whether it is the duplex filter or the Ful-Flo filter which needs cleaning. If the Ful-Flo filter is dirty, all three elements should be replaced and the drain plug at the bottom removed to drain any water accumulation. After reassembling the filter, be sure to vent all air from the chamber through the vent valve on top.

The duplex sintered bronze filter is mounted on the accessory end of the engine and has a handle which can be swung in three positions. With the handle to the left, the left-hand element is cut off. With the handle to the right, the right-hand element is inoperative. With the handle straight down, both elements are filtering. However, the handle should be either to the right or to the left with only one element filtering. The cleaning frequency can be determined best by inspection until a definite schedule can be established. From the duplex bronze filters, the fuel oil goes to the supply header or manifold running to all cylinder heads in both banks of the engine. Just as the fuel oil enters the injector, it is filtered for the fourth time by small thimble-like sintered bronze filters. The purpose of all filtration is solely to protect this finely machined, yet rugged, equipment from getting any dirt inside.

A surplus of fuel oil passes through the injector above that needed for proper power output. This surplus of fuel returns to the main fuel oil tank through the cross-flow lines and 5 pounds sight glass of the relief-valve assembly after cooling and lubricating the precision-made parts of the injector.

In the sight glass and relief-valve assembly the left-hand relief valve is set for 60 pounds, the center relief valve for 100 pounds, and the right-hand relief valve for 5 pounds. The adjustment should not be changed unless it is definitely wrong and then only by a competent maintenance man. The right-hand glass and relief valve passes the return fuel from the engine. A drop in the fuel level or an empty sight glass will indicate that the engine is not receiving its full amount of fuel. Air or gas will show up in the sight glass as bubbles.

The left-hand or sight glass which indicates 60 pounds will show fuel oil passing if either of the filters on the pressure side of the fuel pump collects enough dirt to build up a pressure of 60 pounds, thus causing the fuel to by-pass the filters and the engine and return to the tank. Under such circumstances, close the globe valve between the relief valve and the return line, again forcing the fuel oil through the dirty filters at a higher pressure. This will allow the engine to be operated until the maintenance point is reached and the filters changed or cleaned. If the cleaning of the filters is neglected and the back pressure increases to 100 pounds, the fuel flows through the center sight glass and back to the tank, bypassing the engine and resulting in loss of power, as nothing can be done except clean the dirty filters.

On each side of the unit is located a direct-reading sight-level gauge, 4½ inches from the top. This gauge serves the purpose of checking the fuel level when refueling. The remote type aboard the unit serves the purpose of checking the actual gallonage remaining in the tank. This gauge is actuated by air at reduced pressure, forcing the fuel out of the tank pipe, and exactly balances the weight of the liquid in the tank. The indicator measures the air pressure necessary to balance the weight of the liquid in the tank and is calibrated to read the amount in gallons.

If the fuel pump motor is running but the pump is not delivering fuel up to 5 pounds on the sight glasses, what could be the cause?

Answer: The following are the defects that can and might exist in the fuel system and will interfere with proper operation of the diesel engine:

1. Inadequate supply of fuel in tank.
2. Emergency fuel cutoff valve closed.
3. Broken fuel pump shaft of loose coupling.
4. Leaking supply line.
5. Dirt lodged under relief valve.
6. Defective fuel pump, pump seals, or scores on either the pump or the housing.

If the electric pump is not running, what should be done?

Answer: The following possible defects should be investigated:

1. Burned-out fuses (check with test lamp).
2. Check main battery switch.
3. Check electrical connections at the motor.
4. Check to see if fuel pump switches are on and in good condition.

The following actions and procedures can be helpful when the fuel system fails to operate properly:

1. The fuel pump motor will labor when filters in the delivery line are stopped up.
2. The fuel pump motor will be unloaded with loss of fuel when the emergency fuel cutoff valve is closed; there is no fuel in the tank; there is a leak in the suction line; and with a loss of prime in the fuel pump.
3. A leak in the fuel oil delivery line will be noticeable by the loss of fuel, and if a leak exists between the fuel manifold and the injector connection, lube oil will be diluted with fuel oil. With the engine shut down, operate the fuel pump to find the leak in the fuel line.
4. If air bubbles exist in the 5 pounds sight glass with the fuel pump operating and the engine shut down, check for an air leak in the suction line. If air bubbles occur with the engine operating at maximum power, it indicates that the injector valve is allowing gases to pass into cross-flow line.

   The defective injector check valve can be located by feeling the return line near the injector. If the line is hotter than normal, bar the engine over until the piston of that cylinder is below the scavenger ports. Operate the fuel pump with the injector rack disconnected and in full fuel position. Then inspect the crown of the piston through the liner ports for fuel accumulation.
5. Improper fuel supply to the injector will be indicated by loss of power. Check the servomotor piston at the governor and check the load regulator for correct loading.

What is the arrangement of the lubricating oil system in a diesel electric locomotive?

Answer: The lubricating oil system of a typical diesel power plant (Fig. 30-14) consists of three parts:

1. The main bearing pump, oil switch, and relief valve.
2. The piston-cooling oil pump and "P" pipes.
3. The scavenger-oil pump, standards, and oil separator.

The oil pan of the engine is large enough to hold the main

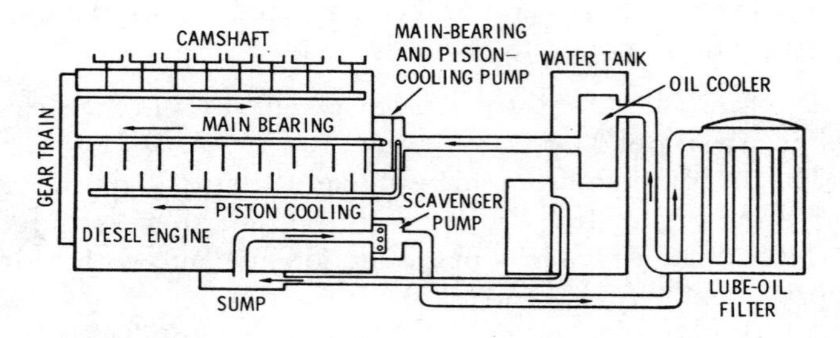

Fig. 30-14. Diagram of the lubricating oil system of a diesel freight locomotive.

portion of the lubricating oil of the oil system, classifying the engine as "wet sump." A bayonet-type oil gauge is used to take oil level readings, which are as follows: System uncharged, system charged, high, and low level. These readings can be taken with engine either running or stopped. The capacity of the sump is 180 gallons. To fill, pour this amount of oil into the filter basket strainer. From here it will fill the oil strainer chamber through the pilot hole. It will also flow into the engine oil pan. As a check on the amount of oil entering the system, the oil level should be at the "uncharged level" on the bayonet gauge when 180 gallons have been added, after the system has been drained.

The systems of lubrication of all diesel engines are fundamentally the same, and whenever pistons, liners, crankshafts, or other parts have been damaged by scoring, the major proportion of such damages, if analyzed correctly, would be assigned to improper maintenance and operation such as:

1. Failure to clean and flush the engine and lubricating oil system properly after dilution (either water or fuel) or after the replacement of scored liners, pistons, bearings, or other internal damage.
2. Operating with too low oil pressure.
3. Water leaks from cooling system washing the lubricant from the piston assembly, resulting, in some instances, in piston seizures.
4. Not cleaning engine lubricating oil system at stated intervals in accordance with manufacturer's recommendation.

5. Failure to handle lubricating oil in clean containers.
6. Operating the engine with either high or low water jacket temperature, high temperature causing oxidation (sludging) of oil, and low temperature causing water, a product of combustion, to wash the lubricant from contact surfaces, resulting in wear at three times the normal rate.
7. The use of reclaimed lubricating oil other than in proportions covered in instructions.

What types of pumps are used in the lubricating oil system?

Answer: The lubricating oil pumps as well as the scavenger pumps are of the positive displacement type and the pump rotors are helical spur gears.

What is the method of mounting the lubricating oil pumps?

Answer: The two lubricating oil pumps (piston cooling and main bearing) are combined in one unit and mounted on the accessory end (or pump end) of the engine, with one suction line connected to the lubricating oil supply.

What is the capacity of the lubricating oil pumps?

Answer: The piston-cooling portion of a typical pump has a capacity of 53 gallons per minute at 800 rpm and delivers lubricating oil through the manifold to the bottom of each cylinder, where a nozzle directs a stream of oil into each piston.

The main-bearing pump has a capacity of 107 gallons per minute and delivers oil to the pressure lubricating system.

What are the mounting method and capacity of the scavenger pump?

Answer: The scavenger pump is mounted on the lower part of the accessory drive housing and has a capacity of 214 gallons per minute at maximum speed. It draws oil through its oil screen and forces it through the filter and oil cooler back to the tank. A constant oil level is provided in the oil cooler on the cooler side of the oil dam. Oil flowing over returns to the oil pan. This keeps the scavenger pump suction pipe submerged in oil and provides a "wet sump."

What method is used for protection against low oil pressure?

Answer: The oil system is protected by a low-oil-pressure switch that consists of a spring-loaded diaphragm actuating a pair of electrical contacts. This low-oil switch should operate at six to nine pounds, that is, it should pick up at nine pounds and drop out at six pounds. When this switch operates, it de-energizes the master magnet valve and drops the engine speed to idle position.

The engine should not be run under such circumstances until the trouble is located and corrected. To test this switch with the engine shut down, place the isolation switch on the engine instrument panel in the "run" position. If the low-oil light comes on and the gong sounds, the switch is operating correctly.

If the main bearing oil pressure drops below 20 pounds (at 800 rpm for hot oil), stop the engine and investigate. Look for a stopped-up suction strainer in the tank, a stuck relief valve, broken oil lines, low oil viscosity, worn oil pump, loss of prime in pump, suction lines leaking and destroying the vacuum, or low oil.

What is the function of the oil-pressure relief valve?

Answer: The function of the relief valve is to keep the oil as it enters the bearing manifold at constant pressure. This spring-loaded relief valve is normally set at 45 pounds and cannot be adjusted while the engine is running.

An accurate check of its setting can be made by attaching a pressure gauge on the oil manifold on the opposite side of the engine. The main bearing pressure gauge is connected to the oil line at the blower end (rear) of the engine. Therefore, with the engine warmed up and operating at 800 rpm, this gauge should register 30 pounds, as there is a 15 pounds drop through the engine.

The piston cooling system has no relief valve or low-oil-pressure switch so pressure depends upon oil temperature and engine rpm.

Usually the pressure will be 20 to 30 pounds at 800 rpm and 4 to 8 pounds at idling speed with a warm engine. If the oil pressure drops below 15 pounds (at 800 rpm for hot oil), stop the engine and investigate the possible causes.

Where is the oil separator located?

Answer: The oil separator is mounted over the main generator. Air is drawn through the oil separator from the crankcase to the air-box blower intake. It is provided with a metal screen which condenses any oil from vapor which is taken from the crankcase. The condensed lubricating oil then returns to the oil pan or sump.

Where is the lubrication oil cooler located?

Answer: The lubrication oil cooler is located in the water tank. It is located there in order that the oil temperature will be reduced to nearly that of the jacket water which is about 163° Farenheit. Therefore if the water jacket temperature is properly controlled and the oil coolers are clean and functioning properly, there should be no breakdown of the oil, causing sludge formation or other oxidation products which are so detrimental to diesel engine operation.

Describe the engine cooling water system.

Answer: The cooling system of the various engines in the locomotive are independent of each other. They are similar to each other in every respect except that the first section (A-units) have heaters to heat the control room. Water is drawn from the water tank by centrifugal pumps located on the accessory end of the engine. The pump impellers are of the single-suction type and operate at 2,440 rpm with an engine speed of 800 rpm, and the combined capacity of the two pumps is 214 gallons per minute at that speed of the locomotive engine.

These pumps draw the water from the tank, which has a capacity of 245 gallons, and circulate it through two tubular passages, one in each bank of the crankcase, which connect to the water spaces between the lower deck plates. From this section the cooling water enters the lower end of the cylinder liner water jacket and flows through the liner and through the cylinder head. From there it discharges into a water space along the upper section of each bank of the engine. The water is then discharged from this chamber into the water outlet elbow connections to the four radiator sections located in the roof as shown in Fig. 30-15.

The four cooling fans in the roof are driven from two speed increasers, one located at each end of the engine. The temperature of the water is controlled by the fans drawing cool air from the outside through the shutters. Closing the shutters restricts the air flow through the radiators and allows the water to heat up. Opening the shutters gives more air flow and causes the water to get cooler. The speed increasers are driven from the main shaft by friction clutches. These clutches are not to be engaged above idling speed; to do so will seriously damage the driving mechanism. In cold water, when the engine temperature is difficult to control, it may be necessary to operate with one clutch disengaged. If this is done, it should be on the accessory end. On the generator end the fan circulates air through the high-voltage cabinet. After the water is cooled in the radiators, it flows down and through the oil cooler, then into the water tank for reuse, thus completing the cycle.

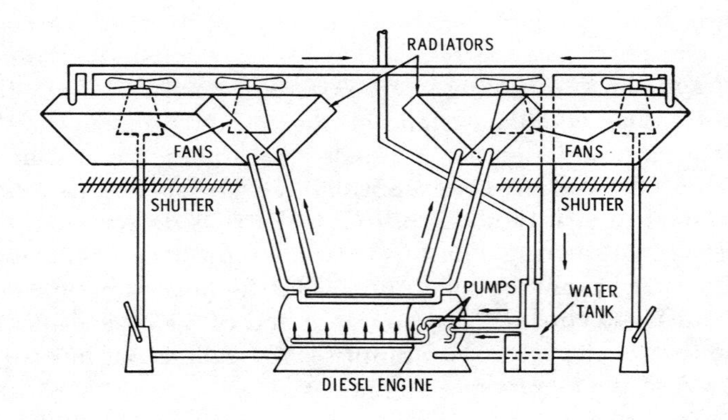

Fig. 30-15. Diagram of the cooling system of a diesel freight locomotive.

There is a water-level gauge located on the water tank. This should be full when operating. The engine should not be operated below the mark indicated on the gauge.

If the engine consumes water and there is no evidence of leaks in the cooling system, it may be that a cylinder lower seal (either rubber or steel) is leaking or a head-to-liner gasket is leaking. In either event, the water will be going out with the exhaust in the form of vapor when the engine is running. Such leaks may be located by removing the cylinder test-valve packing with the engine working. Water will leak to the outside of the engine and will be visible at the cylinder test valve.

What are the customary methods employed when overhauling a diesel engine?

Answer: The detailed procedure to follow depends upon a number of factors, such as the size and type of engine, the facilities available, and the duty and the condition of the engine.

When dismantling an engine, the logical procedure is to start at the top of the engine and work downward, removing the parts one at a time as they become accessible. The diesel engine manufacturer's recommendations and instructions should be followed in dismantling and repairing the engine and its auxiliaries.

Regardless of the make and type of engine, certain general considerations have a bearing on the matter. When an engine is in need of extensive repairs, unit replacement may be made on such items as cylinder heads, pistons, liners, water pumps, and bearings.

Although the procedure may vary from plant to plant, the general routine may be as follows:

1. Remove cylinder heads, dismantle, and clean. Reface valve seats, or if worn to limit, install new inserts or weld in new seats, preferably of specially hardened material, and grind to contour. Grind the valves and if the stems are worn sufficiently to require attention, replate to original diameter or turn down to an undersize dimension and use special valve guides. Renew or build up and remachine the valve guides. Inspect valve levers, springs, spring retainers, bushings, etc. Check the gaskets or seal surface.

2. Remove the pistons and dismantle from the rods. Remove and discard worn rings, thoroughly clean all parts, and inspect for cracks and worn grooves. If the grooves are worn excessively, machine for oversize rings (if a satisfactory method has been developed, machine worn grooves, weld in replacement material, and remachine grooves to standard dimensions). Install new piston rings. Inspect the pins or attachments to the rods.

3. Cylinder liners or sleeves should be inspected and replaced if necessary. Remove any ridge appearing at the top of the piston ring travel. If necessary, lightly hone the interior surface for quick seating of new piston rings. Check the gasket

or seal surface and install all new seal rings if the liner is removed.

4. Inspect the connecting rod and check for straightness and parallelism. Magnaflux for detection of cracks. Clean out oil passages. Inspect, repair, or replace big-end bearing shells, and reinstall with proper clearance.
5. Inspect, repair, or replace main bearing shells and note the condition of the journals. Check the crankshaft with a dial indicator at each crank throw for distortion. Inspect each journal for condition of the surface and measure for out-of-round, blow out, and clean oil passages. Clean the entire crankcase interior thoroughly.

NOTE: If the pins or journals are badly worn or scored, it may be possible to restore them to standard conditions or dimensions by machining, metal spraying, or grinding. This will, of course, involve removal of the crankshaft from the engine, in which instance detailed inspection, including magnetic test, can be made. If wearing or scoring has taken place to a point under the minimum allowable diameter, then the crankshaft of the engine must be replaced.

6. All gear train or trains should be checked as far as possible for backlash. If the crankshaft must be removed, then a more detailed check can be made together with necessary repairs or replacement, such work including camshaft bearings and cams, cam followers, push rods, etc.
7. Superchargers or the blower should be removed for inspection, cleaning, and necessary repairs. Inspect the blower drive and make any repairs necessary. Clean the air intakes and air passages, cleaning or renewing all filter elements. Inspect the exhaust piping and the muffler and clean them.
8. If the condition of the engine requires, remove, dismantle, and repair lubricating oil pumps and inspect the drives. Remove and clean all lubricating oil piping and clean or renew the filter elements on the engine. Renew flexible-hose connections. Check the setting of the regulating valve.
9. Remove, dismantle, and inspect water pumps and check drives. Check and clean all water piping on the engine and renew the flexible-hose connections.

10. Remove the governor, inspect the drive, and install a reconditioned and readjusted unit. Remove the throttle operator, clean, and repair. Check all linkages. Inspect over-speed shutdown devices.
11. Remove all fuel injection equipment (nozzles, high-pressure pumps, injectors) and install either new or factory-reconditioned parts. Check high- and low-pressure piping on the engine and inspect the fittings and clamping. Clean or renew all filter elements. Inspect the fuel pump drive assembly or fuel injector cams and levers and make necessary repairs or renewals.
12. Remove the water and oil radiators, clean thoroughly, and test. Remove the heat exchanger, clean, and test. Remove the locomotive lubricating oil piping and clean. Clean or renew locomotive filter elements. Remove and clean all regulating or relief valves and adjust for proper settings. Clean the interior of the expansion tank and then paint with special mixture. Clean the locomotive cooling water piping.
13. Inspect and make necessary repairs to radiator, fan, and drive, including pulleys, idlers, belts, gear box, angle drive, and fan. Inspect radiators, shutters, and operating mechanisms. Clean the fan, fan housing, shutters, etc.
14. Remove, inspect, clean, and calibrate all indicating gauges, including fuel and water-tank-level indicators.
15. Inspect the fuel transfer pump and motor and make necessary repairs or replace them with a reconditioned unit. Clean and adjust pressure regulating valve. Clean or renew all strainer elements.
16. Drain the fuel tank, flush, and clean.
17. Reassemble the engine, restoring clearances and fits to manufacturer's tolerances and standards as closely as possible. It is not considered good practice to be economical with gaskets and seals. These should be renewed whenever disturbed or where deterioration is known to take place. All bolts and nuts should be tightened and checked with good judgment after the engine has been running for a while. Use either new or reclaimed crankcase oil. Run-in the engine thoroughly.

What type of testing facilities should be provided for testing diesel engines?

Answer: Testing facilities for any diesel engine repair shop should consist of such items as: diesel engine exhaust-gas pyrometer, maximum-pressure indicator, tachometer, rheostat (water or resistor type); for artificial power plant loading, suitable electric instruments, and meter.

What is the general procedure followed in the test on diesel engines?

Answer: It is of the utmost importance that a diesel engine be carefully broken in after extensive repairs have been made. Combined with the running-in process should be a determination of compression and ignition pressure in each cylinder, exhaust gas temperature for each, and the noting of engine speeds, lubricating oil pressures and temperatures, cooling water temperatures, supercharger and blower pressures, horsepower output under varying conditions of current and voltage as indicated by the main-generator characteristic curve, individual notch speed and loading, color and quality of exhaust, general sound and action of the engine under load, and acceleration, etc.

What procedure should be followed preparatory to applying full load to an engine?

Answer: The following general procedure will be useful and should be followed in each instance.

1. After first starting the engine and while idling, determine that conditions are normal in all respects, including proper connecting-rod and main bearing temperatures. Idle the engine for eight hours if main bearings have not been disturbed or for sixteen hours if the main bearings have been removed or replaced.

   A close watch should be kept on the engine during this time and checks of bearing conditions should be made at increasing intervals during the entire eight or sixteen hours of idling.
2. When the idling period is completed, load the engine to the proper degree and run for approximately half an hour in each notch until the highest notch is reached, after which the full

load test can be completed in accordance with load requirements.

How may the horsepower of a diesel locomotive be calculated?

Answer: The horsepower at the rail may be expressed by the following formula:

$$\text{horsepower} = \frac{\text{TE} \times \text{speed}}{375}$$

in which TE equals pounds tractive effort, and speed is given in miles per hour.

This formula may be further simplified to express engine horsepower. Approximately 5 percent of the engine output will be required for driving the auxiliaries such as the compressor, fans for the lighting load, and control load. A reasonable generator efficiency of 94 percent may be assumed, and 90 percent is a reasonable traction-motor efficiency in the normal operating speed range. The transmission losses between the generator and the traction motor may be ignored as they are small. The engine horsepower then may be expressed by the formula:

$$\text{engine hp} = \frac{\text{TE} \times \text{speed}}{375\ (1 - 0.05) \times 0.94 \times 0.90}$$

or

$$\text{engine hp} = \frac{\text{TE} \times \text{speed}}{300}$$

CHAPTER 31

Mercedes 5-Cylinder Turbo Diesel Engine

When was the 5-cylinder turbo diesel introduced?
Answer: The 5-cylinder engine OM617 was introduced in 1974.

Why was the 5-cylinder turbo diesel introduced?
Answer: The 5-cylinder turbo diesel was important for its cylinder dimensions, which represented the best compromise in terms of weight, bulk, and production cost, compared to a 6-cylinder engine.

What type of crankcase ventilation system is used?
Answer: A positive crankcase ventilation system is used (Fig. 31-1).

What type of piston is used in the 5-cylinder turbo diesel engine?
Answer: A cylindrical piston with a circular oil groove has been incorporated on the inside of the piston crown to provide oil

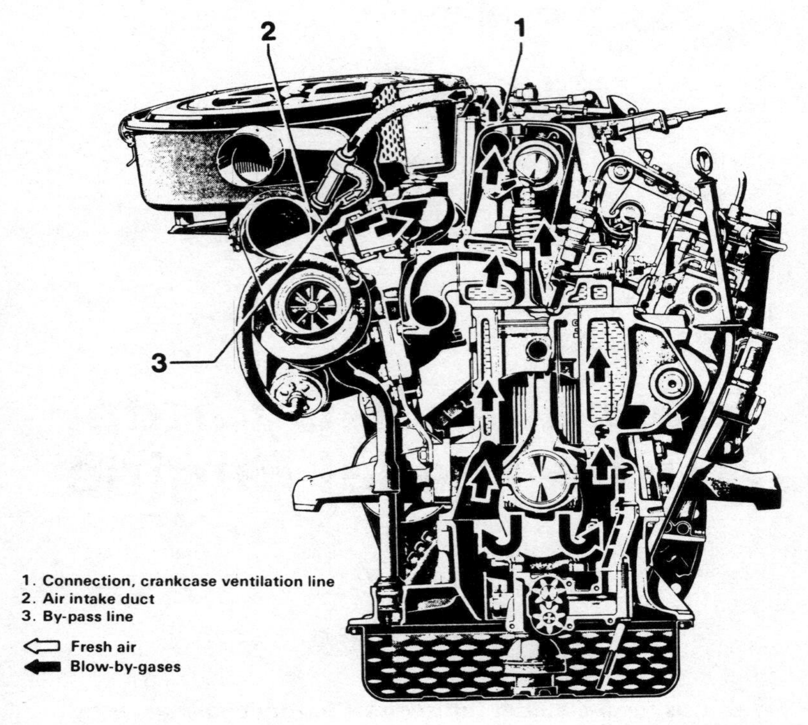

Fig. 31-1. Cross section of Mercedes 5-cylinder turbo diesel crankcase ventilation system.

cooling. This counters the increased thermostress on the pistons (Fig. 31-2).

What rings are used on the turbo diesel engine?

Answer: Two rectangular compression rings with inside chamfers and the oil ring, which is found on the bottom of piston. This twin-scraper ring contains a circumferential coil expander (Fig. 31-3).

How is the piston cooled?

Answer: An oil spray nozzle is used (Fig. 31-4).

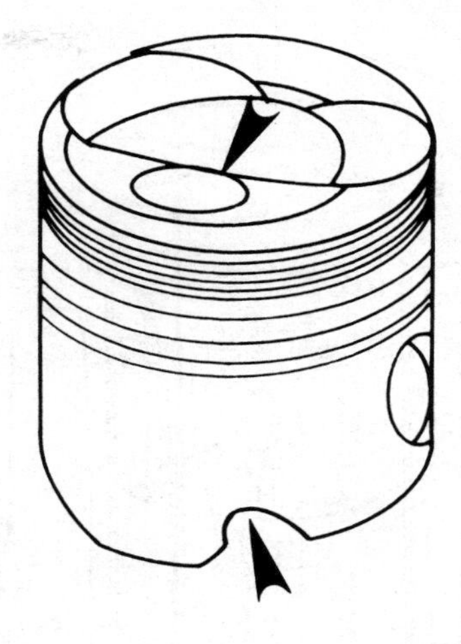

Courtesy Mercedes-Benz

Fig. 31-2. Top view of piston arrows shows piston crown.

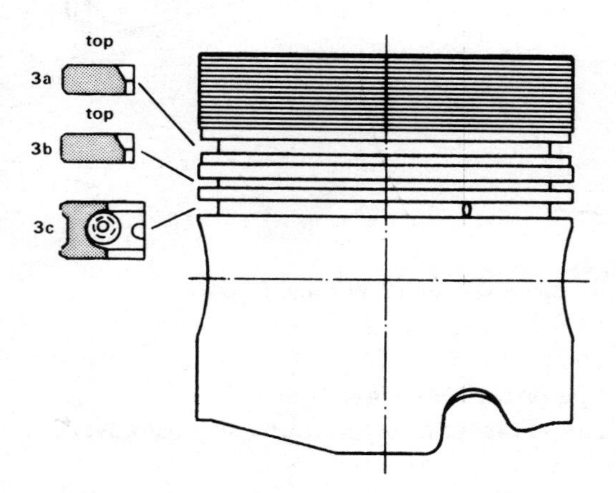

3a. Rectangular ring with inside chamber (3mm thick)
3b. Rectangular ring with inside chamber (2mm thick)
3c. Twin scraper ring with circumferential coil expander (4mm thick)

Courtesy Mercedes-Benz

Fig. 31-3. Cross section of piston.

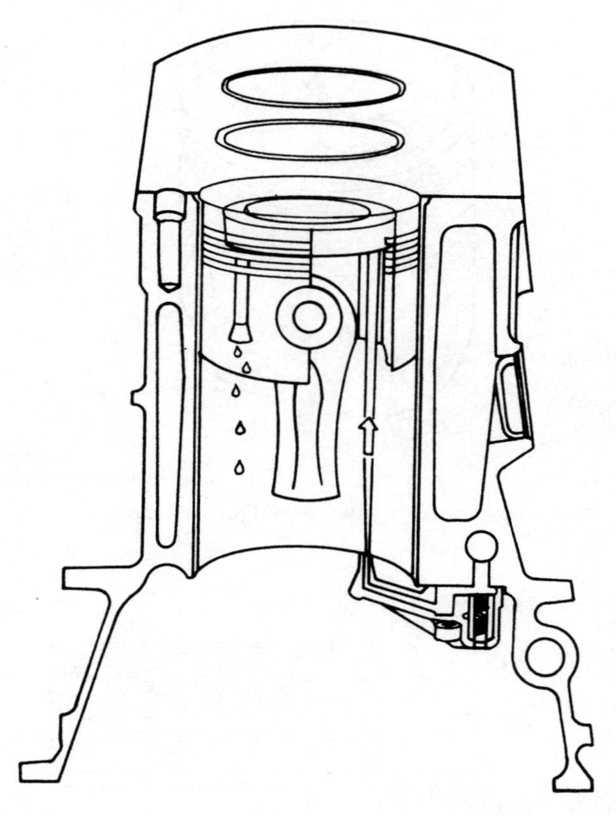

Courtesy Mercedes-Benz

Fig. 31-4. Cross section of oil spray nozzle.

What type of turbocharger is used?

Answer: This engine contains an exhaust gas turbocharger (Fig. 31-5).

How does the turbocharger actually work?

Answer: This turbocharger works by using the aerodynamic energy of the exhaust gases to drive a centrifugal compressor, which in turn delivers high-pressure air to the cylinders of the diesel engine (Fig. 31-6).

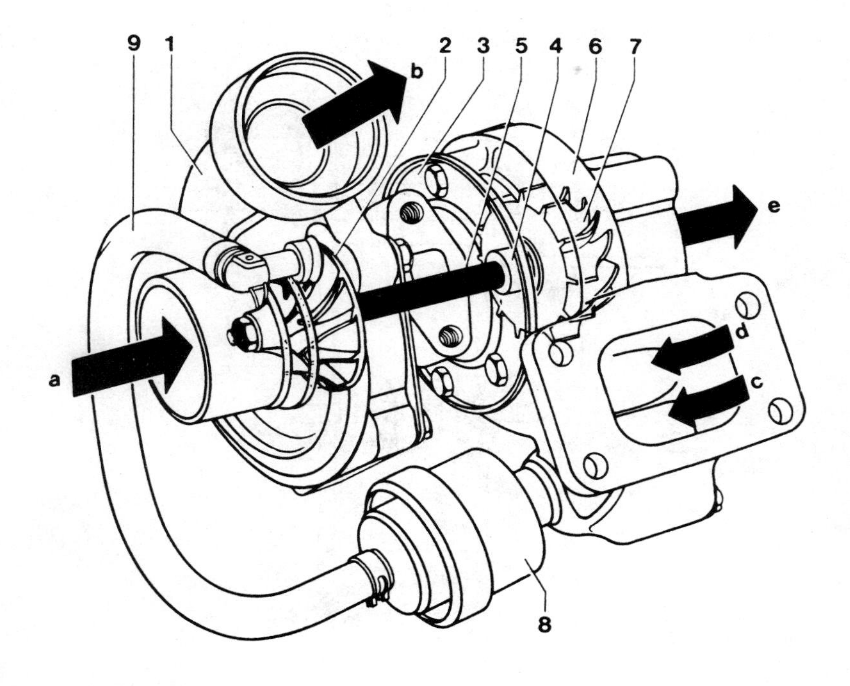

1. Compressor housing
2. Compressor wheel
3. Center housing
4. Bearings
5. Shaft
6. Turbine housing
7. Turbine wheel
8. Boost pressure control valve
9. Connecting hose

a. Compressor intake (fresh air)
b. Compressor discharge (compressed air)
c. Exhaust gas to bypass canal
d. Exhaust gas to turbine wheel
e. Exhaust gas discharge

Courtesy Mercedes-Benz

Fig. 31-5. Diagram of exhaust gas turbocharger.

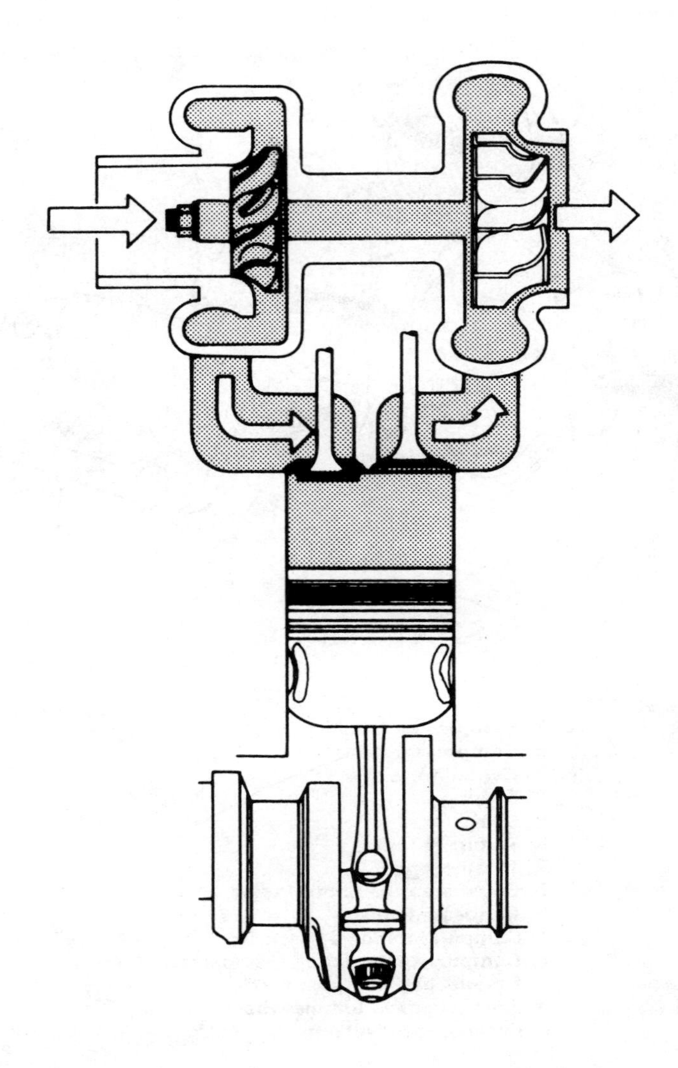

Courtesy Mercedes-Benz

Fig. 31-6. Diagram of fresh-air and exhaust gas flow entering and expelling through exhaust gas turbocharger.

What type of fuel injection pump is used?

Answer: Except for the aneroid compensator and the modulator vacuum control valve, the fuel injection pump is a Bosch-type pump (Fig. 31-7).

Where is the oil pump located?

Answer: The oil pump is fastened with three screws to the first crankshaft-bearing cap (Fig. 31-8).

What type of engine lubrication system is used?

Answer: Since the exhaust gas turbocharger and the piston cooling system have been connected to the engine lubrication system, the oil-filling capacity of the engine and the delivery capacity of the oil pump have had to be increased. Due to high thermoloads, engine oil of viscosity classification 10W must not be used. See Fig. 31-9.

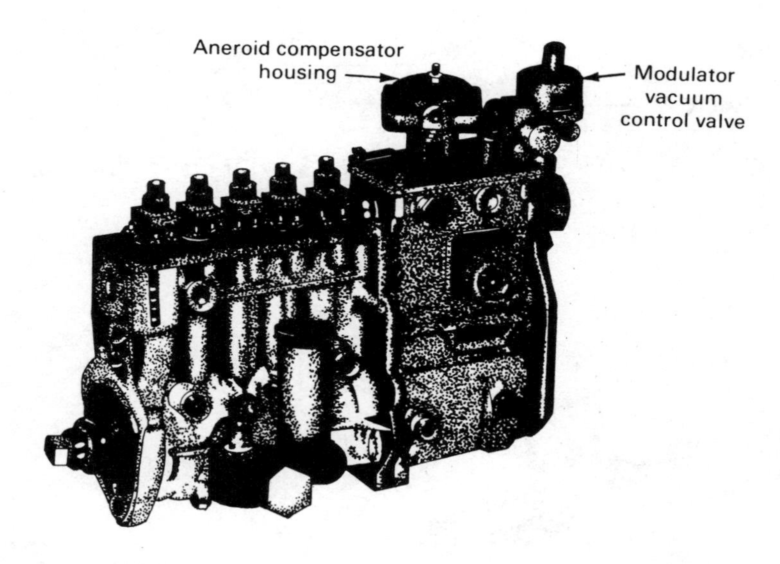

Fig. 31-7. Side view of Bosch injection pump.

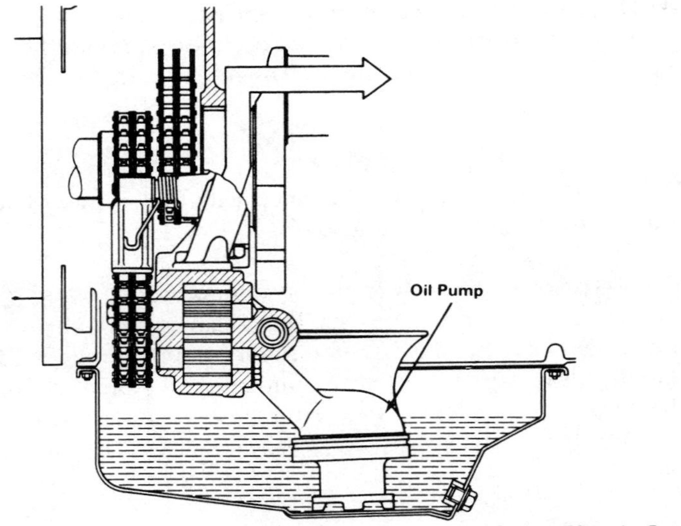

Courtesy Mercedes-Benz

Fig. 31-8. Cross section of oil pump.

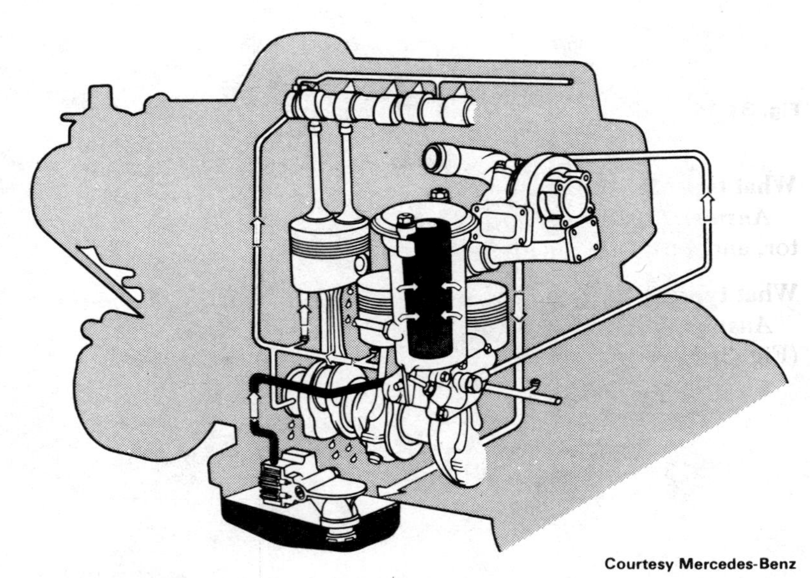

Courtesy Mercedes-Benz

Fig. 31-9. Cross section of oil circulation.

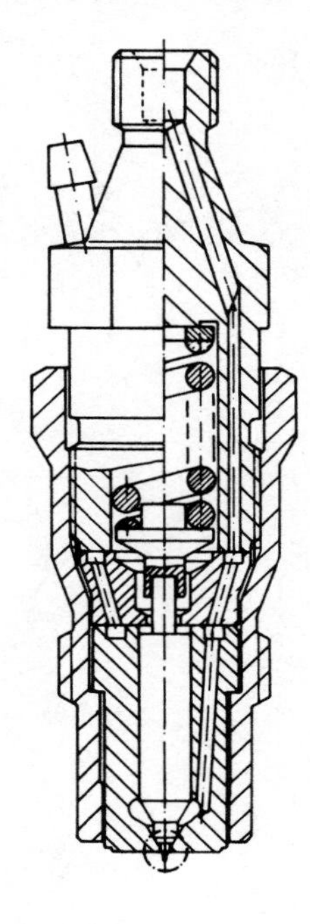

Courtesy Mercedes-Benz

Fig. 31-10. Cross section of Bosch injection nozzle.

What type of cooling system is used?

Answer: A water pump, temperature-control visco-fan, radiator, and a 6-blade fan are used.

What type of injector is used in the Turbo 617 diesel engine?

Answer: A Bosch injector with a central hole in the pintle is used (Fig. 31-10).

CHAPTER 32

Caterpillar D6 Tractor

What is the Caterpillar D6 tractor?

Answer: The Caterpillar D6 tractor is a bulldozer that moves dirt.

What type of engine is used in the Caterpillar D6 tractor?

Answer: The engine that powers the D6 tractor is a 6-cylinder, four-stroke-cycle, valve-in-head, Caterpillar diesel engine.

What type bore does this engine contain?

Answer: This engine contains a 4½-inch bore in the later, or 8U and 9U series, tractors and a 4¼-inch bore in the earlier, or 4R and 5R series, tractors (Fig. 32-1).

What type of injection pump is used?

Answer: A Caterpillar injection pump is used on this diesel. It contains six fuel lines to supply fuel to the engine. Pressure is the

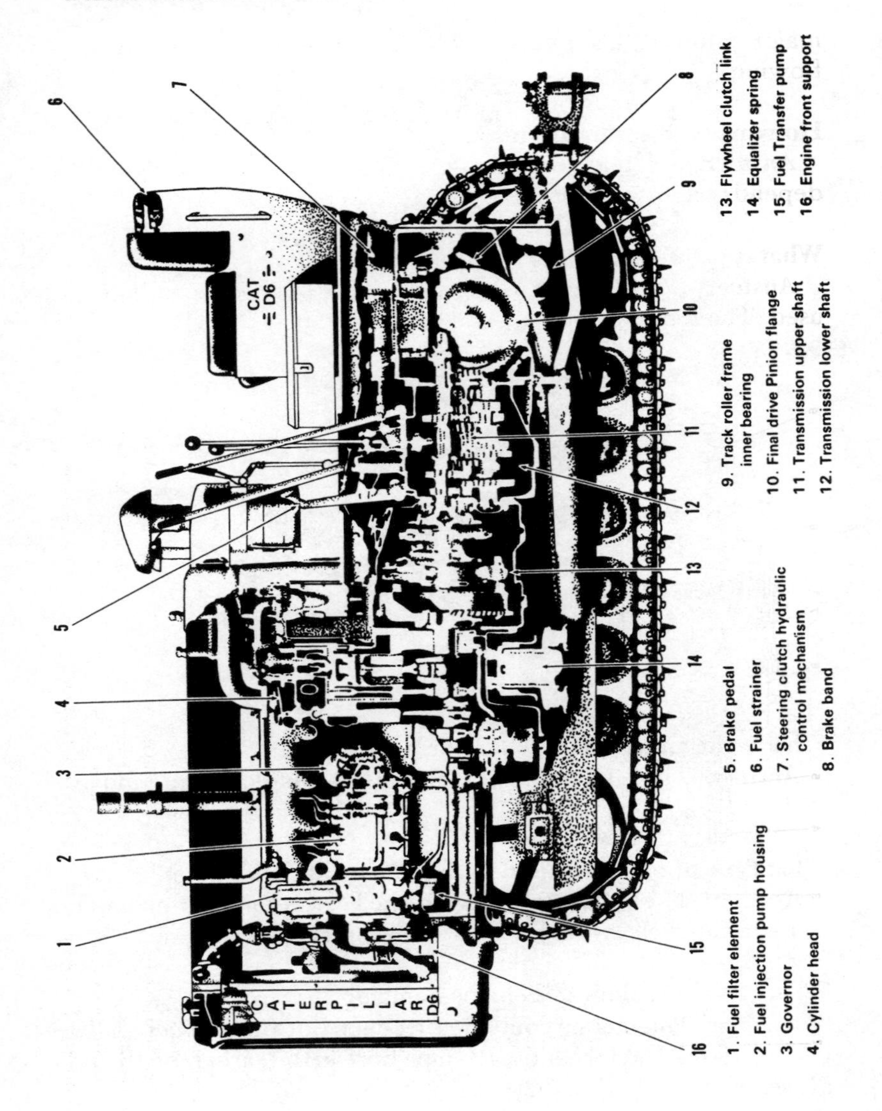

Courtesy Caterpillar Tractor Co.

Fig. 32-1. Cross section of late-model Caterpillar D6 tractor.

major job for this pump; without pressure, fuel would stop flowing.

How many injectors are used?

Answer: Six injectors are used. Each of the six cylinders depends on an injector for vaporized fuel.

What type of lubricating system is used?

Answer: An oil cooler, which is mounted next to the radiator, is used. The radiator's job is to cool the oil cooler.

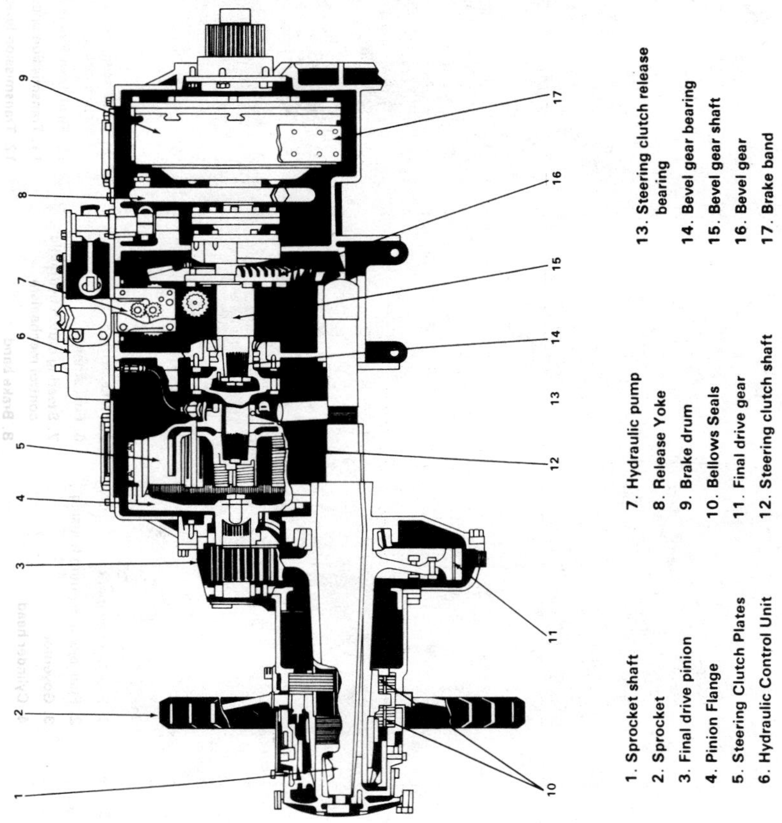

Courtesy Caterpillar Tractor Co.

Fig. 32-2. Cross section of late-model transmission unit.

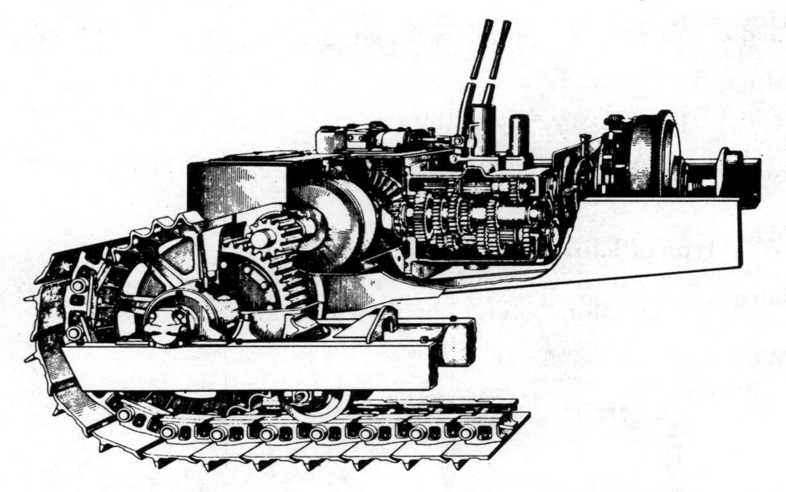

Courtesy Caterpillar Tractor Co.

Fig. 32-3. Cross section of early-model Caterpillar D6 transmission unit.

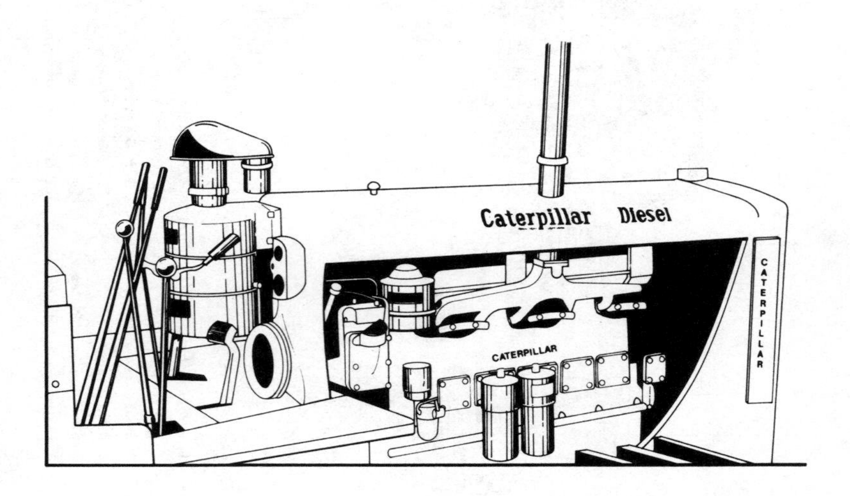

Courtesy Caterpillar Tractor Co.

Fig. 32-4. Side view of hood and dash assembly.

How is the water pump mounted on the D6 tractor?

Answer: The water pump is basically one-unit construction. Mounted in front of the engine, the unique part of the water pump is that it can be removed and replaced by just removing the fan belt and bolts from the engine. There is no need for total overhauling.

What type of transmission is used?

Answer: Steering clutch and final drive mechanism are used in later models of tractors (Fig. 32-2).

What transmission units were used on earlier tractors?

Answer: The power transmission units on earlier tractors con-

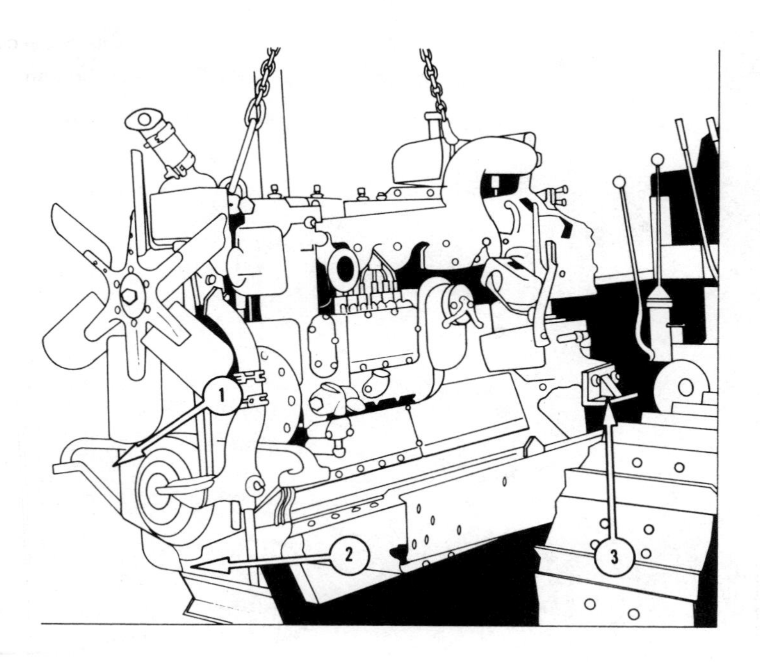

Courtesy Caterpillar Tractor Co.

Fig. 32-5. Front view of engine.

sisted of a flywheel clutch, transmission, bevel gear, steering clutch assemblies, and final drives (Fig. 32-3).

What is the first step in engine removal?

Answer: The hood and dash must be removed first (Fig. 32-4).

How many engine mounts hold the D6 tractor to the frame?

Answer: Three mounts hold the engine in place—one in front of the engine and two on each side of the engine (Fig. 32-5).

What are the most important steps before overhauling and engine removal?

Answer: First, drain all oil. Second, drain all radiator coolant. Third, disconnect all batteries and power sources.

CHAPTER 33

Diesel Engine Service

What parts of the diesel engine require servicing?

Answer: The reciprocating and rotating parts of the diesel engine requiring servicing are:

1. Pistons.
2. Connecting rods.
3. Crankshafts.
4. Valves.
5. Valve gear.
6. Camshafts.

How can the cylinder walls be reconditioned?

Answer: Preliminary to servicing pistons, the cylinder walls should be placed in proper condition. Cylinders that are not badly scored but need cleaning up to bring them within satisfactory working limits may be reconditioned with a hone.

How can the cylinder bore be checked?

Answer: The bore should be checked with an accurate gauge to determine whether or not it is either out of round or tapered.

A used cylinder bore should have an out-of-round measurement of not more than .0005 inch and not more than .0005-inch taper. Maximum allowable out-of-round measurement is .0004 inch and .010 inch for maximum taper.

What should be done if the cylinder walls are badly scored, tapered, or out of round?

Answer: They should first be bored with a reliable boring tool and then polished with a hone.

When reconditioning cylinders, the crankshaft and main bearings should be covered to prevent cuttings and abrasives from getting in bearings. After completing the reconditioning, the engine should be thoroughly cleaned before fitting the pistons.

How is correct fit of a piston determined?

Answer: Correct piston fit is determined by placing a long, 1/2-inch wide feeler gauge along the side of the piston at 90° around the piston from the piston pin and inserting the piston and feeler gauge into the cylinder bore.

A typical piston should push through the cylinder with light pressure when a .002-inch feeler gauge is used and should lock with a .003-inch feeler gauge.

Permanently mark the piston for the cylinder to which it has been fitted and proceed to hone the cylinders and fit the remaining pistons.

What precautions should be taken?

Answer: Handle the pistons with care and do not attempt to force them through the cylinder until the cylinder has been bored to correct size as some pistons can be distorted through careless handling.

What factors should be considered in fitting aluminum-alloy pistons?

Answer: Aluminum-alloy pistons will contract at low temperature, especially after the engine remains idle for a period of time and the oil film drains from between the moving parts. A perceptible sound of pistons and valve mechanism when starting a cold engine is therefore normal and should cause no harm or damage and does not indicate improper piston fitting, provided the sound is not heard after the engine has reached normal engine operating temperature.

Describe the proper fitting of aluminum pistons.

Answer: Pistons should be fitted to the cylinder bore with the greatest of accuracy and care. The clearance between the high spot of the piston-skirt contour at 1/2 to 3/4 inch from the bottom of the piston and the cylinder wall should be .0008 inch (.00025 to .00075 inch with steel-strut pistons) for a typical piston clearance.

The fitting should be made at normal room temperature (70F.) before the piston has been assembled to the rod. This clearance can be checked by using .002-inch feeler stock (.0015-inch feeler stock with steel-strut pistons), 1/2 inch wide and long enough to extend down the bore the full length of the piston.

The piston should be completely inserted upside down in the cylinder bore.

The feeler stock should be placed between the piston and the cylinder wall on the high spot of the piston contour. Draw out the feeler stock with a spring scale.

The amount of pull required to withdraw the feeler stock should be between 5 and 7 pounds (10 to 17 pounds with steel-strut pistons).

For checking sizes of stock pistons, measurements should be made with micrometer calipers across the high points of the piston-skirt contour. The above clearance values are for a typical engine. The manufacturer's specifications should always be followed for a specific engine.

What fact should be noted when fitting semifinished pistons?

Answer: They should have an elliptical shape when proper cam-grinding equipment is used.

The process of cam-grinding a piston creates an elliptical shape. When finished, the diameter of the piston at the wrist-pin holes is less than the diameter across the thrust faces.

Describe the operation of cam-ground pistons.

Answer: At operating temperatures, expansion forces the pin piers away from each other, causing the pistons to assume a shape more nearly round; therefore, semifinished pistons must be finished elliptically (cam-ground) to allow for expansion and to prevent their filling the cylinder bore completely.

Tell how to fit the piston rings properly.

Answer: Piston-ring gap should be measured with the ring about two inches from the bottom of the cylinder bore to which it is fitted. Be sure the piston ring is square in the cylinder bore.

Measure the ring end clearance. Measure the ring side clearance with a feeler gauge. Piston rings should be installed on the piston by means of a piston-ring tool. Rings should be installed in the piston grooves in their proper positions.

What should be done when installing a new set of rings without reconditioning the cylinder bore?

Answer: Always remove the top ridge of the cylinder bore with a reliable ridge reamer.

What precuations should be taken in removing the top ridge of the cylinder bore?

Answer: Care must be taken not to cut below the top of the upper ring position in the cylinder bore and thus cut out the ridge before removing the piston assemblies. Keep the tops of the pistons covered to prevent the cuttings reaching the bearings, crankshaft, etc.

How should the wrist pins (piston pins) be fitted?

Answer: The piston should be assembled to the rod so that the slotted side of the piston is opposite the oil hole in the connecting rod. When the piston and the connecting rod assembly are installed in the cylinder bore, the slotted side of the piston should be farthest from the valve side of the engine. Wrist-pin fit in the connecting-rod bushing should be a tight thumb-push fit.

Heat the piston to 160°F. before inserting the wrist pin so that the piston skirt will not be distorted by driving the wrist pin in place. Wrist pins are available in standard and also the following oversizes: .003 inch and .008 inch.

Describe the removal of the connecting rod and the piston from the cylinder bore.

Answer: Some engines have cylinders large enough to permit removal of the piston and the connecting rod from the top, while in other engines, the unit must be removed from the bottom.

Correct alignment of the connecting rods is important. Whether new connecting rods are being used or the old ones are reinstalled, they should be checked for alignment, so that the crankshaft and crank throw journals will not be subjected to undue stresses and strains.

Describe connecting-rod alignment for a typical General Motors engine.

Answer: General Motors recommends the following procedure for a typical truck connecting rod-alignment:

1. Place the piston pin in the eye of the connecting rod and tighten the clamp volt.
2. Place the connecting rod on the arbor of the aligning fixture J-874-C and tighten the connecting-rod bolts.
3. Place the V-block on the piston pin and move the connecting rod arbor toward the faceplate until the pins on the V-block just engage the faceplate.
4. If either the top or bottom two pins touch the faceplate, but if the other two do not, the rod is bent.
5. If only the two pins on the front, or the two on the back side of the V-block touch the faceplate, the rod is twisted.
6. The fixture is sufficiently strong to hold the connecting rod for straightening.

   Place a bending bar on the rod and twist or bend the rod as required and recheck. Continue this operation until all pins just touch the faceplate.
7. Place the V-block on the piston pin so that the V-block rests against the outside edge of the connecting rod. Move rod and V-block toward the faceplate until all four pins touch.
8. Place the index on the bottom of the fixture, so that it touches the large end of the connecting-rod bearing.

   Remove the connecting rod from the arbor and turn it around.
9. Assemble it again to the arbor and place the V-block on the piston pin in the same place as when checking the other side.

Move the rod and V-block toward the faceplate until either the index touches the bearing or the pins touch the faceplate.

10. If the index does not touch the rod bearing with the four pins touching the faceplate, check the distance between the rod bearing and the index with a feeler gauge. If this distance is greater than .025 inch, the rod should be straightened until the pins touch the faceplate and the index touches the rod bearing within .025 inch.

Describe the reassembly of the connecting rod to the piston in a typical General Motors diesel engine.

Answer: Proceed as follows:

1. Place the piston in piston vise J-1218. Assemble the rod to the piston and install the piston pin. Before tightening the clamp screw, center the piston pin in the piston and the connecting rod in the center of the two piston pin bosses.
2. Tighten the clamp screw to 25 = 35 ft.-lbs. and move the piston on its pin from side to side, checking to see that the piston pin does not extend beyond the outside of the piston.
3. Assemble the remaining rods to the pistons.

What precaution should be taken and why?

Answer: The connecting rod should never be clamped in a bench vise when installing the piston on it, since tightening the clamp screw will likely twist the rod.

Describe the alignment of the connecting rod and the piston for reassembly.

Answer: Proceed to align the piston and connecting rod for reassembly as follows:

1. Assemble the piston and connecting-rod assembly to the alignment fixture and check with the V-block resting against the piston skirt to see that the rod and the piston are in alignment.

   Both pins on the V-block should rest against the face of the plate on the fixture. The piston should be in the same alignment as the connecting rod when this check is made.

2. A quick check of a piston and connecting-rod assembly for both cock and twist can be made without disassembling the rod from the piston.

   This method saves considerable time on any repair operation that does not normally require the removal of the connecting rod from the piston.
3. To make this check, the connecting-rod and piston assembly is mounted on the alignment fixture and the piston is set in line with the connecting rod. Then place the V-block on the piston skirt. If both pins on the block contact the faceplate, the rod is not cocked.
4. Then, with the V-block on the piston skirt and the pins against the faceplate, tip the piston first in one direction and then in the other.

   If the pins on the block remain against the faceplate, there is no twist in the connecting rod.
5. If one pin leaves the faceplate while the piston is being tipped in one direction and the other pin leaves the faceplate while the piston is being tipped in the other direction, the connecting rod is twisted and should be straightened until both pins follow the faceplate.

Describe the crankshaft assembly of a typical diesel engine.

Answer: Measure the crankshaft journals with micrometers. If the crank pins or journals are out of round or tapered in excess of the worn limits (.005 inch for the crankshaft) the crankshaft should be reground to the next undersize bearing available.

What sizes of undersize bearings are available?

Answer: Undersize bearings are available in the following sizes:

.002, .010, .020, and .030 inch

The .002-inch undersize bearing is generally used when the crankshaft shows slightly under size but not enough to warrant regrinding.

How is lengthwise alignment checked?

Answer: Set the crankshaft in V-blocks, and place the dial indicator over the center journal. When the crankshaft is rotated, the indicator will register the amount the crankshaft is out of line.

Move one of the V-blocks to the center journal. Place the dial indicator over the end journal and rotate the crankshaft. The dial indicator will register the number of thousandths that the crankshaft is out of line.

If the crankshaft is out of line in excess of .005 inch at the center or the end, the crankshaft should be straightened in a straightening press.

How are main bearing bores checked for alignment?

Answer: Clean the bearing bores thoroughly. Use an aligning bar ground to the bore diameter and long enough to extend through all the bores.

Install the bearing caps and retainer bolts and tighten down to full tension with a torque wrench. Consult a torque-wrench table.

The bearing bores are in alignment when the alignment bar can be turned by hand without the aid of a wrench or bar.

If the bearing bores have warped or sagged so that the aligning bar cannot be turned by hand, precision-type bearings should not be installed, if bearing life is expected.

Describe reaming of the camshaft bearings in a typical General Motors engine.

Answer: The special camshaft-bearing line reamer has all four cutters mounted on one bar so that all bearings will be in perfect alignment after the reaming operation.

For a typical General Motors engine the instructions for reaming are as follows:

1. Pass the reamer through the first, second, and third bearings. Start the reamer cutters into all four bearings; turn the reamer slowly until the cutters have passed through the bearings.

   While the bearings are being reamed, a liberal supply of kerosene should be used to wash out all metal cuttings.
2. Remove the reamer by pulling it back through the bearings, at the same time turning the reamer slowly in the same direction as when reaming the bearings.
3. Blow all cuttings from the bearings with compressed air.

Install the camshaft and check all bearing clearances with a narrow feeler gauge.

The proper clearance is from .002 to .004 inch. Install the expansion plug in the back end of the cylinder block at the rear camshaft bearing.

Describe the procedure for reconditioning valve-seat inserts in a typical engine.

Answer: Because of the hardness of the special inserts used in the exhaust-valve seats, it is impossible to recut them. They must be reground.

The following should be considered when reconditioning valve inserts:

1. Valve guides must be clean.
2. The upper end of the valve guide must be chamfered.
3. The valve-guide pilot must fit snugly in the valve guide and be tightened securely in place.
4. The grinding stone must be trued for concentricity, on a diamond dressing tool.
5. The finished seats should not exceed .0005-inch runout, when checked with an indicator.
6. Do not use valve-grinding compound on valve-seat inserts. Valve-seat inserts of .010-inch oversize are available for service replacements.

Describe grinding intake-valve seats in an engine.

Answer: The intake-valve seats in the cylinder block should be recut with a suitable valve-seat cutter. The seats should be cut only enough to remove pits or other depressions.

Then grind a new surface on the valve head with a valve-grinding machine. When new seats are finished, the valves and seats in the cylinder block should be lightly lapped together with suitable valve-grinding compound to ensure a tight seat. The valve heads have plain surfaces and the valve grinder may be oscillated by means of a rod fitted with a vacuum cup and operated either by machine or by hand.

What precautions should be taken?

Answer: Considerable care should be taken to make certain

that all grinding compound is removed from the valve, valve seat, intake port, and cylinder block.

Describe the adjustment of the valve tappets.

Answer: If a tappet face scores from too close adjustment, or if breakage results, servicing consists of replacing such a defective tappet with a new one.

The adjusting screw clearance for both the intake and the exhaust valves on the engine is .014 inch with the engine cold. Tappet screws should not be set closer than indicated because if this should be done, when the engine has become heated and normal expansion takes place, the valves will hold open, with the result that tappet and cam faces become badly scored or cut, the head of the valve becomes warped, and in some instances actually badly burned.

If the tappet face becomes scored, there is nothing to do but to replace it. If the damage is not too bad, marks on the cams of the camshaft can often be smoothed out by honing them with an oilstone.

Proper setting of the valves in relationship to the crankshaft (valve timing) is one of the most important phases of successful

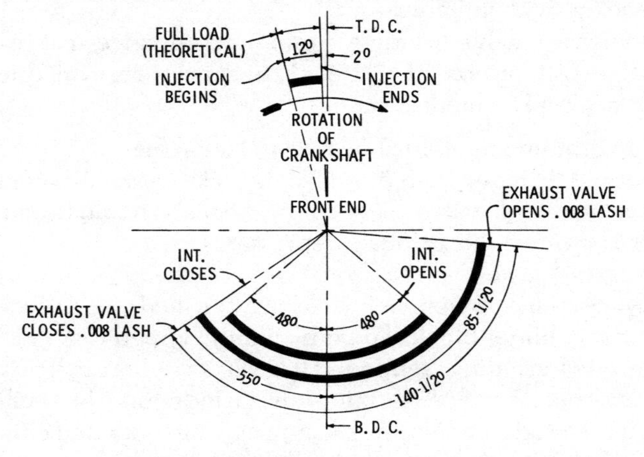

Fig. 33-1. Timing for a typical diesel truck engine. If the engine is timed properly, intake ports and exhaust valves must open and close at the proper time. Also, fuel injection must occur at the correct point in the cycle.

motor operation. Obviously, the engine manufacturer has conducted exhaustive tests to determine proper timing, and for this reason the mechanic should never attempt to alter the original factory setting. This phase of the timing will be correct if the marked tooth on the crankshaft gear is properly meshed with the two correspondingly marked teeth of the camshaft gear.

APPENDIX

Diesel Terms and Tables

For purposes of clarity and uniformity, the following definitions are supplied for use in discussions concerning diesel engines.

acceleration—The average rate of change of an increasing velocity or speed.

API—American Petroleum Institute. An arbitrary scale used to designate the specific gravity of fuel oils. Diesel fuels usually range from 24 to 36 API gravity.

atomize—To reduce to extremely fine particles.

axially—Parallel to the center line of a cylinder or shaft.

axis—A center line. A line about which something rotates or about which it is evenly arranged.

Babbitt—A soft, antifriction bearing metal.

back pressure—The result of resistance to the normal flow of gases and liquids.

bore—The interior diameter of an engine cylinder.

British thermal unit (Btu)—A unit for measuring heat. One British thermal unit equals 778 foot-pounds of work (or energy). It represents the heat required to change the temperature of one pound of water through one degree Fahrenheit.

by-pass—A separate passage that permits a liquid or gas to take a course other than normally used.

caloric value—Heat value of fuel measured by the metric system; calorie equals 0.003968 Btu or 1 Btu equals 252 calories.

carbon residue—The carbon remaining after evaporating off the volatile portion of a fuel oil by heating it in the absence of air under controlled test conditions. It is an indication of the amount of carbon that may be deposited in an engine.

centrifugal force—The force acting on a rotating body that tends to throw it farther from the axis of its rotation.

cetane rating—A system of numbers for indicating the ignition quality of diesel fuels.

chamfer—A beveled corner or edge.

check valve—A device that permits passage of a fluid or gas in one direction only. It stops (or checks) reverse flow.

clearance—The space between a moving part and a stationary part. Clearance is usually allowed between two surfaces to provide for expansion and contraction and for lubrication.

clearance volume—The amount of space confined within the engine cylinder and related parts when the piston is at its top-dead-center position.

coefficient—A ratio, a known factor or quantity that is always constant.

combustion chamber—The space within a cylinder in which the fuel mixture is burned.

compression—The act or result of pressing a substance into a smaller space. One of the events of an internal-combustion engine cycle.

compression pressure—The amount of pressure resulting from the compression stroke of a piston when it has reached top dead center.

compression ratio—A ratio expressing the extent to which a fuel or air charge is compressed. It is a relationship between clearance and displacement volumes and is found as follows:

$$\frac{\text{piston displacement} + \text{clearance volume}}{\text{clearance volume}} = \text{compression ratio}$$

concentric—Having a common center.

condensation—The process by which a vapor is reduced to a liquid.

contraction—Becoming smaller in size—usually, in metals, as a result of cooling or lowering of temperature.

cycle—A series of events, operations, or movements that repeat themselves in an established sequence.

cylinder, in-block—A group of cylinders cast as one piece.

detonation—The result of violent uncontrolled burning of a fuel in the combustion chamber.

diesel cycle—A cycle of events that occurs in diesel engines similar to gasoline (Otto cycle) engines except that air without fuel is compressed. Fuel is injected and burns immediately.

distillation—Separation of the more volatile parts of a petroleum oil from those less volatile by vaporization and subsequent condensation.

dribbling—A characteristic of an injection nozzle in which the fuel seeps slowly from the injection nozzle tip.

eccentric—A circle not having the same center as another within it. A device mounted off-center for converting rotary motion into reciprocating motion.

efficiency, mechanical—The ratio between the brake horsepower (bhp) and the indicated or total horsepower (ihp).

efficiency, volumetric—The ratio of the volume of air or fuel mixture actually taken into the cylinder to the volume of the piston displacement.

energy—Capacity for doing work.

engine—A machine which produces power to do work, particularly one that converts heat into mechanical power. The term "engine" should be used in referring to the power plant of a motor vehicle, and the term "motor" should be used in connection with electric motors.

fit—A fit may be considered as the desired clearance between the surfaces of machine parts.

friction—The action between two bodies at the surfaces of contact, which resists motion.

fuel pump—A small pump for delivering fuel to the engine. A diesel fuel pump operates at a fixed ratio to engine speed and may be either of the constant-volume or of the metering type.

fulcrum—The support on which a lever turns.

governor—A device used to control speed (rpm) of an engine.

heat—A form of energy.

heat units—The unit of heat (one British thermal unit—Btu).

helical—A term used to describe gear teeth shaped like a helix.

helix—A line shaped like a screw thread.

horsepower—A unit for measuring power. It is the rate at which work is done. One horsepower equals 33,000 foot-pounds per minute. (See brake horsepower, friction horsepower, and indicated horsepower).

brake horsepower (bhp)—Amount of net available power produced by an engine as measured at the crankshaft.

friction horsepower (fhp)—The horsepower consumed by the engine in running itself; that is, the power lost within the engine due to its internal friction.

indicated horsepower (ihp)—Total power developed by the engine; or brake horsepower added to friction horsepower.

hunting—Erratic engine operation; caused by the inability of a governor to respond accurately to changes in engine speed.

hydraulics—The science of using liquids under pressure to do work.

idling—Engine running at lowest speed possible, without stalling.

impeller—The rotating part of a blower or pump that imparts motion to air or a liquid by forcing it outward from the center of the part.

inertia—The property of a body that causes it to persist either in a state of rest or in uniform motion.

injection—The forcing of fuel oil into the combustion chamber of a diesel engine by means of high pressure.

integral—The whole made up of parts; constituting a part of a whole necessary to completeness.

intermittent—Occurring at intervals.

jet—A metered opening in an air or fuel passage to control the flow of air or fuel.

journal—The finished part of a shaft that rotates in or against a bearing.

laminated—Made of thin layers, such as "laminated shims," "laminated cores."

lean mixture—A mixture in which the proportion of air to fuel is greater than the ideal.

motor—Technically applied to an electric motor.

mph—Miles per hour.

needle valve—A small plain or threaded rod having a conical or tapered joint, operated within a jet to vary the flow of fuel through the jet.

Otto cycle—A cycle of four events that occurs in a gasoline engine in order: intake, compression, power, and exhaust.

poppet valve—A valve opened by the action of a cam and closed by a strong spring. This type of valve is used almost exclusively in the automotive industry.

port—An opening, hole, or passage.

pour point—The lowest temperature at which fuel oil will just flow under controlled test conditions. It is an indication as to the suitability of the fuel for cold weather operation.

power—The capacity to do work.

radial—Radiating from a common center, as the spokes of a wheel.

reciprocating—A back-and-forth (or up-and-down) linear motion, such as the action of pistons in the engine.

rectilinear motion—Motion in a straight line.

rotary—Revolving or circular. Rotary motion is considered the opposite of linear reciprocating (up-and-down or back-and-forth) motion in power transfer.

rpm—Revolutions per minute.

seat, valve—That part of the valve mechanism upon which the valve face rests to close the port.

servomotor—A small hydraulic motor for operating heavy control mechanisms.

stability—Ability of lubricating oil to stand up without physical change under severe operating conditions.

stress—The forces exerted on, within, or by a body during either tension or compression. The opposing reaction of the interior elements of a body against forces tending to deform them.

stroke—The distance a piston travels up or down inside a cylinder.
temperature—The intensity (or degree) of heat.
thermodynamics—The theory of changing heat into mechanical work.
thrust—A stress or strain tending to push anything out of alignment.
torque—A twisting or wrenching effort. Torque is the product of force multiplied by the distance from the center of rotation at which it is exerted. For example: A force of 40 pounds applied on the end of a 1-foot pipe wrench would be 40 pounds × 1 foot, or 40 foot-pounds of torque. Similarly, 40 pounds of force exerted on the end of a 2-foot pipe wrench would be 40 pounds × 2 feet, or 80 foot-pounds of torque. This indicates why it is easier to unscrew a pipe coupling with the 2-foot wrench than with the 1-foot wrench, the torque incident to the 2-foot lever (wrench) being greater.
torsion—The deformation in a body caused by twisting.
vacuum—Result of reducing atmospheric pressure.
venturi—A tube with a narrowing throat or constriction to increase the velocity of the gas or fluid flowing through it.
viscosity—Internal resistance to flow. The fluid body of a liquid.
volatility—Ability of a liquid to vaporize or turn into gas.
work—The use of energy to overcome resistance.

Fractions and Decimals of Inches and Millimeter Equivalents

Inches			Inches			Inches		
Fractions	Decimals	mm	Fractions	Decimals	mm	Fractions	Decimals	mm
	.0039	.10	11/64	.1719	4.37	21/64	.3281	8.33
1/64	.0156	.40		.1772	4.5		.3347	8.5
	.0197	.50				11/32	.3438	8.73
1/32	.0313	.79	3/16	.1875	4.76		.3543	9.
	.0394	1.00		.1969	5.	23/64	.3594	9.13
3/64	.0469	1.19	13/64	.2031	5.16		.374	9.5
	.0591	1.50		.2165	5.5			
			7/32	.2188	5.56	3/8	.375	9.53
1/16	.0625	1.59	15/64	.2344	5.95	25/64	.3906	9.92
5/64	.0781	1.98		.2362	6.		.3937	.10
	.0787	2.00					.3937	(1 cm)
3/32	.0938	2.38	1/4	.25	6.35	13/32	.4062	10.32
	.0984	2.50		.2559	6.5		.4134	10.5
7/64	.1094	2.78	17/64	.2657	6.75	27/64	.4219	10.72
	.1181	3.00		.2756	7.		.4331	11.
			9/32	.2813	7.14			
1/8	.125	3.18		.2953	7.5	7/16	.4375	11.11
	.1378	3.5	19/64	.2969	7.54		.4528	11.5
9/64	.1406	3.57				29/64	.4531	11.51
5/32	.1563	3.97	5/16	.3125	7.94	15/32	.4688	11.91
	.1575	4.		.3150	8.		.4724	12.
31/64	.4844	12.3		.689	17.5	59/64	.9219	23.42
	.4921	12.5	45/64	.7031	17.86		.9252	23.5
				.7087	18.	15/16	.9375	23.81
1/2	.5	12.7	23/32	.7188	18.26		.9449	24.
	.5118	13.		.7284	18.5	61/64	.9531	24.21
33/64	.5156	13.1	47/64	.7344	18.65		.9646	24.5
17/32	.5313	13.49		.748	19.	31/32	.9688	24.61
	.5315	13.5	3/4	.75	19.05		.9843	25.
35/64	.5469	13.89	49/64	.7656	19.45	63/64	.9844	25.
	.5512	14.		.7677	19.5			
			25/32	.7813	19.84	1"	1.	25.4
9/16	.5625	14.29		.7874	20.	1 1/4	1.25	31.75
	.5709	14.5	51/64	.7969	20.24	1 1/2	1.5	38.1
37/64	.5781	14.68		.8071	20.5	1 3/4	1.75	44.45
	.5906	15.				2	2.	50.8
19/32	.5938	15.08	13/16	.8125	20.64	3	3.	76.2
39/64	.6094	15.48		.8268	21.		3.937	100.
	.6102	15.5	53/64	.8281	21.03			10cm
			27/32	.8438	21.43			1dm
5/8	.625	15.88		.8465	21.5	6	6.	152.4
	.6299	16.	55/64	.8594	21.83	1ft.	12.	
41/64	.6406	16.27		.8661	22.	2ft.	24.	
	.6496	16.5				3ft.	36.	
21/32	.6563	16.67	7/8	.875	22.23		39.37	1000mm
	.6693	17.		.8858	22.5			100cm
43/64	.6719	17.07	57/64	.8906	22.62			10dm
				.9055	23.			1m
11/16	.6875	17.46	29/32	.9063	23.02			

Areas and Circumferences of Circles

Diameter	Circum-ference	Area	Diameter	Circum-ference	Area
1/16	0.1964	0.00307	1 13/16	5.6941	2.58016
1/8	0.3927	0.01227	1 7/8	5.8905	2.76117
3/16	0.5890	0.02761	1 15/16	6.0868	2.94831
1/4	0.7854	0.04909	2	6.2832	3.1416
5/16	0.9818	0.07670	2 1/16	6.4795	2.3410
3/8	1.1781	0.11045	2 1/8	6.6759	3.5466
7/16	1.3744	0.15033	2 3/16	3.8722	3.7584
1/2	1.5708	0.19635	2 1/4	7.0686	3.9761
9/16	1.7671	0.24850	2 5/16	7.2649	4.2
5/8	1.9635	0.30680	2 3/8	7.4613	4.4301
11/16	2.1598	0.37122	2 7/16	7.6576	4.6664
3/4	2.3562	0.44179	2 1/2	7.8540	4.9087
13/16	2.5525	0.51849	2 9/16	8.0503	5.1573
7/8	2.7489	0.60132	2 5/8	8.2467	5.4119
15/16	2.9452	0.69029	2 11/16	8.4430	5.6727
1	3.1416	0.78540	2 3/4	8.6394	5.9396
1 1/16	3.3379	0.88664	2 13/16	8.8357	6.2126
1 1/8	3.5343	0.99402	2 7/8	9.0321	6.4918
1 3/16	3.7306	1.10753	2 15/16	9.2284	6.7772
1 1/4	3.9270	1.22718	3	9.4248	7.0686
1 5/16	4.1233	1.35297	3 1/16	9.6211	7.3662
1 3/8	4.3197	1.48489	3 1/8	9.8175	7.6699
1 7/16	4.5160	1.62295	3 3/16	10.0138	7.9798
1 1/2	4.7124	1.76715	3 1/4	10.2102	8.2958
1 9/16	4.9087	1.91748	3 5/16	10.4066	8.6179
1 5/8	5.1051	2.07394	3 3/8	10.6029	8.9462
1 11/16	5.3014	2.23654	3 7/16	10.7992	9.2807
1 3/4	5.4978	2.40528	3 1/2	10.9956	9.6211
3 9/16	11.1919	9.9678	4 9/16	14.3335	16.3492
3 5/8	11.3883	10.3206	4 5/8	14.5299	16.8002
3 11/16	11.5846	10.6796	4 11/16	14.7262	17.2573
3 3/4	11.7810	11.0447	4 3/4	14.9226	17.7206
3 13/16	11.9773	11.4160	4 13/16	15.1189	18.19
3 7/8	12.1737	11.7933	4 7/8	15.3153	18.6655
3 15/16	12.3701	12.1768	4 15/16	15.5116	19.1472
4	12.5664	12.5664	5	15.7080	19.6350
4 1/16	12.7628	12.9622	5 1/8	16.1007	20.6290
4 1/8	12.9591	13.3641	5 1/4	16.4934	21.6476
4 3/16	13.1554	13.7721	5 3/8	16.8861	22.6907
4 1/4	13.3518	14.1863	5 1/2	17.2788	23.7583
4 5/16	13.5481	14.6066	5 5/8	17.6715	24.8505
4 3/8	13.7445	15.0330	5 3/4	18.0642	25.9673
4 7/16	13.9408	15.4656	5 7/8	18.4569	27.1084
4 1/2	14.1372	15.9043	6	18.8496	28.2744

Caloric Values of Fuels

Degrees Baumé	Specific Gravity	Lb. per gal.	Btu per lb.	Btu per gal.	Remarks
14	.9722	8.10	18810	152361	Mexico, California, Texas and Kansas Crudes, Fuel Oil
15	.9655	8.05	18850	151743	
16	.9589	7.99	18890	150931	
17	.9523	7.94	18930	150304	
18	.9459	7.88	18970	149484	
19	.9395	7.83	19010	148848	
20	.9333	7.78	19050	148209	
21	.9271	7.73	19090	147506	
22	.9210	7.68	19130	146918	
23	.9150	7.63	19170	146267	
24	.9090	7.58	19210	145612	
25	.9032	7.54	19250	145145	
26	.8974	7.49	19290	144482	
27	.8917	7.44	19330	143815	
28	.8860	7.39	19370	143144	Kansas, Indian Territory and Illinois Crudes; Penna. Fuel; California Refined Fuel Oil
29	.8805	7.34	19410	142469	
30	.8705	7.29	19450	141790	
31	.8695	7.25	19490	141303	
32	.8641	7.21	19530	140811	
33	.8588	7.16	19570	140121	
34	.8536	7.12	19610	139623	
35	.8484	7.07	19650	138926	
36	.8433	7.03	19690	138421	
37	.8383	6.99	19730	137913	Ohio, Penna., and West Va. Crude; California and Kansas Refined
38	.8333	6.95	19770	137402	
39	.8284	6.91	19810	136887	
40	.8235	6.87	19850	136370	
41	.8187	6.83	19890	135849	
42	.8139	6.80	19930	135524	
43	.8092	6.76	19970	134997	
44	.8045	6.72	20010	134467	
45	.8000	6.68	20050	133934	
46	.7954	6.64	20090	133398	Kerosene and Gasoline
47	.7909	6.60	20130	132858	
48	.7865	6.57	20170	132517	
49	.7821	6.53	20210	131971	
50	.7777	6.49	20250	131423	

Useful Formulas For 4-Stroke-Cycle Engines

Piston area = $.7854D^2$ sq. in.
Piston velocity = $^2/_6$ LN ft. per min.
Piston displacement = .7854 Ld^2 cu. in. per cylinder
Engine displacement = .7854 Lnd^2 cu. in.

S.A.E. horsepower = $\frac{nD^2}{2.5}$

Brake horsepower = = $\frac{DPN}{792,000} = \frac{NT}{5252}$

Brake torque = $\frac{5252\ bhp}{N}$ ft.-lb.

Brake mean effective pressure = $\frac{792,000\ bhp}{DN}$ psi

Brake specific fuel consumption = $\frac{\text{lbs. fuel per hour}}{\text{bhp}}$

Note: In the above formulas:
- D = piston diameter, inches.
- L = length of stroke, inches.
- N = engine speed, revolutions per minute.
- n = number of cylinders.
- D = engine displacement, cubic inches.
- P = mean effective pressure, pounds per square inch.
- T = brake torque, foot-pounds.

Engine Data

$$\text{bmep} = \frac{33{,}000 \times \text{bhp}}{\text{LAN} \times \text{No. Cyls.}}$$

bmep . . . . brake mean effective pressure
bhp . . . . brake horsepower
L . . . . length of stroke in feet
A . . . . area of piston in square inches
N . . . . number of power strokes per min.

N . . . . $\frac{\text{rpm}}{1}$ for two-stroke-cycle engines
$\frac{\text{rpm}}{2}$ for four-stroke-cycle engines

1 KWH . . . . 3413 Btu
1 hp hr . . . . 2545 Btu
Torque in lb.-ft. . . . . $\frac{\text{hp} \times 5250}{\text{rpm}}$
1 atmosphere . . . . 14.69 lbs. per sq. inch
29.92 in. of mercury
33.90 ft. of water
1 inch of mercury . . . . 0.491 lbs. per sq. inch
1 foot of water . . . . 0.434 lbs. per sq. inch
Speed droop . . . . decrease in steady speed of an engine caused by increase in load from No Load to Full Load without change in govenor adjustment. This decrease in speed is expressed as a per cent of mean speed or:

$$\frac{(\text{no load speed} - \text{full load speed}) \times 100}{\text{mean speed}}$$

Index

A

B

C

The Audel® Mail Order Bookstore

Here's an opportunity to order the valuable books you may have missed before and to build your own personal, comprehensive library of Audel books. You can choose from an extensive selection of technical guides and reference books. They will provide access to the same sources the experts use, put all the answers at your fingertips, and give you the know-how to complete even the most complicated building or repairing job, in the same professional way.

Each volume:

- **Fully illustrated**
- **Packed with up-to-date facts and figures**
- **Completely indexed for easy reference**

APPLIANCES

HOME APPLIANCE SERVICING, 3rd Edition
A practical book for electric & gas servicemen, mechanics & dealers. Covers the principles, servicing, and repairing of home appliances. 592 pages; $5\frac{1}{2} \times 8\frac{1}{4}$; hardbound.
Price: $12.95

REFRIGERATION AND AIR CONDITIONING LIBRARY—2 Vols. Price: $21.95

REFRIGERATION: HOME AND COMMERCIAL
Covers the whole realm of refrigeration equipment from fractional-horsepower water coolers, through domestic refrigerators to multi-ton commercial installations. 656 pages; $5\frac{1}{2} \times 8\frac{1}{4}$; hardbound. **Price: $12.95**

AIR CONDITIONING: HOME AND COMMERCIAL
A concise collection of basic information, tables, and charts for those interested in understanding troubleshooting, and repairing home air conditioners and commercial installations. 464 pages; $5\frac{1}{2} \times 8\frac{1}{4}$; hardbound. **Price: $10.95**

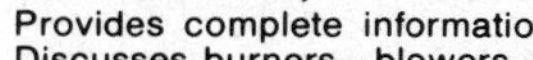

OIL BURNERS, 3rd Edition
Provides complete information on all types of oil burners and associated equipment. Discusses burners—blowers—ignition transformers—electrodes—nozzles—fuel pumps—filters—controls. Installation and maintenance are stressed. 320 pages; $5\frac{1}{2} \times 8\frac{1}{4}$; hardbound.
Price: $9.95

Use the order coupon on the back of this book.
All prices are subject to change without notice.

AUTOMOTIVE

AUTOMOBILE REPAIR GUIDE, 4th Edition
A practical reference for auto mechanics, servicemen, trainees, and owners. Explains theory, construction, and servicing of modern domestic motorcars. 800 pages; $5^1/_2 \times 8^1/_4$; hardbound. **Price: $14.95**

AUTOMOTIVE AIR CONDITIONING
You can easily perform most all service procedures you've been paying for in the past. This book covers the systems built by the major manufacturers, even after-market installations. Contents: introduction—refrigerant—tools—air conditioning circuit—general service procedures—electrical systems—the cooling system—system diagnosis—electrical diagnosis—troubleshooting. 232 pages; $5^1/_2 \times 8^1/_4$; softcover. **Price: $7.95**

DIESEL ENGINE MANUAL
A practical guide covering the theory, operation and maintenance of modern diesel engines. Explains diesel principles—valves—timing—fuel pumps—pistons and rings—cylinders—lubrication—cooling system—fuel oil and more. 480 pages; $5^1/_2 \times 8^1/_4$; hardbound **Price: $12.95**

GAS ENGINE MANUAL, 2nd Edition
A completely practical book covering the construction, operation, and repair of all types of modern gas engines. 400 pages; $5^1/_2 \times 8^1/_4$; hardbound. **Price: $9.95.**

BUILDING AND MAINTENANCE

ANSWERS ON BLUEPRINT READING, 3rd Edition
Covers all types of blueprint reading for mechanics and builders. This book reveals the secret language of blueprints, step by step in easy stages. 312 pages; $5^1/_2 \times 8^1/_4$; hardbound. **Price: $9.95**

BUILDING MAINTENANCE, 2nd Edition
Covers all the practical aspects of building maintenace. Painting and decorating; plumbing and pipe fitting; carpentry; heating maintenace; custodial practices and more. (A book for building owners, managers, and maintenance personnel.) 384 pages; $5^1/_2 \times 8^1/_4$; hardbound. **Price: $9.95**

COMPLETE BUILDING CONSTRUCTION
At last—a one volume instruction manual to show you how to construct a frame or brick building from the footings to the ridge. Build your own garage, tool shed, other outbuildings—even your own house or place of business. Building construction tells you how to lay out the building and excavation lines on the lot; how to make concrete forms and pour the footings and foundation; how to make concrete slabs, walks, and driveways; how to lay concrete block, brick and tile; how to build your own fireplace and chimney. It's one of the newest Audel books, clearly written by experts in each field and ready to help you every step of the way. 800 pages; $5^1/_2 \times 8^1/_4$; hardbound. **Price: $19.95**

Use the order coupon on the back of this book.
All prices are subject to change without notice.

GARDENING & LANDSCAPING

A comprehensive guide for homeowners and for industrial, municipal, and estate groundskeepers. Gives information on proper care of annual and perennial flowers; various house plants; greenhouse design and construction; insect and rodent controls; and more. 384 pages; $5^1/_2 \times 8^1/_4$ hardbound. **Price: $9.95**

CARPENTERS & BUILDERS LIBRARY, 4th Edition (4 Vols.)

A practical, illustrated trade assistant on modern construction for carpenters, builders, and all woodworkers. Explains in practical, concise language and illustrations all the principles, advances, and shortcuts based on modern practice. How to calculate various jobs. **Price: $39.95**

Vol. 1—Tools, steel square, saw filing, joinery cabinets. 384 pages; $5^1/_2 \times 8^1/_4$; hardbound. **Price: $10.95**
Vol. 2—Mathematics, plans, specifications, estimates. 304 pages; $5^1/_2 \times 8^1/_4$; hardbound. **Price: $10.95**
Vol. 3—House and roof framing, layout foundations. 304 pages; $5^1/_2 \times 8^1/_4$; hardbound. **Price: $10.95**
Vol. 4—Doors, windows, stairs, millwork, painting. 368 pages; $5^1/_2 \times 8^1/_4$; hardbound. **Price: $10.95**

CARPENTRY AND BUILDING

Answers to the problems encountered in today's building trades. The actual questions asked of an architect by carpenters and builders are answered in this book. 448 pages; $5^1/_2 \times 8^1/_4$; hardbound. **Price: $10.95**

WOOD STOVE HANDBOOK

The wood stove handbook shows how wood burned in a modern wood stove offers an immediate, practical, low-cost method of full-time or part-time home heating. The book points out that wood is plentiful, low in cost (sometimes free), and nonpolluting, especially when burned in one of the newer and more efficent stoves. In this book, you will learn about the nature of heat and its control, what happens inside and outside a stove, how to have a safe and efficient chimney, and how to install a modern wood burning stove. You will learn about the differnt types of firewood and how to get it, cut it, split it and store it. 128 pages; $8^1/_2 \times 11$; softcover. **Price: $7.95**

HEATING, VENTILATING, AND AIR CONDITIONING LIBRARY (3 Vols.)

This three-volume set covers all types of furnaces, ductwork, air conditioners, heat pumps, radiant heaters, and water heaters, including swimming-pool heating systems. **Price: $38.95**

Volume 1
Partial Contents: Heating Fundamentals . . . Insulation Principles . . . Heating Fuels . . . Electric Heating System . . . Furnace Fundamentals . . . Gas-Fired Furnaces . . . Oil-Fired Furnaces . . . Coal-Fired Furnaces . . . Electric Furnances. 614 pages; $5^1/_2 \times 8^1/_4$; hardbound. **Price: $13.95**

Volume 2
Partial Contents: Oil Burners . . . Gas Burners . . . Thermostats and Humidistats . . . Gas and Oil Controls . . . Pipes, Pipe Fitting, and Piping Details . . . Valves and Valve Installations. 560 pages; $5^1/_2 \times 8^1/_4$; hardbound. **Price $13.95**

Volume 3
Partial Contents: Radiant Heating . . . Radiators, Convectors, and Unit Heaters . . . Stoves, Fireplaces, and Chimneys . . . Water Heaters and Other Appliances . . . Central Air Conditioning Systems . . . Humidifiers and Dehumidifiers. 544 pages; $5^1/_2 \times 8^1/_4$; hardbound. **Price: $13.95**

HOME MAINTENANCE AND REPAIR: Walls, Ceilings, and Floors

Easy-to-follow instructions for sprucing up and repairing the walls, ceiling, and floors of your home. Covers nail pops, plaster repair, painting, paneling, ceiling and bathroom tile, and sound control. 80 pages; $8^1/_2 \times 11$; softcover. **Price: $6.95**

HOME PLUMBING HANDBOOK, 2nd Edition

A complete guide to home plumbing repair and installation. 200 pages; $8^1/_2 \times 11$; softcover. **Price: $7.95**

Use the order coupon on the back of this book.

All prices are subject to change without notice.

MASONS AND BUILDERS LIBRARY—2 Vols.

A practical, illustrated trade assistant on modern construction for bricklayers, stonemasons, cement workers, plasters, and tile setters. Explains all the principles, advances, and shortcuts based on modern practice—including how to figure and calculate various jobs. **Price $17.95**

Vol. 1— Concrete Block, Tile, Terrazzo. 368 pages; $5\frac{1}{2} \times 8\frac{1}{4}$; hardbound. **Price: $9.95**
Vol. 2—Bricklaying, Plastering, Rock Masonry, Clay Tile. 384 pages; $5\frac{1}{2} \times 8\frac{1}{4}$; hardbound. **Price: $9.95**

PLUMBERS AND PIPE FITTERS LIBRARY—3 Vols.

A practical, illustrated trade assistant and reference for master plumbers, journeymen and apprentice pipe fitters, gas fitters and helpers, builders, contractors, and engineers. Explains in simple language, illustrations, diagrams, charts, graphs, and pictures the principles of modern plumbing and pipe-fitting practices. **Price $29.95**

Vol. 1—Materials, tools, roughing-in. 320 pages; $5\frac{1}{2} \times 8\frac{1}{4}$; hardbound. **Price: $10.95**
Vol. 2—Welding, heating air-conditioning. 384 pages; $5\frac{1}{2} \times 8\frac{1}{4}$; hardbound. **Price: $10.95**
Vol. 3—Water supply, drainage, calculations. 272 pages; $5\frac{1}{2} \times 8\frac{1}{4}$; hardbound. **Price: $10.95**

PLUMBERS HANDBOOK

A pocket manual providing reference material for plumbers and/or pipe fitters. General information sections contain data on cast-iron fittings, copper drainage fittings, plastic pipe, and repair of fixtures. 288 pages; 4×6 softcover. **Price: $9.95**

QUESTIONS AND ANSWERS FOR PLUMBERS EXAMINATIONS, 2nd Edition

Answers plumbers' questions about types of fixtures to use, size of pipe to install, design of systems, size and location of septic tank systems, and procedures used in installing material. 256 pages; $5\frac{1}{2} \times 8\frac{1}{4}$; softcover. **Price: $8.95**

TREE CARE MANUAL

The conscientious gardener's guide to healthy, beautiful trees. Covers planting, grafting, fertilizing, pruning, and spraying. Tells how to cope with insects, plant diseases, and environmental damage. 224 pages; $8\frac{1}{2} \times 11$; softcover. **Price: $8.95**

UPHOSTERING

Upholstering is explained for the average householder and apprentice upholsterer. From repairing and regluing of the bare frame, to the final sewing or tacking, for antiques and most modern pieces, this book covers it all. 400 pages; $5\frac{1}{2} \times 8\frac{1}{4}$; hardbound. **Price: $12.95**

WOOD FURNITURE: Finishing, Refinishing, Repairing

Presents the fundamentals of furniture repair for both veneer and solid wood. Gives complete instructions on refinishing procedures, which includes stripping the old finish, sanding, selecting the finish and using wood fillers. 352 pages; $5\frac{1}{2} \times 8\frac{1}{4}$; hardbound. **Price: $9.95**

ELECTRICITY/ELECTRONICS

ELECTRICAL LIBRARY

If you are a student of electricity or a practicing electrician, here is a very important and helpful library you should consider owning. You can learn the basics of electricity, study electric motors and wiring diagrams, learn how to interpret the NEC, and prepare for the electrician's examination by using these books.

Use the order coupon on the back of this book.

All prices are subject to change without notice.

Electric Motors, 3rd Edition. 528 pages; 5½×8¼; hardbound. **Price: $12.95**

Guide to the 1981 National Electrical Code. 608 pages; 5½×8¼; hardbound. **Price: $13.95**

House Wiring, 5th Edition. 256 pages; 5½×8¼; hardbound. **Price: $9.95**

Practical Electricity, 3rd Edition. 496 pages; 5½×8¼; hardbound. **Price: $13.95**

Questions and Answers for Electricians Examinations, 7th Edition. 288 pages; 5½×8¼; hardbound. **Price: $9.95**

ELECTRICAL COURSE FOR APPRENTICES AND JOURNEYMEN

A study course for apprentice or journeymen electricians. Covers electrical theory and its applications. 448 pages; 5½×8¼; hardbound. **Price: $11.95**

RADIOMANS GUIDE, 4th Edition

Contains the latest information on radio and electronics from the basics through transistors. 480 pages; 5½×8¼; hardbound. **Price: $11.95**

TELEVISION SERVICE MANUAL, 4th Edition

Provides the practical information necessary for accurate diagnosis and repair of both black-and-white and color television receivers. 512 pages; 5½×8¼; hardbound. **Price: $11.95**

ENGINEERS/MECHANICS/ MACHINISTS

MACHINISTS LIBRARY

Covers the modern machine-shop practice. Tells how to set up and operate lathes, screw and milling machines, shapers, drill presses and all other machine tools. A complete reference library. **Price: $29.95**

Vol. 1—Basic Machine Shop. 352 pages; 5½×8¼; hardbound. **Price: $10.95**

Vol. 2—Machine Shop. 480 pages; 5½×8¼; hardbound. **Price: $10.95**

Vol. 3—Toolmakers Handy Book. 400 pages; 5½×8¼; hardbound. **Price: $10.95**

MECHANICAL TRADES POCKET MANUAL

Provides practical reference material for mechanical tradesmen. This handbook covers methods, tools equipment, procedures, and much more. 256 pages; 4×6; softcover. **Price: $8.95**

MILLWRIGHTS AND MECHANICS GUIDE

Practical information on plant installation, operation, and maintenance for millwrights, mechanics, maintenance men, erectors, riggers, foremen, inspectors, and superintendents. 960 pages; 5½×8¼; hardbound. **Price: $19.95**

POWER PLANT ENGINEERS GUIDE

The complete steam or diesel power-plant engineer's library. 816 pages; 5½×8¼; hardbound. **Price: $16.95**

QUESTIONS AND ANSWERS FOR ENGINEERS AND FIREMANS EXAMINATIONS, 3rd Edition

Presents both legitimate and "catch" questions with answers that may appear on examinations for engineers and firemans licenses for stationary, marine, and combustion engines. 496 pages; 5½×8¼; hardbound. **Price: $10.95**

WELDERS GUIDE

This new edition is a practical and concise manual on the theory, practical operation and maintenance of all welding machines. Fully covers both electric and oxy-gas welding. 928 pages; 5½×8¼; hardbound. **Price: $19.95**

Use the order coupon on the back of this book.

All prices are subject to change without notice.

WELDER/FITTERS GUIDE

Provides basic training and instruction for those wishing to become welder/fitters. Step-by-step learning sequences are presented from learning about basic tools and aids used in weldment assembly, through simple work practices, to actual fabrication of weldments. 160 pages. 8½×11; softcover. **Price: $7.95**

FLUID POWER

PNEUMATICS AND HYDRAULICS, 3rd Edition

Fully discusses installation, operation and maintenance of both HYDRAULIC AND PNEUMATIC (air) devices. 496 pages; 5½×8¼; hardbound. **Price: $10.95**

PUMPS, 3rd Edition

A detailed book on all types of pumps from the old-fashioned kitchen variety to the most modern types. Covers construction, application, installation, and troubleshooting. 480 pages; 5½×8¼; hardbound. **Price: $10.95**

HYDRAULICS FOR OFF-THE-ROAD EQUIPMENT

Everything you need to know from basic hydraulics to troubleshooting hydraulic systems on off-the-road equipment. Heavy-equipment operators, farmers. fork-lift owners and operators, mechanics—all need this practical, fully illustrated manual. 272 pages; 5½×8¼; hardbound. **Price: $8.95**

HOBBY

COMPLETE COURSE IN STAINED GLASS

Written by an outstanding artist in the field of stained glass, this book is dedicated to all who love the beauty of the art. Ten complete lessons describe the required materials, how to obtain them, and explicit directions for making several stained glass projects. 80 pages; 8½×11; softbound. **Price: $6.95**

Use the order coupon on the back of this book.

All prices are subject to change without notice.

BUILD YOUR OWN AUDEL DO-IT-YOURSELF LIBRARY AT HOME!

Use the handy order coupon today to gain the valuable information you need in all the areas that once required a repairman. Save money and have fun while you learn to service your own air conditioner, automobile, and plumbing. Do your own professional carpentry, masonry, and wood furniture refinishing and repair. Build your own security systems. Find out how to repair your TV or Hi-Fi. Learn landscaping, upholstery, electronics and much, much more.

HERE'S HOW TO ORDER

1. Enter the correct title(s) and author(s) of the book(s) you want in the space(s) provided.
2. Print your name, address, city, state and zip code clearly.
3. Detach the order coupon below and mail today to:

Theodore Audel & Company
4300 West 62nd Street
Indianapolis, Indiana 46206
ATTENTION: ORDER DEPT.

All prices are subject to change without notice.

ORDER COUPON

Please rush the following books(s).

Title ______________________________

Author ______________________________

Title ______________________________

Author ______________________________

NAME ______________________________

ADDRESS ______________________________

CITY ____________________ STATE __________ ZIP ______

☐ Payment enclosed ______________
(No shipping and handling charge) Total

☐ Bill me (shipping and handling charge will be added)

Add local sales tax where applicable.

Litho in U.S.A.